Electricity generation and
the ecology of natural waters

Electricity generation and the ecology of natural waters

T. E. LANGFORD

Central Electricity Research Laboratories
Marine Biological Laboratory
Fawley, Hampshire, England

Liverpool University Press

Published by
LIVERPOOL UNIVERSITY PRESS
Press Building, Grove Street, Liverpool L7 7AF

First published 1983

British Library Cataloguing in Publication Data

Langford, T.E.
    Electricity generation and the ecology of natural waters.
    1. Water-power electric plants
    I. Title
    574.5′222   TK1081
    ISBN 0-85323-334-9

Text set in 10/12 pt Times and printed and bound
by the Alden Press in the City of Oxford

TO MY WIFE, JEAN,
MY PARENTS,
AND LATE
PARENTS-IN-LAW

Preface

Early in 1978, when I was half-way through the the second draft of this work, I was at a conference in the U.S.A. In his introduction, one of the organizers stated that 'it is obvious now, that except possibly for very small areas, thermal discharges have no effect on the ecology of aquatic habitats'.

This short phrase dismissed (or some might say summarized) many hundreds of millions of dollars' worth of research completed, in progress, and planned by universities, the electricity industry, and consultancies throughout much of the civilized world. If true, it also contradicted the expert predictions of many noted scientists and made nonsense of some complex and expensive legislation in the U.S.A. and many other countries.

In continuation, the speaker stated that 'impingement of animals on intakes and entrainment of organisms through power stations were the current areas of ecological concern'. It occurred to me at that point to try to predict how long I might have to wait before another phrase at another conference would summarize and dismiss the results coming from the vast array of research projects at present in progress on this topic. The problem is to determine the truth of such statements, if there is a truth. If I needed any stimulus to complete this work, it was the idealistic and naïve determination to put the ecological truths into perspective without reference to politics or public relations. To this end, the book is, I hope, an objective reference for ecologists with similar aims.

Much of the literature on the ecological consequences of hydro-electricity and thermal generation has come from ecologists or conservationists initially committed to the belief that that such huge operations *must* have dramatic deleterious ecological effects. Opposing viewpoints, usually from the industry, have argued that hydro-electricity generation causes no harm because it only holds back natural water. Also, thermal discharges must be beneficial as 'it is well known that fish grow faster in warm water and fish and fishermen congregate around discharges'. Even though I have been employed by the electricity industry for 15 years, I have tried not to be biased and I hope I have assessed the scientific data as objectively as is possible. The best result would be that each 'side' believes that I have favoured the other.

There are some thousands of papers now published which deal with the effects of power developments on aquatic habitats. I have read over 3,000 papers and have tried to select references because I felt that they illustrated the particular points better than other papers on the same subject or, in many cases, because they include comprehensive bibliographies from which the reader interested in more detail, could begin a literature survey.

The very extensive Russian literature is mentioned and referred to where English translations are available but, because of my own shortcomings in the language, the numbers of references are small in relation to those published. Those papers listed in the bibliography may help the interested reader to begin further research. Usually, however, the Russian work has been paralleled by work in other countries. I have numbered the references, in addition to arranging them alphabetically, so that cross-referencing is easier for those papers published in proceedings or symposia, or books of collected papers. In addition, where papers have more than two authors, I have usually listed only the first, followed by *et al.*

I have received a great deal of help and advice from many experts in various fields over the past 2 years, and I would like to thank everyone who has assisted in any way. Most of my colleagues have begged me to return scientific papers, books, and other documents which I have kept for long periods. I have to thank them for their patience, especially Drs. R. J. Aston and D. J. Brown, Mr. A. G. P. Milner, and Miss D. J. Foster of the C.E.R.L. Freshwater Biology Unit at Nottingham. Also, Messrs. J. Coughlan, J. F. Spencer, M. H. Davis, R. Holmes, and J. M. Fleming, Drs. R. Bamber, A. Turnpenny, N. J. Utting, C. H. Dempsey, and P. A. Henderson of the C.E.R.L. Marine Biological Laboratory, who have supplied information, discussed points, criticized chapters, and brought literature to my notice. For this continued cooperation, I am grateful. Mr. J. W. Whitehouse, of C.E.R.L. Leatherhead, has also patiently forwarded material for which he knew he would have to request return in due course.

Dr. J. S. Forrest F.R.S., former Director of the Central Electricity Research Laboratories, gave me permission to begin the work and Dr. P. F. Chester, the present Director, has allowed me to continue and utilize a little time and material on occasions. Mr. R. S. A. Beauchamp and Mr. F. F. Ross, both now retired from the C.E.G.B., personally provided me with a great deal of information about chlorination and legal actions respectively, and for this I am in their debt. Mr. F. B. Hawes of C.E.G.B. Planning Dept. very kindly sent details of E.E.C. and European work, provided me with much basic information, and gave valuable criticism on chapters 8 and 9. The libraries of the Central Electricity Research Laboratories and the C.E.G.B. have also been extremely helpful. Data have also come from E.N.E.L. (Italy), E.D.T. (France), and the Belgian Electricity Industry.

I would like to thank all the authors whose work I have quoted and whose figures I have included and my hope is that I have represented all their results and conclusions as accurately as possible.

Finally, individuals without whom I may never have completed the work.

Dr. Gwyneth Howells, head of Biology at C.E.R.L., has allowed me to use some facilities and given the benefit of her advice and help. Mrs. Audrey Betteridge did the major work on the typescript and for her time and expertise I will never be able to thank her enough. Additional typing was done by Mrs. Melanie Malcolm, Mrs. Susan Williams, and Mrs. Beth Paterson. Dr. J. W. Eaton of the Department of Botany at the University of Liverpool generously accepted the onerous task of criticizing and correcting the draft manuscripts. For his advice and constructive comments I am most grateful.

Professor H. B. N. Hynes, Head of the Department of Biology, University of Waterloo, Ontario, was kind enough to take precious time from his own research to read two successive drafts and make valuable and stimulating comments at a point when the task of bringing this project to fruition seemed too great.

Mr. J. G. O'Kane, the former Secretary of Liverpool University Press was, as I am sure he was with most of his authors, able to stimulate my efforts many times and his encouragement and help, apart from the immense amount of work he did at various stages in the project, are very much appreciated. I regret that with his retirement in September 1981, he did not have the opportunity of seeing the work through to publication. But, I am grateful to the present Secretary, Rosalind Campbell, who has supervised the final stages. My wife, Mrs. Jean Langford, made the terrible mistake one weekend of asking if there was anything she could do to help. She has since done her own responsible job, and looked after the household while living, sleeping, and eating references in her 'spare' time. All the good work on the bibliography, and first-order corrections is entirely due to her and my thanks can only be an understatement. Any errors are mine.

As the book includes power station terminology which may not be familiar to ecologists I would recommend for reference the *Glossary of Power Station Terms* published by the Education and Training Department of the C.E.G.B. London (see Bibliography under C.E.G.B. 1963).

A common difficulty for authors is the dilemma of the use of different terminology in different countries—e.g. 'power plant' in the U.S.A. is the equivalent of 'power station' in the U.K. Where possible I have used British terminology and I hope this does not confuse readers in other countries too much. Any real or implied criticisms of research and any mistakes or misrepresentation of results are my responsibility and not that of my colleagues or employers.

TERRY LANGFORD

Wainsford, Hampshire
England
September, 1980

Contents

List of illustrations

List of tables

Chapter 1 Development of the industry, water use, and ecological concern

1.1. INTRODUCTION

The development of electricity generation has been one of the major factors determining the progress of civilization during the twentieth century. From being an unknown power source in the 1850s, electricity has become one of the essential features of modern life in developed countries. In less-developed countries a cheap, continuous supply of electricity is probably vital to future social and industrial progress. Since the beginning of large-scale generation, water has been a major resource, either for supplying the motive power directly in a hydro-electricity power station, or for producing and cooling the high pressure, high-temperature steam used to drive the huge turbo-generators in a thermal power station.

Today, the electricity industry is the largest single user of water in all the developed regions of the world (Van der Leeden, 1975). In England and Wales, the fresh-water used for electricity generation represents almost 50 per cent of the total demand for all uses and 75 per cent of the total industrial use. It also represents over 50 per cent of the average freshwater run-off from England and Wales and exceeds the total dry-weather flow of all major rivers (Hawes, 1970; Water Resources Board, 1973a,b). Cooling-water may be extracted from fresh or tidal water, but almost all the water used for hydro-electricity generation is from rivers or lakes (Table 1.1).

In the United States, where the amount of power generated in megawatts per square kilometre is only about one-ninth of that in Britain with rivers much larger (Hawes, 1970), the electricity industry accounts for almost a third of the total water use (Federal Power Commission, 1969). About 20 per cent of the freshwater run-off is used for cooling purposes in power stations (Krenkel and Parker, 1969a).

This cooling water, unlike that used in other industrial or in domestic processes, is mostly returned to its source near the point of abstraction with few obvious changes in quality. However, the use of these vast amounts of water from rivers, lakes, estuaries, and the sea has resulted in many physical, chemical, and biological changes. The massive dams necessary to impound rivers and provide the hydrostatic head for hydro-electricity developments have blocked the passage of migrating fishes and created long, deep stratified lakes where once flowed fast, shallow, and turbulent streams. Lakes used as

Table 1.1. *Sources and uses of cooling-water by the C.E.G.B. (U.K.), 1971 and 1977* (in $m^3 \times 10^6$ day^{-1})

	Year	
Use and source	1971	1977
Once-through cooling		
Tidal waters	47	46
Freshwater	18	16
Recirculated (mainly fresh)	105	110
Evaporative loss	0·4	0·5
Average freshwater run-off 190 $m^3 \times 10^6$ day^{-1} in England and Wales		

N.B. In comparison the daily cooling-water circulation in all U.S. power plants for 1971 averaged 7,433 $\times$ 10^6 m^3 s^{-1} of which 5,245 $\times$ 10^6 were from fresh waters and 2,188 $\times$ 10^6 from tidal waters.

(From Pipe, 1972.)

reservoirs in pumped-storage schemes now have daily water-level fluctuations over many metres where previously a stable shoreline existed. Water abstracted for cooling purposes and returned at a higher temperature changes the temperature regime of the receiving water. All of these physical changes can affect the survival, life cycles, and distribution of resident animals and plants acclimatized over thousands of years to the seasonal cycles in natural waters. There are no major chemical processes involved in electricity generation, with the exception of antifouling or anticorrosion procedures, though some changes in the chemistry of receiving waters may be caused indirectly as a result of the physical alterations in the water body.

Whether the physical and chemical changes in water bodies always result in significant biological changes in ecosystems has been the subject of considerable research, discussion, and plain conjecture during the past two decades. Concern has arisen with the rapid expansion of the electricity industry, particularly of nuclear power generation with its problems of radio-active waste disposal and its greater demand for cooling water than in conventional thermal generation plants. Before discussing the ecological aspects it is essential that the basic processes of generation are outlined.

1.2. ELECTRICITY GENERATION AND WATER USE

Electricity is generated on a commercial scale by one fundamental process. Energy is used to rotate a turbine which in turn rotates a dynamo or generator from which the electric current is produced. There are two major alternative means of providing the energy, each of which requires large quantities of water. In a *hydro-electricity* installation (Plate 1.1) the potential energy of water stored at a high level is converted to kinetic energy by allowing the water to fall to a lower level, turning the turbines as it falls. In the second process,

Plate 1.1. Hydro-electricity station, showing dam and turbine house, Pitlochry, Scotland.

usually known as *thermal* or *steam* generation, water is heated to produce steam at high pressure and high temperature, and this steam is fed as the driving force to the turbine. The fuel used to provide the heat may be coal, oil, natural gas, or more recently, some radio-active element like Uranium 235, which gives out heat through nuclear fission.

1.2.1. Hydro-electricity generation, pumped storage, and tidal power

There are two main types of hydro-electricity installation, classified according to the method of water storage.

The *pure* or *run-of-river* installation is found where a natural stream or river has been dammed to create the reservoir (Fig. 1.1). Water levels and flows are controlled at the dam to provide energy for the turbines and to maintain stream flow below the dam. If the rivers are large enough and the gradients steep, a number of dams and hydro-stations can be installed in sequence. For example the River Tummel in Scotland and its tributaries are harnessed by nine power stations and dams to provide a total power output of 250 MW. On

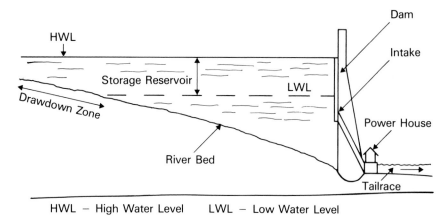

HWL − High Water Level   LWL − Low Water Level

Fig. 1.1. Diagrammatic section of a conventional run-of-river hydro-electricity scheme.

a larger scale, the Columbia River and its tributaries on the border of the United States and Canada are impounded by more than fifty major dams, most of which involve hydro-electricity plant generating thousands of megawatts (Trefethen, 1972). In the second type of hydro-electricity installation, i.e. the *pumped storage* scheme, water is pumped from a lower source, using electricity at times of low demand (off-peak), to a high-level reservoir. It is then allowed to fall back to the lower source, via the turbines, producing electricity on demand (Fig. 1.2). In some schemes pumped-storage and run-of-river installations are used in combination, to maximize water use. In many countries pumped-storage schemes are used increasingly to augment other types of generation and to meet peak power demands. The proportion of pumped-storage to other types of generation will, however, probably not exceed about one-sixth because of power demand for pumping and peak demand characteristics (Birkett, 1979). Hydro-electricity installations are further classified according to their head height (i.e. approximately the vertical distance between the upper reservoir and the turbine house). 'High-head' schemes exceeding 160 m and up to 1,000 m are typical of Alpine regions. Medium-head schemes, i.e. 30–160 m, are commonly found in Britain and the U.S.A., while low-head schemes, of under 30 m head, are typical of regions with low relief such as Sweden (*Modern Power Station Practice*, 1971).

   Since the potential water power of any given site is proportional both to the size of the stream discharge *and* to the available head height, the world's best hydro-electricity sites are, as might be anticipated, in regions of heavy rainfall and high relief. There are a number of advantages, however, which, owing to the patterns of electricity demand and consumption, necessitate the inclusion of hydro-electricity generation in the power supply system of most countries, even when the rainfall or relief are not especially favourable. For example, hydro-generators can be started very quickly, often in a few seconds, whereas

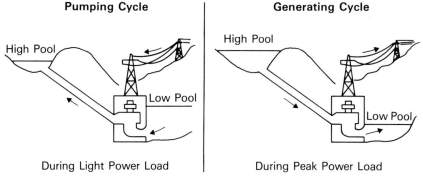

Fig. 1.2. Diagrammatic representation of the operation of a typical pumped-storage installation, redrawn after Hauck and Edson (1970).

thermal generators may require from 20 minutes to several hours to run up a full load. Very fast starts are often necessary, to meet sudden peaks in demand. Further advantages of hydro-generation are that maintenance costs are lower than for thermal plant, installations have a life of up to 50 years compared with 30 years for thermal installations, and there are no handling or storing costs for fuel. The main disadvantages are the high capital cost of construction and the siting limitations in countries with low or medium relief.

A recent variation of the hydro-electricity installation is the *tidal storage* scheme, where the rise and fall of tides are used, either to drive turbines directly, or to fill an upper-level reservoir, from which water is released to the turbines as the tide ebbs. The world's first large-scale operating tidal-power scheme began operating at La Rance, near St. Malo, in France, during 1966, and a small experimental scheme was completed in the U.S.S.R. in 1968 (Gordon and Longhurst, 1979).

Cotillon (1974) listed the three basic requirements of a suitable tidal-storage site as:

(a) A large tidal range to provide head.

(b) A large-capacity reservoir for adequate storage.

(c) A short dam, to keep construction costs down.

La Rance met all three requirements, having the right geographical features and an equinoctial spring tidal range of 13·5 m. The available output in 1974 was about 65 MW.

This type of installation is, of course, best suited to regions with a high tidal amplitude, such as the Severn Estuary in Britain (Glendenning, 1977) or the Bay of Fundy, both with tidal ranges of over 14 m (Shaw, 1971). In this latter, the Passamaquoddy project in Maine (U.S.A.) could become operative in the 1980s* (Gordon and Longhurst, 1979). Hubbert (1971) suggested that about 2

* The large Passamaquoddy project may not now proceed though a small scale trial project is in progress.

per cent of the total hydro-electricity in the world could originate from such sources, but the economic advantages of tidal power are, as yet, uncertain. For example, though the Severn Estuary is possibly the best site in the world for a tidal storage scheme, potentially producing up to 7,000 MW, there is no economic justification for the proposal, based solely on electricity generation (Shaw, 1975; Leason, 1976).

Whichever type of installation is used, hydro-electricity developments always involve physical alterations in the relevant aquatic habitats. In many cases these changes may lead to social benefits which can offset any deleterious ecological effects.

1.2.2. Thermal power generation

The basic process for generating electricity in thermal and 'steam' power stations is the same, whichever fuel is utilized.

In *fossil-fuelled* plant, coal, oil, natural gas, or in rare instances, peat is burned to provide the heat input (Plate 1.2).

In *nuclear-fuelled* plant, the energy derived from the fission of the radio-active elements is used as the heat source (Plate 1.3). Although nuclear power stations are a relatively recent development, the percentage of electricity generated from such installations is increasing rapidly throughout the world though in the next two decades the rate of increase is expected to be somewhat slower.

Ross (1959) divided the water uses in thermal power stations into two categories:

(a) Major uses: viz. cooling, ash disposal, boiler water (Table 1.2).

(b) Minor uses: viz. domestic sewage, fuel store drainage, flue-gas washings, and boiler water treatment wastes. The second category also includes water containing small amounts of radio-active contaminants from nuclear power stations.

Table 1.2. *Analysis of daily water use in thermal electricity generation processes in British power stations (C.E.G.B. only) during 1971*

Requirement	$m^3 s^{-1}$	%
Condenser cooling	422·1	90·1
Steam raising (boiler feedwater)	22·3	4·8
Auxiliary plant cooling	17·4	3·7
Ash handling	4·8	1·0
Gas washing	1·5	0·3
Personal uses (sewage, laboratory)	0·5	0·1
Total	468·6	100·0

(From Pipe, 1972)

Plate 1.2. Typical coal fired power station, with cooling-towers, High Marnham, Nottinghamshire, U.K.

Plate 1.3. Nuclear power station, Trawsfynydd, North Wales, U.K.

In a few geothermal regions of the world, for example, Italy, New Zealand, Japan, Iceland, the U.S.A., and the U.S.S.R., there are small areas where heat emitted from the earth's magmatic core turns underground water to steam. This steam may be tapped and fed directly to turbines or used to pre-heat water in boilers. Small amounts of cooling water are used, usually in combination with some artificial cooling device.

1.2.3. **Other sources of electricity generation**

Potential large-scale electricity sources for the future are solar power, wind power, tidal power, fuel cells, and magneto-hydrodynamics. Any ecological implications of water use in these are discussed in chapter 9.

1.2.4. **Historical development of electricity generation**

The increase in electricity generation in most countries over the past hundred years has been accompanied by increases in efficiency of water use. Developments in Britain have generally been paralleled in most industrialized countries. The emphasis on hydro-electricity and thermal generation has differed from country to country, mainly as a result of differences in topography, rainfall, and fuel availability. The history of power station development and of the integrated generation and transmission system in Britain illustrates the general progress of the industry (Electricity Council, 1973; Hannah, 1979).

Generators were used to provide electric lighting for British lighthouses in the second half of the nineteenth century. In 1880, Sir William Armstrong used a 6 h.p. hydro-electric installation to light the picture gallery at his home at Cragside, Northumberland. This was probably the first dwelling house in the world to be lit by incandescent electric lamps. In 1881, the first British public supply was produced by a hydro-electric power station at Godalming on the River Wey, and used to light the streets. The Electric Lighting Act (1882) authorized local authorities to produce electricity but only seven licences were issued by 1888. In 1882 Britain's first thermal power stations opened, producing electricity for public and private consumption at Holborn Viaduct, Brighton, Hastings, and Eastbourne. In the same year Edison's Pearl Street power station opened in New York, and the U.S.A.'s first hydro-electric station began operating.

The first turbo-generator, the forerunner of the huge machines used world-wide today, was developed by Sir Charles Parsons in 1884, and in that year a 1,000 kW (1 MW) plant was installed on the Grosvenor Gallery site in London.

In 1914, generating units each producing 11 MW were in operation in Britain using steam at $17 \cdot 57$ kg cm^{-2} (250 lbs in^{-2}) pressure and 380°C. By 1933, generating units of over 100 MW were developed and in 1966, the first 500 MW units were commissioned in Britain at Ferrybridge in Yorkshire using steam at a pressure of $210 \cdot 9$ kg cm^{-2} (3,000 lbs in^{-2}) and a temperature of 565°C. In 1968, thermal power stations each providing 2,000 MW from groups of 500 MW units were completed at Moss Landing (California) at Ferrybridge (Yorkshire) and West Burton (Nottinghamshire).

During the early years, hydro-electric power kept pace with thermal generation in Britain mainly owing to developments in Scotland and North Wales. Before long, however, the phenomenal advances in turbo-generator design and materials allowed thermal generation to outpace hydro-electricity in England and Wales. In 1978, out of a total installed capacity of over 55,000 MW, only about 120 MW were produced from run-of-river hydro-electric stations, and one of the three tiny English installations dated back as far as 1913. Pumped storage produced about 360 MW.

In Scotland, however, the trend was different. By 1965, hydro-electricity accounted for over 60 per cent of the total generation of about 1,700 MW, though the construction of large nuclear and oil-fired power stations in the last decade has now changed this proportion (North of Scotland Hydro-electric Board, 1973).

In the rest of the world the trend towards larger generating units has been similar to that in Britain. Units of 660 to over 1000 MW are now operating in several countries. Where the topography and rainfall are favourable, hydro-electricity generation is still dominant. About 20 per cent of electricity generated in the U.S.A. comes from this source compared with about 3 per cent in England and Wales, and 25 per cent in Europe as a whole. In Switzerland, hydro-generation supplied 100 per cent of electricity used until 1969 when the 300 MW Vouvry thermal plant opened. In Austria hydro-electricity decreased from 75 to 58 per cent of the total generated between 1961 and 1972 (Fry, P., *et al.*, 1971; Bauer and Sterk, 1974).

The size of hydro-electricity generators has also increased over the years. At the turn of the century generators were mostly of under 5 MW output but by 1972 units of up to 500 MW were being commissioned in Canada. At the Churchill Falls project in Labrador a total of over 5,000 MW can be generated from eleven units. The largest hydro-electricity scheme at present is at Krasnoyarsk on the River Yenisey in Siberia, with an installed capacity of over 6,000 MW. In remote areas, tiny hydro-stations with outputs as low as 0·1 MW still supply communities (North of Scotland Hydro-electric Board, 1973, Smil, 1976).

The first stage of the world commercial nuclear power programme was launched in Britain in 1957 and the first full-scale commercial nuclear power station became operational in 1962 at Berkeley on the Severn Estuary. By 1971 the total nuclear capacity in Britain was 4,500 MW and although Hunt (1971) forecast that some 9,500 MW of nuclear plant would be operating in Britain by 1975, this target was not met, mainly for technical, economic, and political reasons. Nuclear generating plant accounted for 7 per cent of the total installed capacity in England and Wales during 1977–8 but actually produced 13 per cent of output (England, 1978).

After a head start, Britain is developing nuclear generation more slowly than other European countries and North America. For example, in Switzerland nuclear power now accounts for about 20 per cent of the total generated, while in the U.S.A. the predicted proportion was 25 per cent by 1980 (Hubbert, 1971).

The development of the pumped-storage scheme in Germany in the 1930s opened the way for new hydro-electric generation. The Dinorwic scheme at present under construction in North Wales will produce about 1,800 MW, 1,320 MW of which can be produced within 10 secs of starting up. Pumped storage schemes are proliferating in many countries (Karadi, *et al.*, 1971).

The rate of growth in electricity output has varied markedly from country to country, but has generally been most rapid since 1945. In the U.K., output increased 4·5 times between 1948 and 1968, with an average annual growth of 7·3 per cent. In recent years, however, this has slowed dramatically and in 1976–7 a reduction in demand was recorded for the first time since the 1930s. As extreme examples, demand in India has increased seven times since 1950 and in Iran the annual growth from 1967 to 1972 averaged over 18 per cent. In 1933, a national network for electricity distribution (National Grid) was opened in Britain, and at the present time, Britain has the largest fully integrated electricity supply in the world, with the possible exception of the U.S.S.R. Along with the growth in output in the U.K. and other countries there has been a marked decline in the number of power stations (Fig. 1.3). In England and Wales, 450 power stations could produce around 6,500 MW in

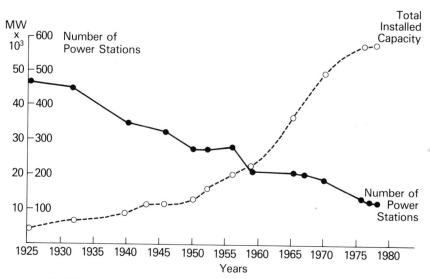

Fig. 1.3. Concentration of generating capacity in Britain from 1925 to 1977.

1932, but by 1977, 58,000 MW could be supplied from only 137 stations. In 1980, this number was reduced to about 120. The trend toward larger thermal and hydro-electric power stations has resulted in greater water requirements at each location.

1.2.5. The development of ecological concern

During the past two decades the increasing concern about the ecological consequences of electricity generation for rivers, lakes, and tidal waters has been reflected in legislation, research, and the media. The fact that obstructions in rivers block the passage of migratory fish has been well known for centuries (Netboy, 1968; Eicher, 1970; Gregory, 1974; Larkin, 1979) but, even so, there is still controversy over dams and weirs in salmon rivers (*Sunday Times*, 1978; Power, 1978). The legal aspects of fish migration and hydro-electricity schemes are discussed in chapter 9. From the late 1940s until today, research on salmonids and the effects of hydro-schemes on migration has been carried out in Scotland (Berry, 1955) and in addition, stocks have been augmented by introductions of artificially reared young fish, paid for by the electricity undertakings. The research and fish rearing were paralleled in several other countries including Sweden, the U.S.A., and U.S.S.R. (see chapter 4). Dams without provision for fish passage were still being built in some countries until quite recent times (Liepolt, 1972). In the U.S.A. and Canada some dams, completed as late as the 1940s and 1950s, had no facilities for allowing free fish migration. Controversy about high dams in Canada and the U.S.A. continues as hydro-electricity schemes proliferate (Geen, 1974, 1975; Power, 1978; Larkin, 1979).

Published research on hydro-electricity dams and their associated lakes has existed for many years (Berry, 1955; Larkin, 1958) but since the late 1960s, there has been a new upsurge of research on fish migration, river faunas, and physico-chemical effects, such as gas-bubble disease, particularly in the U.S.A. and Canada (Lowe-McConnell, 1966; Benson, 1970; Ackermann, *et al.*, 1973; Ward and Stanford, 1979).

There have been marked increases in legislation, research, and press coverage related to thermal power generation in many countries over the last 20 years, coinciding with the rapid growth of this type of generation. The first environmental problem to be noted was the potential effect of heat discharged to natural waters. In fact, the Electricity Supply Act of 1919 in Britain included reference to potential hazards of raised river temperatures.

Under Common Law in Britain, riparian owners have, for hundreds of years, had some protection against pollution from upstream, and a rise in temperature has been considered to be an adverse alteration of water quality (Coulson and Forbes, 1952; Wisdom, 1956). In the River Trent, rises in water temperature near power stations have been recorded for 30 years or more but the first real concern with heat as a pollutant in Britain stemmed from problems in two rivers, namely the Thames and the Derwent, between about

1949 and 1952. In the Thames, increased water temperatures were believed to be the cause of decreasing oxygen concentrations (D.S.I.R., 1964). The Hobday Report on River Pollution Control (Ministry of Health, 1949) included a section on the potential effects of thermal discharges and stated that 'during summer a rise of even 1 °C may have a very adverse effect—either directly or indirectly—on the biota'. The 'Pride of Derby' case (see chapter 9) in 1952 brought 'thermal pollution' to public attention when a heated discharge from the Spondon Power Station was said to have contributed to the destruction of the fishery in the River Derwent, and the Electricity Authority and others paid heavy compensation. Biologists began working on the problems of fish and power station cooling-water discharges in Britain around 1952, and by 1954 the Electricity Authority had established a biological laboratory. At about the same time government research bodies became involved in similar research (Alabaster, 1958). With the proposal to build Bradwell Nuclear Power Station on the Essex coast, a team of marine biologists was established in about 1955 to look at the problems of discharging cooling-water to the sea, and the effects of the cooling-water system on oyster larvae which might pass through it (Hawes, *et al.*, 1975). The effects of radio-nuclides discharged to natural waters were also being monitored by Government scientists and by the industry.

In the U.S.A., publications by Cairns (1956a,b) brought thermal effects to the notice of American scientists and the public. Also research on the Delaware was showing marked ecological changes below a power station cooling-water discharge (Trembley, 1960; Coutant, 1962). After about 1964, research *and* public concern over 'thermal pollution' accelerated until by 1970 there were over 300 separate projects being supported by the Edison Electric Institute in the U.S.A., employing hundreds of research workers. At the same time in the U.S.S.R. many similar projects were in progress and teams were growing in Scandinavia and other European countries. Publications concerned with thermal power station effects grew in number each year from about 5 in 1958 to more than 800 by 1975. A recent bibliography (Hannon E. H., 1978) lists over 1,200 reports relating to biological investigations of nuclear and fossil-fuelled power station cooling-water discharges in the U.S.A. Around 1969, 'thermal pollution' was also a major public concern, particularly in the U.S.A., where the words entered almost every form of public life. The Environmental Protection Agency (E.P.A.) formed in the U.S.A. in 1970 as a result of the National Environmental Policy Act (N.E.P.A.) played a major role in stimulating research and establishing criteria for power station intakes and discharges.

Since the 1950s there has also been concern over the effects of radio-nuclides discharged to natural waters and their potential effects on man through accumulation and transport in food organisms (see chapter 7). These discharges have been monitored for some years in many countries and there are strict international guidelines for controlling them. Recently, however, radio-nuclides have come into more public prominence as a result of the acceleration of nuclear power programmes and building of fuel reprocessing plants. Pressure groups such as the Friends of the Earth and local conservation groups have kept nuclear power in the public eye.

Recently, concern has focused on the effects of aerial emissions on terrestrial and aquatic ecology, particularly on the effects of sulphur and nitrogen oxides on the pH of precipitation and the subsequent reductions in the acidity of lakes and rivers. In parts of North America and in Scandinavia this increased acidity is alleged to have resulted in the destruction of salmon and trout fisheries (Gorham, 1976) (see chapter 7). The ecological concerns have eventually been reflected in legislation, particularly in the U.S.A., where design and operation criteria for cooling-water discharges and intakes are now causing operating difficulties and massive planning and construction costs at established and proposed power station sites.

1.2.6. Nature of discharges from thermal power stations

The use of water in the various processes involved in thermal generation of electricity have already been mentioned briefly (see 1.2.2). The composition of discharges which result from these uses can be complex and although the processes are dealt with separately below, any discharge from a power station site may contain effluents from two or more processes (*Modern Power Station Practice*, 1971).

(a) Heat disposal and thermal discharges

Steam-electric generating plants, whether nuclear or fossil-fuelled, operate through the thermodynamic process known as the *Rankine cycle*. In this cycle, steam produced at high temperature and pressure is fed to a turbine, giving up energy to the turbine rotor, which in turn drives the generator. At the exhaust end of the turbine the steam is condensed and returned to the boiler for a repetition of the cycle. Massive volumes of cooling-water are used to absorb the heat given up in the condensing process. The heat which is eventually discharged into the cooling-water is the residue which is not turned into mechanical work in the turbine or lost in some other way. Cootner and Lof (1965) describe this heat as 'waste' but waste in a *technological* rather than an *economic* sense. The reason for this is the low efficiency of the Rankine cycle itself. The maximum possible efficiency of conversion of fuel to electricity depends on the difference between maximum and minimum steam temperatures. With the steam temperatures currently used (see 1.2.4) the maximum theoretical thermal efficiency is slightly more than 60 per cent. At the present state of technological development however the best overall efficiency which can be achieved is about 40 per cent owing to the thermal, mechanical, and electrical losses in the process. Actual efficiencies in most power stations range from about 30 to 37 per cent. The theoretical energy balance and cooling-water requirements at a station which is 40 per cent efficient are shown in Table 1.3.

Table 1.3. *Typical energy balance in steam-electric power plant (figures for 1 kWh net electrical output)*

Assumed overall efficiency	40%
Assumed generator efficiency	97·5%
Heat equivalent of 1 kWh	3,618 kJ
Fuel energy required	9,045 kJ
Heat losses from boiler furnace at 10 per cent fuel use	904 kJ
Heat loss from electric generator at 2·5 per cent of generator input	92 kJ
Electric generator output	3,618 kJ
Energy required for generator equals energy output from turbine	3,710 kJ
*Energy remaining in steam leaving turbine removed in condenser	4,430 kJ
Total cooling water use for 5·5°C rise	229 l
Total cooling water use	153 l

* For nuclear plant 7,102 kJ (After Cootner and Lof, 1965) (1 Btu ≡ 1·06 kJ (approx)).

Consequently, about 265×10^9 kJ of heat need to be disposed of each day from a 2,000 MW power station and this requires around 65 $m^3 s^{-1}$ of water. This volume is far greater than the average daily dry-weather flow of any British river (Pipe, 1972). Because of the lower steam pressures, differences in heat exchange efficiencies, and no stack losses, the amount of heat to be discharged to the cooling-water from nuclear power stations may be up to 50 per cent greater than in fossil-fuelled stations.

During the past 50 years there have been considerable advances in design which have led to more efficient use of cooling-water (Fig. 1.4). Cootner and Lof (1965) give a detailed account of heat transfer processes in power stations to which the interested reader should refer. Even with these remarkable increases in the efficiency of water use, the total amount of water necessary for electricity generation has increased because of the larger power stations and the increases in output. The development of magneto-hydrodynamic (M.H.D.) generation, which uses 75 per cent less cooling-water, may lead to some reduction (Vielvoye, 1977).

Because of the lack of fresh-water for cooling in Britain and some European countries, and the need for inland siting near centres of population, alternative methods of cooling-water use and recycling have been developed (Leason, 1974). In countries with larger rivers and lakes, constraints were not generally as restrictive and alternative cooling-systems have been slower to emerge. The last power station in Britain to use river water for cooling without provision for extra-cooling capacity was commissioned in 1953.

(b) Cooling-water systems

Several alternative cooling-systems have been developed, either to provide

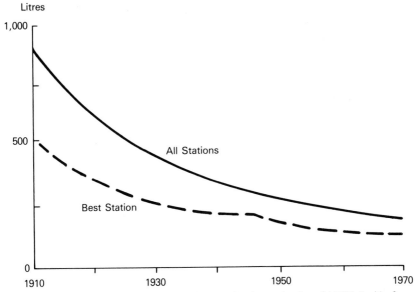

Fig. 1.4. Reduction in cooling water requirement for the production of 1 kWh (unit) of electricity in U.K. thermal power stations between 1910 and 1970, redrawn from Pipe (1972).

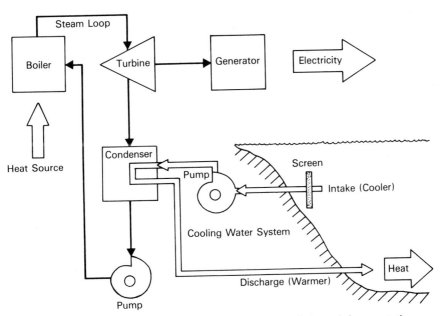

Fig. 1.5. Schematic diagram of the cooling-water system for a direct-cooled power station, redrawn from Schubel and Marcy (1978).

adequate cooling capacity or to alleviate potential thermal effects in the receiving water.

(i) *Direct-cooling*

Direct or 'once-through' cooling systems (Fig. 1.5) are widely used where water supplies are adequate, for example on the coast, in larger rivers and estuaries, or at the smaller inland sites. In such systems the cooling-water is pumped from the source to the condenser system in which the rejected heat is transferred to the cooling-water. This heated cooling-water is discharged directly to the receiving water. Once in the receiving water the heat is dissipated to the air by surface evaporation and to the surrounding water by dilution.

(ii) *Indirect-cooling*

Cooling ponds. Direct-cooling is the least effective method of limiting heat discharge to the receiving water. To reduce this heat discharge the cooling-water may be retained in shallow ponds or lakes where it will lose its heat by evaporation, radiation, and conduction before being discharged to the receiving water. The main disadvantage of cooling ponds is that large areas of land are required to provide an adequate cooling surface. Scott (1973) suggested an area of 485–607 ha (1,200–1,500 acres) for a 1,000 MW fossil-fuelled power station. If the ponds were operated on a closed-cycle basis, i.e. recycling cooled water back to the power station, a 1,000 MW station would need up to 1810 ha (2,000 acres) of pond surface area and a nuclear station 1215 ha (3,000 acres) (Budenholzer, *et al.*, 1971). Natural lakes or reservoirs may also be used as cooling ponds (Burnett, *et al.*, 1974) and recent developments have used pumped storage reservoirs as cooling lake systems for nuclear power stations (Kipelainen, 1971).

Spray ponds. Fine-nozzle sprays which atomize the water into minute droplets as they spray it into the air may be incorporated into the cooling ponds. Heat transfer is facilitated by the increased water surface area and the area of land required for each cooling pond may be reduced by a factor of twenty (Budenholzer, *et al.*, 1971; Scott, 1973). A further modification of this system is the floating spray whereby the pumps and jets are floated on to the surface of the cooling pond.

Cooling-towers. In countries where water for cooling is scarce or where constraints on the temperature of discharges are severe, cooling-towers are used extensively. There are two main types of cooling-tower in use in the world (McKelvey and Brooke, 1959; Gurney and Cotter, 1966).

Natural draught evaporative cooling-towers have been in use at power stations for over 50 years in Britain and currently there are almost 300 of these in operation (Leason, 1974) (Plate 1.2). The operating principle is relatively simple. Water leaves the condensers carrying its maximum heat load. It is pumped to the tower, entering at a point about 10–13 m above the base, from where it is distributed via radial channels to small spray cups or nozzles. These create droplets which then fall over a system of wooden or asbestos slats to a collecting pond at the tower base. In falling, the droplets are in close contact with an up-current of cold air to which the heat is transferred. Between the inlet to the tower and collection in the base ponds the water may be cooled by 8–10°C, depending on the air temperature and the humidity. In the more humid regions of the world, temperature loss may be as little as 3–4°C. During this evaporative heat transfer, most of the waste heat and about 1 per cent of the cooling water are lost to the atmosphere. At the same time, 2 per cent of the recirculating water is purged to the receiving water. This purge and the necessary make-up water abstracted from the source are to prevent the progressive concentration of dissolved solids in the cooling-water system which might encourage deposition of materials in pipes. A deterioration in the quality of water discharged to the receiving water may result from this concentration (see 3.5.2).

Hyperbolic natural draught towers are less common in countries where surface water is plentiful. For example, the first such tower in the U.S.A was built in 1962 at the Big Sandy site in Kentucky, though a number are now in use and more are planned. At some stations in the U.S.A., unlike all those in Britain, cooling towers are used to cool the discharge before it goes to the receiving water. Although these towers produce climatic and visual amenity problems because of their size (approximately 130 m diameter base by 130 m high) the increased efficiency of water use makes them technologically and economically desirable in many countries (Budenholzer, *et al.*, 1971; Leason, 1974). In the U.S.A. towers have been proposed for use at coastal sites using sea-water, but local salt deposition from tower plumes may be a problem. In the U.K. saline water has been used in cooling towers at Fleetwood Power Station for some years with few problems.

Mechanical or forced draught cooling-towers have been widely used at power stations in warmer, humid regions. Large, motor-driven fans are used to move the air past falling water droplets. The cooling principle is similar to that in the natural draught tower but the structures are generally longer and lower (Budenholzer, *et al.*, 1971). As a result plumes of water vapour from the tower can often influence local climatic conditions, more so than plumes from the higher natural draught towers.

Dry cooling-towers are in use at several inland power stations. Here the water is cooled like that in a car radiator, that is, not in direct contact with the air, and there is no evaporative loss to the atmosphere (Leason, 1974; Reymeysen, *et al.*, 1979). The first large dry cooling-tower was constructed at Rugeley Power Station (U.K.), and is a hyperbolic structure some 110 m high by 100 m diameter at the base, operating on natural draught. With this and other dry towers, capital costs and corrosion are major disadvantages (Dutkiewitz, 1974; Belter, 1975).

(iii) *Relative costs of cooling-water systems*
There are considerable costs involved in providing any extra cooling facility at
a power station (Monn, *et al.*, 1979) though the choice of a cooling system may
depend on economic, social, or ecological factors. It is difficult to give any
absolute cost figures, but in Table 1.4 the relative costs of some installations
are shown.

Table 1.4. *Relative costs of cooling water systems at power stations*

Type of system	Investment cost kW^{-1}	
	Fossil-fuelled plant*	Nuclear-fuelled plant*
Once through (direct)†	1·0–1·5	1·5–2·5
Cooling ponds‡	2·0–3·0	3·0–4·5
Wet cooling towers		
Mechanical draught	2·5–4·0	4·0–5·5
Natural draught	3·0–4·5	4·5–6·5

* Unit costs based on systems of over 600 MW.
† Circulation from natural water body.
‡ Artificial ponds designed specifically for the cooling system.
(From Federal Power Commission, 1969)

(c) Non-thermal discharges

This is a generalized description describing all those effluents which do not
originate directly from the cooling process (Zimmerman, *et al.*, 1974).

(i) *Ash disposal and effluents*

The electricity industry is a major consumer of low-grade coal with high ash
content. The coal is burned after being ground or milled to fine particles and
the ash, known as pulverized fuel ash (P.F.A.), or more commonly 'fly ash', is
the residue. From a 2,000 MW installation operating for 24 hours each day
about one million tonnes of ash are disposed of each year (*Modern Power
Station Practice*, 1971). If consolidated, this would cover a site of 11·5 ha (28
acres) to a depth of almost 10 m (32 feet). In the U.S.A. the production of fly
ash was over million tonnes in 1966 and is expected to be up to 45·7 million
tonnes per year by 1980 (Zimmerman, *et al.*, 1974). In the U.S.S.R., 60 million
tonnes of ash and slag are produced annually (Migachev, *et al.*, 1974).

In power stations not using pulverized coal, the ash is normally collected
from the bottom of the furnaces by a water sluicing system (Ross, 1959).
Where pulverized coal is burned, the ash is separated from the waste gases by
electrostatic precipitators (*Modern Power Station Practice*, 1971). Alternatively the gases themselves may be separated out with water. The separated
ash may be transported dry for use in industrial processes such as building
block manufacture or road-building, or it may be mixed with water and

sluiced to settlement ponds or lagoons where it is allowed to settle out. Approximately 18.6×10^3 m^2 of ash pond area is required for a 1,000 MW coal-fired power station (N.T.I.S., 1977). The supernatant liquid contains suspended particles of ash together with various minerals and is usually discharged to a nearby watercourse or to the sea. The normal volume of ash-effluent from a 2,000 MW installation is between 9 and 13 million litres per day.

There are very few published studies of the nature and composition of non-thermal discharges from power stations. Volumes are relatively small in comparison to those of cooling water, but ash effluents generally form the largest proportion of non-thermal effluents. Fly ash from two power stations in Northern England is dumped into the sea for disposal (Bamber, 1978).

(ii) *Boiler cleaning effluents*
In common with coal, fuel-oil leaves a deposit on boilers which reduces efficiency if not regularly removed. Cleaning and descaling is carried out annually, usually using acetic or other weak acids. Boiler-cleaning wastes may be discharged to the cooling-water system or separately. At some sites the acids may be neutralized in the ash-settling lagoons. Acid cleaning discharges contain strong concentrations of metal salts in addition to acids.

(iii) *Fuel store drainage*
Coal-fired power stations often maintain a coal stock representing 50–100 days' use. Thousands of tonnes are usually stored uncovered. Rainfall and the natural water associated with the coal drain through the coal heaps into site drainage systems and eventually into rivers or streams. In many cases drainage is directly into the ground, but after heavy rainfall some inevitably enters the site drainage system. Usually this water contains sulphuric acid, a high concentration of sodium chloride, heavy metal salts, and suspended solids. Discharges are generally intermittent.

Where oil is stored, spills from tankers and leakage from pipes may contaminate site drainage, though effluents are usually passed through a separator to remove the oil before discharge.

(iv) *Water treatment wastes*
The water used for producing the steam in a power station boiler needs to be very pure. Cootner and Lof (1965) give a figure of 99.998 per cent purity, far greater than the raw feedwater which may originate from a river or a municipal water supply. The dissolved minerals and gases are removed by a series of complex chemical processes in a water treatment plant (*Modern Power Station Practice*, 1971). The wastes include some concentration of the natural salts in addition to soda and phosphates used to reduce scale formation. Small amounts of hot boiler water or water from evaporating plant may also be present in some cases. If the raw water is purified by means of

ion-exchange resins, then high concentrations of salts, acids, or alkalis may be included in the waste. Volumes of boiler feedwater may be quite large (Table 1.2), but the water-treatment wastes are of relatively small volume.

(v) Gas washing

The air used for the combustion of the coal, oil, or other fossil-fuel is converted to flue gas, which is a mixture of nitrogen, carbon dioxide, and water vapour, contaminated with sulphur and nitrogen oxides. To reduce atmospheric pollution in urban areas, the flue gases are sometimes 'washed' with water to remove the sulphur and nitrogen oxides (Ross, 1959). The resulting liquid is highly toxic, but is not usually of great volume.

(vi) Domestic sewage

A modern, 2,000 MW power station under construction may have over 1,000 workers on site. When in operation this is reduced to about 250 at any time if the station is on base-load. If the domestic waste can be discharged to a municipal sewer few problems arise in local watercourses, but in most cases sites are so remote that each has a small sewage disposal plant. There are rarely major pollution problems from these small plants, though some do discharge separately into small streams rather than via the cooling-water system or other drains.

(vii) Chemical additives in cooling-water

Biological fouling of pipes and culverts is a problem in power station cooling systems as we will see in chapter 6. Consequently algicides, fungicides, bacteriocides, or 'total' biocides may be added to the cooling water. Anticorrosive agents such as ferrous-sulphate or poly-electrolytes may also be added and these substances may be potentially harmful to the biota of the receiving water. Chlorine, in some form or other, is the most common biocide in both freshwater and salt-water cooling systems. It is dosed either continuously or intermittently. In 1976, 10,000 tonnes of chlorine were used by the C.E.G.B. in the U.K., mainly for biocidal purposes in cooling systems (Coughlan and Whitehouse, 1977).

Dosing rates vary in different countries from 0·5 to 10 p.p.m. and the concentrations of free-chlorine discharged vary with the chemistry of the water being dosed (see chapter 3). Various antifouling methods have been tried, but as yet few are as successful and economic as chlorination. For safety reasons, however, the current trend is away from gaseous chlorine, which requires potentially dangerous large-scale storage, and towards chlorine produced electrolytically from sea-water on the site.

(viii) Radio-active effluents

In nuclear power stations, spent fuel rods are stored in ponds known as ageing ponds. The water in these ponds becomes contaminated with radio-active elements, and overflow from the ponds may be disposed of via the cooling-water system. Additional radio-activity may appear in the cooling-water because of activation or corrosion products in the coolant system, leakage of fission products through cladding, diffusion of tritium through stainless steel cladding, and activation of the cooling-water itself (Eisenbud, 1973). The water is generally decontaminated before discharge, but this is rarely 100 per cent successful. Concentrations in effluents are, however, extremely low (see chapter 7).

(d) **Aerial emissions**
Sulphur and nitrogen oxides discharged from very high power station and factory chimneys may reduce the pH of rain by forming sulphuric and nitric acids. Deposited on soils or snow and subsequently leaching into naturally acidic lakes and rivers these acids have been blamed for losses of fisheries in parts of Scandinavia and North America (Jensen and Snekvik, 1972; Likens and Bormann, 1974; Schofield, 1976; Scriven and Howells, 1977) (see chapter 7).

   In most countries few measures are taken to desulphurize fuels or emissions and if this became necessary, costs would be immense.

1.3. SUMMARY

Water is a vital resource for both hydro-electricity and thermal generation. Volumes used outstrip those in any other industry in most countries. The problems of heat disposal have led to a number of augmenting and alternative recycling methods, most of which also reduce the waste heat load to natural waters. A variety of chemical contaminants may be discharged from power stations though their volumes are relatively small compared to those of cooling-water discharges. Aerial emissions of gaseous oxides are alleged to cause pH reductions in rain and as a result biological problems in rivers and lakes in some regions.

Chapter 2 Effects of electricity generation on the physical characteristics of natural waters

2.1. INTRODUCTION

The composition of plant and animal communities (the biocoenosis) in any aquatic habitat depends upon a complex interrelationship between the physical, chemical, and biological components.

The relationships between aquatic organisms, communities, and their environments are dealt with in a number of excellent books, all of which have extensive bibliographies. These include works on running waters (Hynes, 1970), estuaries (Perkins, 1974), the sea (Kinne, 1970), sea-shores (Newell, 1970), and lakes (Macan and Worthington, 1951; Welch, 1952; Hutchinson 1957).

Both Macan (1963) and Hynes (1970) emphasize the significance of water movement and temperature in the formation of freshwater ecosystems. Macan (1963) stresses that 'water movement, whether it be the flow in rivers, or the action of waves beating upon a lake shore, is one of the main factors shaping biocoenoses, because of its effect on the substratum'. In tidal waters and in lakes, the action of waves on open shores is vital in forming the shore topography and substratum and hence the composition of the flora and fauna. In estuaries the scouring of the substratum by tidal currents is a vital factor. In most aquatic habitats fast water currents or wave action carry away fine particles of silt, leaving heavier particles, such as coarse sand, gravel, or rocks. All particles are not, however, free to move because of compaction or accretion. As a general rule, Hynes (1970) suggested that below current speeds of about 20 cm s^{-1}, silt and mud will settle out, while at velocities between 20 and 40 cm s^{-1}, stream beds would be expected to include a considerable amount of sand, possibly among stones and gravel.

In tidal situations the movement and settlement of substrate material depends on the angle of collision between coastline and waves, as well as the depth of water and the geology of local rocks. Usually on open shores waves strike obliquely, moving finer materials along the coast. Sand and finer material are deposited in sheltered bays or inlets where waves tend to hit the shore at right-angles (Yonge, 1949; Newell, 1970). In all waters silt and mud are deposited as water movement lessens. In rivers, fast waters with 'eroding' substrata and slow waters with 'depositing' substrata have characteristic plant and animal communities, many species of which are physiologically, physically, or behaviourally modified to survive in that habitat (Hynes, 1960, 1970). Similarly the various substrata in lakes and tidal waters have their characteristic and highly adapted communities (Macan, 1963; Newell, 1970).

Water temperature is also vitally important in determining the distribution, metabolism, and life-histories of aquatic organisms (Hynes, 1970; Kinne, 1970; Newell, 1970). Significantly the physical characteristics of any aquatic habitat most altered by electricity generating installations are usually water movement and temperature. Turbidity may also be altered with consequent effects on light penetration.

2.2. ELECTRICITY GENERATION AND WATER MOVEMENT

2.2.1. Stream discharge and current velocities

(a) Natural

Hynes (1970) states that 'the one thing which is certain about stream discharge is that it is irregular, and that any regularity of pattern it does show is largely a statistical phenomenon'. Although all types of natural rivers show widely fluctuating discharge patterns, small headwater streams are generally less stable in the short-term than large rivers. Also, spates in temperate rivers usually occur more frequently in spring and winter than in summer. Given the different climatic conditions, timing, and great variations in the magnitude of discharge, similar patterns of fluctuation occur in most rivers, though today it is hardly possible to find any river in its natural state.

The magnitude of discharge which has to be considered by power station design engineers varies enormously from country to country and region to region. As extreme examples, the discharge in the middle reaches of the River Trent, one of the large British rivers, ranges from about 140 to 560 $m^3\ s^{-1}$, while the mean low discharge for the Austrian part of the Danube is about 800 $m^3\ s^{-1}$ (Ottendorfer, 1975), and the Columbia River discharge ranges from about 1,400 to 18,400 $m^3\ s^{-1}$ (Trefethen, 1972). Some smaller rivers in Britain on which thermal power stations are sited have recorded dry-weather discharges as low as 0–0·2 $m^3\ s^{-1}$ (Langford, 1972; Langford, et al., 1979). Baxter (1961) showed that mean annual discharge in some British rivers was exceeded only for short periods in any one year. The volume and duration of the dry-weather flows are critical ecological factors where cooling-water is abstracted and discharged. In most natural streams and rivers, the flora and fauna have to withstand periods of low flow lasting up to several months, interspersed with spates which may last from periods of a few hours in an upland stream to weeks or even months in very large rivers. The seasonal discharge patterns vary with many factors. For example high mountain streams have low winter flows because of icing. Maximal flows occur in spring and early summer. In contrast rivers in temperate and sub-tropical regions have low-flow periods generally in summer as a result of low precipitation.

Maxima occur in winter and spring. In many streams the wetted area of substrate is drastically reduced during low flows, thus reducing the size of the available habitat. For example, at about 12 per cent of the mean discharge only 30–50 per cent of a small stream bed may be covered with water (Baxter, 1961). In larger rivers, however, the percentage of uncovered bed may be small even at 12 per cent of the mean discharge. In lowland rivers with U-shaped channels little exposure of river bed occurs even at zero flows (Langford and Bray, 1969).

Hynes (1970) discussed the various factors which affect current velocity in natural stream channels, including roughness of the bed, depth of water, and gradient. He states that 'natural stream flow is a complex process and in natural water bodies is always turbulent'. Current velocities in rivers can vary from nil up to over 600 m s^{-1} (Leopold, *et al.*, 1964; Hynes, 1970). For this discussion the most important fact is that, in general terms, current velocity is related to the discharge. Also the current velocity on the stream bed is usually lower than at the surface. Low discharge may lead to desiccation, to critically low current velocities, and to silt deposition which are all unfavourable to obligatory fast-water species on clean substrates (see chapter 4).

(b) Effects of hydro-electricity generating stations

Run-of-river hydro-electricity generation usually has marked effects on river flow and current velocities both upstream and downstream of the installation. The most obvious upstream effect is the alteration of a free-flowing riverine habitat into a lacustrine habitat and this will be referred to many times in later chapters. Downstream, the artificially controlled discharge patterns may be very different from those of the pre-impoundment regime. Several major symposia and books have dealt with the many hydrological and biological aspects of impoundments and the reader is referred to these for details not included below (Lowe-McConnell, 1966; Obeng, 1969; Hall, 1971; Acker-mann, *et al.*, 1973; Rzoska, 1976; Mordukhai-Boltovskoi, 1979; Ward and Stanford, 1979).

(i) *Dams*

The major structure in any impoundment scheme is the dam, either interrupting the natural flow of a river or controlling the outlet from an already existing lake. Man-made dams are not new phenomena in rivers. Hammerton (1972) reports that the first large dam on the Nile was constructed by Menes, the first king of Egypt, over 5,000 years ago, to allow him to build the city of Memphis. Today the Nile is almost fully controlled by a series of dams of which the Aswan High Dam is the most recent and ecologically the most controversial (Rzoska, 1976).

In North America, the Columbia River and its main tributaries are controlled by more than fifty major dams, though the first large dam was not built until 1927. Most major river systems in the world are impounded to a greater or lesser extent either for flood control, water conservation, navigation hydro-electricity, or some combination of these. Dams built specifically for electricity generation are, however, relatively recent. None of the major English rivers have mainstream hydro-electricity stations, though some existed in the late nineteenth and early twentieth centuries. Scotland, however, has some fifty-six hydro-electricity dams registered in the World Register of Large Dams (North Scotland Hydro-electric Board, 1973).

(ii) *Lake formation upstream of dams*

The closure of a dam across a river begins the process of lake formation. Depth of water increases as the dam is approached from upstream, and current velocities are reduced. Where many dams are installed on a river, the change to lake conditions may be almost total.

Apart from short reaches immediately below each dam the normal riverine habitat may be replaced by a series of lakes. In Sweden this occurs in most rivers as almost every available metre of head has been utilized for electricity generation (Pyefinch, 1966). In the Danube thirteen dams and hydro-electric stations in the Austrian section alone have reduced average current velocities from 2·5 to 0·05 m s^{-1} (Ottendorfer, 1975).

One of the main effects of the reduced current velocity is the increased deposition of finer materials, for example, mud, silt, and organic solids, on and around the normal stony or gravelly river beds. These silt deposits accumulate and flushing facilities are often incorporated to remove such deposits from behind the dam (see 2.2.3).

The physical effects are not restricted to the immediate vicinity of the dam. In many regions, lakes extend for long distances upstream, flooding land bordering the previous river channel. In upland Britain with its smaller rivers, hydro-dam lakes are rarely longer than 5–10 km and small in area, but in the larger river valleys of the world, dam lakes may extend for much greater distances and cover huge areas (see Lowe-McConnell, 1966; Ackermann, *et al.*, 1973; Mordukhai-Boltovskoi, 1979). For example, the Mica Dam on the Columbia created a lake 67 km long and the proposed Moran Canyon Dam on the Fraser River in Canada could create a reservoir extending 270 km upstream (Geen, 1974). Lake Volta in Ghana, formed by damming the Volta River, has an area of 8,482 km^2 and a shoreline of over 4,800 km. Once such a lake is formed, water movements differ completely from the previous riverine regime in that currents are limited to stratification and density currents, or those caused by winds, and these are generally of much lower velocities than in free-flowing systems (Viner, 1970a). On the shore itself wave action mainly determines the extent of erosion and deposition of sediments.

(iii) *Downstream discharge and velocities*

The effects of generation on river discharge and current velocities downstream of any hydro-electric power station depend on many factors including the daily pattern of generation, natural river flows, diversion schemes, and the amount of water allowed to by-pass or spill over to maintain a flow in the river below the dam, i.e. compensation water. Where hydro-electricity provides the basic daily supply (for example, in Switzerland, Norway, Austria) generation continues for 16–24 h each day. Where hydro-electricity only supplies power at peak demand periods generation may last for only 2–5 h daily, and not necessarily every day. In the former situation, downstream discharges can vary markedly but regularly, with low flows lasting only a few hours. In the second situation, low flows may last for several days or even weeks in periods of low power demand. In some schemes the minimal river discharge below a hydro-power dam may be *greater* than the minimum natural discharge prior to construction of the dam, because of enhanced compensation flows. Arrangements may also be made to vary compensation water releases seasonally to match migratory fish movements (Pyefinch, 1966), and other artificial releases may be made for dilution of pollutants or flushing of sediments.

In general, the long-term effect of well-planned hydro-electricity developments downstream is to suppress the natural extremes of spate and drought and make the *annual* flow patterns more uniform. Short-term fluctuations may, however, be quite violent, to the extent that anglers fishing downstream may be in danger when generation begins. In other situations, compensation flows may be so low that the downstream reaches are almost completely dry during non-generating periods. For example Trotzky and Gregory (1974) made detailed studies of discharge, velocities, and the fauna of the Kennebec River in Maine, U.S.A., below the Wyman hydro-electric power station. Discharge rates of less than $8 \cdot 5 \text{ m}^3 \text{ s}^{-1}$ occurred most nights for 1–6 hours. In contrast, high daytime discharge, during generation, reached $170 \text{ m}^3\text{s}^{-1}$. Current velocities varied diurnally from $0 \cdot 1 \text{ ms}^{-1}$ up to $0 \cdot 9 \text{ ms}^{-1}$. Approximately 25 per cent of the river bed was uncovered for up to 6 h each day. The river was described, from an ecological standpoint, as essentially two rivers *daily*, one consisting of riffles and pools at night, the other a long uninterrupted stretch of fast-flowing water during generation. The Kennebec may not be the most typical case, but given differences in timing, magnitudes of discharge, and current velocity, the pattern is similar in other situations. Compensation flows are often fixed by law or agreement at most hydro-electric power stations today to try to minimize the daily fluctuations (see chapter 9).

(iv) *Diversions and water transfers*
In some hydro-electricity schemes, water storage is enhanced by diversions from one lake or drainage-basin to another. Where a river is diverted, flows downstream of the diversion dam are usually drastically reduced, while in the receiving stream flows are increased. For example, plans to divert the Churchill River to the Nelson River in Canada would reduce the discharge in

the former by 80 per cent (Geen, 1975) and several such schemes in operation in Canada and other countries have already caused significant flow reductions.

2.2.2. Lake water levels

(a) Natural
The water level in most natural temperate lakes varies annually over a small range because of the temporary imbalance between inflow, evaporation, and outflow. Changes in water level are usually seasonal and slow and 'the fall or rise no greater than the resident animals can keep up with' (Macan, 1963). The area of lake bed exposed during low water levels depends on the slope of the shore, shallower slopes having greater exposure than steeper ones.

Lake Winnipeg in northern Manitoba has an annual range of about 2·5 m (Dickson, 1975). Tropical lakes can fluctuate much more as a result of long periods of evaporation and sudden heavy rains. Records of seasonal fluctuations of 5–7 m are not unusual. In most temperate lakes, levels usually vary seasonally over about 1·0–2·0 m.

(b) Effects of hydro-electricity generation
Where natural lakes are regulated for hydro-electric schemes the effect is usually to *increase* fluctuations in water level by drawing heavily on the water during dry periods and refilling during wet periods. The consumption of water from the lake (or reservoir) is known as drawdown. The rate at which the level falls depends upon the storage capacity of the lake in relation to generation requirements. Diversions and augmentation may allow the lake to fill completely between generating periods. In the extreme situation large areas of shore may be exposed for several months each year, or even for several years in very dry regions.

Lake regulation schemes may in some situations actually *reduce* seasonal fluctuations. For example, the regulation of Lake Winnipeg will raise the minimum water level by 1·5 m, but reduce the annual fluctuation from 2·5 m to 1·3 m (Dickson, 1975). Similarly, the closing of the W.A.C. Bennett Dam will reduce fluctuations in Lake Athabaska in Canada (Fig. 2.1). Drawdown in run-of-river reservoirs can vary from 1 or 2 m up to perhaps 20–40 m, depending upon the ratio of water demand to storage capacity.

(c) Effects of pumped storage schemes
Pumped-storage schemes, in their simplest form, involve two 'reservoirs' at different altitudes (Fig. 1.2). Either or both may be natural lakes. Where either lake has a small holding capacity in relation to generating requirement, water levels fluctuate *daily* over a wide range. In many cases the fluctuation may be more than 12 m each day, instead of the *annual* range of 2 m or less commonly found in natural lakes. In many areas, pumped storage schemes may also be combined with run-of-river hydro-electricity schemes, though Velz (1971)

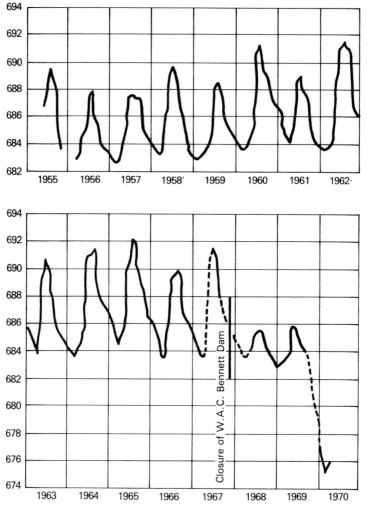

Fig. 2.1. Water levels in Lake Athabasca, Canada, before and after regulation, redrawn from Blench (1972).

suggested that the diurnal pulsation of flow in the lower reservoir could be detrimental to river flow. However this water could be used to augment cooling-water at thermal power stations.

2.2.3. Sediment transport and effects of hydro-electricity stations

Transport of sediments and silt is of considerable importance to bank erosion and rebuilding in rivers and to alluvium deposition in deltas and riverside washes (Einstein, 1972; Fraser, 1972; Hammerton, 1972; Bardach and Dussart, 1973; Glymph, 1973). The Nile, for example, carries 100×10^6

tonnes of silt each year. In the slower currents upstream of dams the finer silt and mud settle out prematurely at considerable loss to the deltas and alluvial plains. The floods and the resulting irrigation and deposition of alluvial soils were at one time so important to crop production in the Nile washlands that agricultural taxes were based on the height of the annual flood, and were waived if this was below critical levels (Hammerton, 1972). Although dams may destroy the annual flood patterns, the stored water and electricity supply do allow enhanced irrigation and cultivation over wider areas.

The turbidity of water spilling over dams or coming from turbine outlets is generally lower than in uncontrolled rivers except where phytoplankton blooms cause it to rise (Hynes, 1970; Bardach and Dussart, 1973). However the flushing out of sediment accumulated above a dam creates excessive turbidity and silting downstream. Nisbet (1961) described how the Rhone was badly affected by sediment flushing from a hydro-electricity dam and Gross, *et al.* (1978) showed that transport of sediment from upstream of dams during floods in the Susquehanna River was similarly increased. Hynes (1970) recommended that such flushing practices be discouraged, especially where pollution increases the amount of organic matter in accumulated sediments. As a peculiar example, damming of the Alpine reaches of the Danube has prevented the downstream movement of large rocks and boulders which used to be a feature of spates in the region (Liepolt, 1972).

Although it has been suggested that reduced sediment transport may be detrimental to estuaries and deltas, Perkins (1974) shows that in some British estuaries river-transport of silt is small compared with that carried by tidal and wave movements. Blench (1972) and Geen (1974, 1975), however, both consider that reduced sediment and nutrient transport have drastically reduced productivity of the marsh ecosystem of the Peace-Athabaska river deltas in Canada since the construction of the W.A.C. Bennett Dam.

2.2.4. **Ice-formation and hydro-electricity**

Reduced current velocities upstream and downstream of dams may enhance ice-formation. Liepolt (1972) states that surface ice is found more often on the Danube *upstream* of dams than in the free-flowing reaches. Gas exchange between the water and atmosphere may also be significantly reduced by prolonged icing, and deoxygenation may occur (see chapter 3). Where current velocities are reduced in shallow reaches *downstream* of dams, ice may form on the substratum in winter which may impede the flow and also scour the substrate when it melts and breaks up, removing large numbers of bottom-living animals in the process (Hynes, 1970; Fraser, 1972).

2.2.5. **Tidal movements and currents**

(a) **Natural**

Water movements in tidal waters are highly complex and influenced by many factors, including tidal amplitude, tidal phase, shore topography, winds, fresh-water inflows, and wave motion (Newell, 1970; Dyer, 1973; Perkins, 1974).

Maximum tidal amplitudes vary globally, from less than 0·5 m in some parts of the Mediterranean to 14–15 m in areas like the Bay of Fundy. The speed of tide currents up and down any shore varies mainly with the slope, i.e. faster movements over very shallow shores. These movements will also be modified by wind speed and direction.

Tidal currents, averaged over many tidal cycles, leave a residual horizontal drift, varying from 0·02 to 0·2 m s^{-1} in Britain depending upon location, time of year, and wind conditions (Spurr and Scriven, 1975).

In estuaries the direction and velocities of the main currents vary cyclically with the tides. This, together with the lateral vectors caused by incoming and outgoing tides makes the prediction of water velocity and direction at any one point a complex exercise. As examples of over-all velocities, the means for one area of Southampton Water (U.K.) range from 4·28 to 10·10 cm s^{-1}. As a more extreme example surface velocities of 45–120 cm s^{-1} and deep water velocities of 15 cm s^{-1} have been recorded in the Knight Inlet in British Columbia (Dyer, 1973). For further details the reader should refer to Sverdrup, *et al.* (1963) and Dyer (1973).

(b) Effects of tidal power schemes

There is, as yet, little global experience of the effects of operating tidal power stations on water movements. The physical effects of any one scheme can be generally predicted given the design of the scheme (Shaw, 1975). Normal tidal cycles and amplitudes will almost certainly be disturbed, at least 'inland' or 'upstream' of any dam or barrage.

The littoral zone in the reservoir will be reduced because the fill and empty regime will not have the normal daily tidal movement. Wave motion may be less as the dam will absorb much of the energy. In areas with wide expanses of mud flats, exposure may be reduced from twice daily (Shaw, 1971). Silt deposition will be enhanced and scour reduced upstream of a barrage or dam on an estuary. Downstream, increased current speeds near the power station outlets will increase scour. If the turbines generate on both incoming and outgoing tides scour will increase on both sides of the dam. At La Rance for example, increased current velocities near the outlets have prohibited shipping from using some areas and some sandbanks have been completely removed by scour (Cotillon, 1974).

2.2.6. Thermal generation and water movements

The abstraction and discharge of cooling-water usually causes changes in water currents in the vicinity of both intake and outfall. The extent of the changes and the current velocities depend upon the design and siting of the intake or outfall and the amount of water abstracted. Intake and outfall channels may have water velocities of up to 3 m s^{-1}, and be several kilometres long at some power stations. Often these high water velocities are produced in lakes or tidal waters where they would not normally occur.

The abstraction of water can result in flow reversal in rivers. For example, in some British rivers the cooling-water requirement at power stations may be equal to, or even exceed extreme dry weather flow. The discharge may therefore be drawn upstream to the intake, and recirculation occurs. Thus river currents may be in an upstream direction for up to several months in any year, in that particular reach (Langford, 1970, 1971b, 1972). In the larger rivers of the world recirculation is of little ecological significance.

2.3. ELECTRICITY GENERATION AND WATER TEMPERATURE

There have been many studies of water temperature in recent years either as backgrounds to ecological studies or for the planning of power station sites. These studies have varied in Britain from relatively simple thermometer or thermograph records (for example, Macan, 1958; Gameson, et al., 1959; DeTurville and Jarman, 1965; Edington, 1966; Crisp and Le Cren, 1970; Langford, 1970; Spencer, 1970, 1977; Smith, K., 1975; Maddock and Swann, 1977) to the more expensively instrumented studies (for example, Burnett, et al., 1974; Spurr and Scriven, 1975). In North America and Europe vast amounts of temperature data have accrued from studies prior to power station siting and studies of operating power station discharges. References to many studies are given in proceedings of several symposia, notably are Parker and Krenkel (1969); Krenkel and Parker (1969b); I.A.E.A. (1971, 1975a); E.D.F. (1977); Gallagher (1974); Gibbons and Sharitz (1974); and Esch and McFarlane (1976); and in reports such as those by Edinger and Geyer (1965) and Brady, et al. (1969).

The brief summary given below is intended mainly to illustrate the types of temperature changes caused by power developments, in relation to regional climatic and seasonal variability in natural temperature patterns.

2.3.1. Natural temperatures

The natural temperature of any water body depends on a wide variety of factors including geographical location (and hence climate), altitude, size, depth, amount of shade, and source of feedwater (Hutchinson, 1957; Macan, 1963; Hynes, 1970; Smith, K., 1972; Walker and Lawson, 1977). Cold and warm currents also alter temperature patterns in some tidal waters (Yonge, 1949; Kinne, 1970; Perkins, 1974).

(a) Running water

The temperature regimes of the larger rivers on which thermal power stations are sited are generally more stable than those of the smaller, shallower, and usually higher altitude streams which feed hydro-electricity reservoirs. Running-water temperatures ranges are approximately 0–15°C in sub-arctic regions, 0–25°C in temperate regions, 5–30°C in the sub-tropics, and 10–35°C in the tropics, though the ranges may be modified by altitude. In geothermal springs, natural temperatures reach 100°C and streams originating in such springs may run at 50–70°C for some distance (Brock, 1975).

In any one region or even locality, annual maxima and minima can vary from year to year, particularly in temperate zones. For example studies of the middle reaches of the River Severn in the English Midlands showed that in 5 years out of 10 the natural water temperature exceeded 21°C. Summer maxima varied from 19·6 to 22·8°C (Langford, 1970, 1971a). As an extreme example a difference of 12°C in river temperature was recorded between similar dates on successive years at the same place. On 10 June 1970 the natural temperature was 22°C. On the same date in 1971 it was only 10°C. Such differences, usually of smaller magnitude, occur in most years in most rivers as a result of localized climatic conditions. Annual ranges in large temperate rivers are normally between 20 and 30°C while in Arctic and tropical zones the range may be as little as 5°C. In the short-term, running-water temperatures can fluctuate widely. Diurnal ranges of 6–14°C have been recorded in small streams (Sprules, 1947; Macan, 1958; Mackichan, 1967). In larger rivers, however, diurnal ranges are much smaller. In the Severn, for example, the maximum daily range was about 3°C but on many days this was suppressed by high flows and changeable weather (Fig. 2.2) (Langford, 1970). Similar phenomena were shown by Smith, K. (1975) in the River Tees. The greatest rate of change in the Severn was about 0·5°C h^{-1} over about 4 h, but the average was usually less than 0·08°C h^{-1} (Langford, 1970). In small streams, rates of 1·0–1·5°C h^{-1} have been recorded (Sprules, 1947; Macan, 1958). In general, water temperature varies along valleys downstream of the source so that, at least in summertime, the increase in temperature is roughly proportional to the logarithm of the distance from the source (Hynes, 1970). Larger rivers are usually at more or less mean monthly air temperature at the point of measurement (Hynes, 1970). Spring-fed streams may be at a constant temperature and warmer at source in winter than in the lower reaches (Illies, 1952; Minckley, 1963).

Vegetation and shading can affect running-water temperature and rivers are often cooled during passage through tunnels, culverts, and forests (Macan, 1958; Edington, 1966; Brown and Krygier, 1967). Deforestation has been shown to cause increases of maximum temperature of 6·5–8·8°C in some streams (Brown, G., 1970; Brown and Krygier, 1970; Brown and Brazier, 1972).

Therefore, although there are some underlying patterns, it is difficult to predict stream temperatures at any point in space and time very accurately, as so many factors are involved. Given certain climatic conditions, however, it is possible to make approximate predictions using basic heat exchange data (Edinger and Geyer, 1965; Edinger, et al., 1968; Walker and Lawson, 1977). In free-flowing shallow streams there is usually little vertical or horizontal

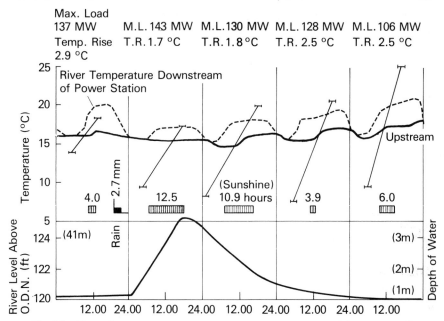

Fig. 2.2. Effects of a summer spate on water temperature upstream and downstream of Ironbridge power station, cooling water outfalls, River Severn (June 1966), redrawn from Langford (1970).

temperature stratification though there may be slight differences between top, bottom, and sides in larger rivers and streams (Neel, 1951; Shadin, 1956).

(b) Natural lakes

The temperature regimes of many natural lakes differ from those of fast-flowing waters in that vertical stratification occurs at least for some part of the year. Thus surface and bottom temperatures may differ by as much as 20°C (Welch, 1952; Hutchinson, 1957). As with rivers, however, lake temperatures vary with respect to region, climate, and altitude. Maximum surface temperatures in temperate lakes reach 20–24°C, while in the warmer regions, surface temperatures can exceed 30°C. In contrast Arctic lakes may never exceed 4°C at the surface at any time of year. Altitude affects temperature maxima. Pesta (1929), cited in Macan (1963), recorded maxima ranging from 3·2 to 17·5°C in various Alpine lakes, while Dussart (1955) showed that altitude differences of 200 m may result in temperature differences of up to 8°C in different lakes on the same day.

Thermal stratification in lakes (and in some tidal waters) is, according to Welch (1952), 'so profound and so far reaching in its influence that it forms directly and indirectly the substructure upon which the whole biological framework rests at least in the temperate zone'. Surface temperatures may show diurnal fluctuations of 2–3°C, though in some tropical lakes dense mats of plants inhibit heat exchange and suppress such short-term change (Dale and Gillespie, 1976).

In deeper temperate lakes a temperature profile in winter will show cold surface water and a slight increase in temperature with depth, usually to around 4°C in deeper parts (Hutchinson, 1957). As the surface warms during spring, the density of surface and deeper water becomes equilibrated and the lake becomes homeothermous as a result of wind mixing. Subsequent warming of the surface layers increases the density difference between top and bottom until even strong winds cause little mixing. The bottom water tends to remain at or around 4–6°C. This stratification persists through the summer until cooling occurs in the autumn and the lake mixes or 'overturns' again.

The thickness of the strata, viz. the *epilimnion* (surface), the *thermocline* (a transitional stratum usually with temperature decreasing at least 1°C for each 0·3 m depth), and the *hypolimnion* (bottom stratum), depends very much on the climate and physiography of the lake (Hutchinson, 1957). The development of stratification may be much more complex in some lakes where one, two, or no overturns may occur in any year. In larger lakes, such as the Great Lakes of North America, thermoclines occur intermittently and locally. Larger tropical lakes do not always stratify though smaller lakes show distinct stratification (Welch, 1952). Also stratification tends to be more consistent in eutrophic and mesotrophic lakes than in oligotrophic lakes. Under natural conditions a heat balance is established in water bodies, matching incoming solar energy with outgoing losses through evaporation, radiation, conduction, and convection (Sverdrup, *et al.*, 1963; Edinger and Geyer, 1965; Spurr and Scriven, 1975). The major natural heat input is from the sun.

Natural heat budgets are important to the design of systems for heat disposal or dispersal, especially where the receiving water body is not very large in relation to the effluent volume (Burnett, *et al.*, 1974).

(c) Coastal waters and the sea

Minimum water temperatures in the sea are lower than in fresh-water, reaching −2 or −3°C (Sverdrup, *et al.*, 1963; Crisp, 1964). In open oceans the annual range globally is from about −2 to 30°C while in inshore waters the range is −2 to 43°C. In polar regions annual fluctuations may be as low as 0·1°C while in temperate waters the range is about 16–18°C. In general, annual temperature ranges decrease with increasing depth (Kinne, 1970). Ocean surface waters show *diurnal* ranges varying from about 0·2–1·9°C depending on solar input. In extreme cases diurnal ranges of up to 3·0°C may occur (Kinne, 1970). As with lakes, seas show thermal stratification in summer, with the thickness of the surface stratum varying with region and physiography. At the equator, the temperature at the surface is approximately 26°C, while at 2,000 m depth, it is 3·3°C (Sverdrup, *et al.*, 1963). In other waters, the epithalassa is quite thin and overlies a thick (several thousand metres) layer of cold water. The surface layer (epithalassa) of the English Channel in summer is about 15 m thick and the thermocline about 2–3 m (Harvey, 1945). Close inshore, there may be no stratification because of turbulence and tidal currents (Perkins, 1974), though some nearshore areas and sheltered bays may be stratified in summer (Parkhurst, *et al.*, 1962). Most power stations extract their cooling water from, and discharge it into relatively shallow coastal waters. These waters usually have large annual temperature ranges, perhaps up to 20°C in some regions (Adams, 1969).

The intertidal zone shows the greatest annual seasonal and diurnal fluctuations as the substrate is exposed to the extremes of air temperature twice each day (see 2.3.1.e).

(d) Estuaries

As with all the other water bodies, estuary temperatures depend very much on geography, topography, hydrography, and climate (Dyer, 1973). In estuaries with large areas of exposed sand or mud the temperature at the edge of the incoming tide may be significantly altered by direct heat transfer to and from the substrate (Spencer, 1970). Temperature gradients can develop from inshore to offshore and the shallows may be several degrees warmer in summer and colder in winter than the deeper water. In some sheltered estuaries, temperature stratification occurs, much as in other waters (Dyer, 1973). In other estuaries, reverse stratification may occur when cold, less dense fresh-water floats on denser warm incoming saline water (Raymont and Carrie, 1964). The heat contribution from freshwater flows is generally relatively low in estuaries. For further details the reader is referred to Dyer (1973) and Perkins (1974).

(e) Substrate temperatures

For many organisms living on and in the substrata of rivers, lakes, and tidal waters, the temperature of the substrate may be more important than that of the overlying water and in fact may well be different. Temperatures ranging from −20°C to over 35°C have been recorded on some sea-shores during a year. Further, Fig. 2.3 shows that on a late summer day the temperature range can be 13°C in an intertidal rock pool and over 14°C on an exposed rock surface. Offshore the mean range was usually less than 5°C in the open water. Temperatures in temperate shores have smaller annual and diurnal ranges (Newell, 1970).

Diurnal temperature ranges decrease with depth in most substrata. For example Spencer (1970) showed that in the Blackwater estuary (U.K.), littoral substrate temperatures increased on a sunny day by up to 6°C at 1 cm depth. At 15 cm the rise was only 1·5°C. Johnson (1965) showed similar differences in a sand flat and concluded that animals living in the upper 1 cm were exposed to a daily range three times greater than that at 10 cm depth. In intertidal areas, substrate temperatures also vary annually over a much larger range than

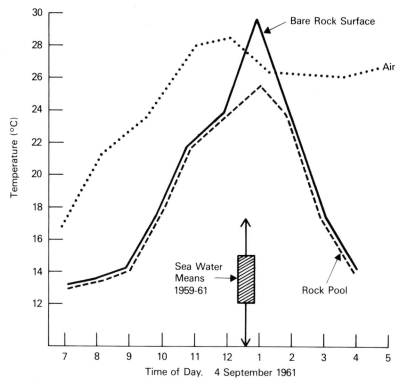

Fig. 2.3. The temperature range in the *Endocladia* habitat at Pacific Grove on 4 September 1961 compared with the sea-water means and extremes for Monterey, redrawn from Strickland (1969).

water temperatures owing to direct heat exchange with the air. Thus temperatures as low as $-20°C$ may be reached on Arctic shores while the upper limits may well be over $40°C$ in tropical regions.

In temperate lakes, temperatures tend to be around 4 to $6°C$ near the substrate surface for much of the year owing to stratification in the water (see 2.3.1.b), while in fast-flowing rivers the uppermost substrate temperature will be very near that of the water and have much wider annual and diurnal ranges than in lakes. Generally, any diurnal fluctuations which occur will be greater at the substrate surface than at depth.

2.3.2. Effects of hydro-generation

(a) Upstream of dams

The main temperature change upstream of a hydro-dam is from a riverine to a lacustine state, that is, the uniformly distributed but diurnally fluctuating regime is usually replaced by seasonal stratification (Fig. 2.4). The stability of the stratification depends on the regularity of flushing in the lake or reservoir, but in the larger manmade lakes of Europe, North America, and Africa, it is

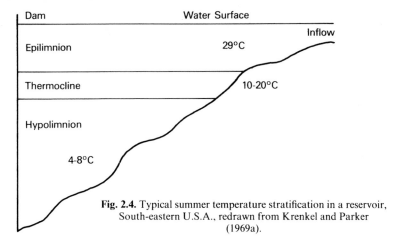

Fig. 2.4. Typical summer temperature stratification in a reservoir, South-eastern U.S.A., redrawn from Krenkel and Parker (1969a).

usually very stable for several months. Whatever the region, the previous temperature regime of the free-flowing river and its substrata is changed. In a temperate region it would be reduced from an annual range of 0–23°C to 4–8°C in the deeper parts of the reservoir. Near-shore and inshore temperatures may continue to approximate those of the river before impoundment. There is little doubt, however, that the original fauna of the main river channel would experience dramatic changes in temperature, i.e. warmer in winter and colder in summer.

The operation of hydro-electricity turbines influences stratification markedly, depending upon the depth of the turbine inlets (Wünderlich, 1971). These inlets are often in the hypolimnion in older stations, though later

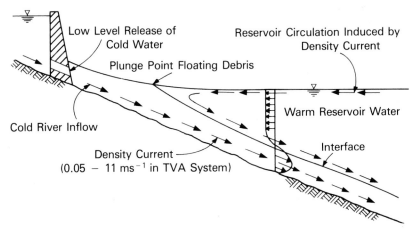

Fig. 2.5. Density currents due to low level (hypolimnial) release from an upstream to a downstream reservoir, redrawn after Elder from Kenkel and Parker (1969a)

developments have incorporated inlets and outlets at different depths mainly to minimize temperature and chemical changes downstream of the dam. The effect of a hypolimnial release from a stratified reservoir is to remove the colder water, expand the depth of the epilimnion, and lower the thermocline, as might be predicted (Orlob, 1969). With outlets at different levels, water may be abstracted from all the strata without disturbing the stratification if this is required. Hypolimnial depletion may also be alleviated by releases of cold hypolimnial water from reservoirs upstream (Fig 2.5).

The exposure of reservoir shores to the air during drawdown causes much greater fluctuations in substrate temperature than would occur in a lake with no drawdown. Prolonged summer drawdown leads to desiccation in the surface and deeper layers of the substrate, while winter drawdown results in substrates being frozen for some depth, particularly in colder regions (Quennerstedt, 1958; Andrew and Geen, 1960; Grimas, 1961, 1964; Geen, 1974, 1975; Kaster and Jacobi, 1978).

(b) Downstream of dams

In natural lakes, overspill water usually originates from the epilimnion and is usually at or around the temperature of local natural streams. Hypolimnial releases from a reservoir therefore alter the temperature regime of the downstream river considerably, both in the long- and short-term. In summer they usually cause a reduction in downstream river temperature, while in winter they may cause an increase of 1–5°C. The duration of the temperature change depends on the generating schedule at each power station. The magnitude and distance of the change depend on downstream dilution (Orlob, 1969; Jobson, et al., 1979).

The effects of a hypolimnial release decrease with distance downstream of the dam. For example, Ward (1974) showed that normal diurnal temperature ranges were reinstated in the South Platte River, Colorado, some 5 km downstream of the Cheeseman Dam, although the maximum fluctuation at the outlets was only 1–2°C. The first 9 km below the dam was, however, ice-free throughout the winter.

Reduced flow caused by hydro-dams may enhance heating by the sun resulting in much higher summer temperatures than would normally be experienced (Trotzky and Gregory, 1974). Jaske and Goebel (1967), however, found that the low-head dams on the Columbia River did not affect *average* annual water temperatures though ranges were reduced and the period of maximum temperatures shifted from August to September.

The effect of hydro-generating stations with hypolimnial outlets is therefore to reduce the annual downstream temperature range near the dam from say 24°C in temperature regions to 8°C or less though the effect will depend on dilution. In the short-term, diurnal fluctuations may be very complex, with short periods of lower or higher temperatures (depending on the season) during generating periods, interspersed with periods when natural water temperature is related to air temperature, i.e. when generation ceases (Jobson, et al., 1979).

Banks (1969) concluded that the building of any river dam will always affect the temperature regime downstream if the impoundment is large and deep enough to stratify, and the deeper the impoundment the more drastic the changes will be.

The incorporation of multi-depth inlets and outlets together with surface overspill has however minimized downstream temperature changes at the more recent power dams.

Management of the hydro-electricity generating station and the balance between epilimnial and hypolimnial releases are also highly significant factors in reducing the downstream temperature shocks. In Japan, cold-water releases from dams may affect the rice crop. Temperature balance here is therefore of both ecological and economic significance (Lavis and Smith, 1972). Thermal discharges from power plants may partly restore temperatures in rivers cooled by hydro-generation (Jobson, et al., 1979).

2.3.3. Effects of tidal power generation

The changes caused by impounding tidal areas are not fully known. However, it is likely that instead of a twice daily exposure, substrates 'upstream' of dams would be exposed less often and therefore not be heated and cooled as frequently as in the natural system (see 2.3.1.e). In the short-term, temperatures would therefore fluctuate less. In an impounded estuary, stratification may develop upstream of a barrage though it may be complicated by freshwater flows and the restriction of the incoming saline water.

2.3.4. Effects of pumped storage

There are energy losses in both run-of-river hydro-generation and pumped storage generation, and some of these are generally passed to the lower reservoir as heat, to be dissipated by radiation and evaporation. In most cases heating effects are not significant owing to the massive dilutions involved. However, where the ratio of installed generating capacity to reservoir area is high, temperature rises in lower reservoirs may be larger. For example, in the Cruachan pumped storage scheme in Scotland, the ratio is $0 \cdot 1$ MW ha^{-1} and water temperature rises are hardly measurable. In the Ffestiniog lower reservoir (9 MW ha^{-1}) average temperature rises of $1-2°C$ are recorded, but in the proposed Dinorwic scheme where the ratio is 27 MW ha^{-1} average water temperature rises in the lower reservoir could be more than $3 \cdot 5°C$ above ambient, with extreme maximum temperatures possibly reaching $30°C$. Downstream river temperatures would rarely rise by more than $1-2°C$ owing to mixing with cold compensation water.

Temperature effects of pumped storage generation have usually been considered insignificant, though obviously at Dinorwic this is not so. The short-term exposure of the reservoir shore (littoral) substrates during drawdown will also expose organisms to greater ranges in temperature, though there is less likelihood of long-term desiccation or freezing than in run-of-river reservoirs, if operation is regular and frequent.

2.3.5. Effects of thermal generation

The temperature rise in cooling-water (ΔT) between intake and outfall at British power stations is usually in the range 8–12°C but in other countries ΔT may be up to 15 or 16°C (Schubel and Marcy, 1978). Generally, however, efficient operation of power plants dictates that ΔT should be 15°C at maximum. The effect of the cooling-water discharge on the receiving water depends on many factors, including dilution, water currents, shape, size, and construction of the outfall, topography of the bed, and salinity gradients. Added to this complexity are the plant-operating conditions, which can vary enormously depending on power demand and maintenance programmes (Langford, 1972, 1974). In temperate regions, maximum electricity demand tends to be in winter, but in sub-tropical regions such as the southern parts of North America, peak demand is in summer because of air-conditioning equipment (Langford, 1972). In tropical regions, demand probably remains more or less constant throughout the year. Plant maintenance also occurs at different times in different climates. In Britain, the machinery is overhauled in summer, resulting sometimes in less efficient and older plant coming into operation. Heat disposal at these tends to be greater than in the more modern stations. In sub-tropical regions, in contrast, overhauls occur mainly in the cooler seasons. Thus, at any one site, annual operating and maintenance schedules can greatly influence the amounts of heat being rejected to the aquatic environment.

(a) Discharge temperatures and ΔT

Maximum discharge temperatures at thermal power stations depend mainly on the intake temperature, assuming full output conditions. In Britain outfall temperatures exceeding 30°C have been recorded on many occasions. In larger rivers such as the Severn and Trent high ambient temperatures (21–24°C) have resulted in outfall temperatures of 31–33°C. In smaller rivers temperatures of 32–34°C have been recorded because of recirculation. The highest published record is 36·5°C at the Castle Donington outfall on the River Trent under experimental conditions during the lowest flows for 200 years (Alabaster, 1969).

In other countries temperatures of 45–50°C have been recorded near the Savannah River Nuclear Plant in Georgia (U.S.A.) (Gibbons, *et al.*, 1975), and Mesarovic (1975) quotes maximum summer discharge temperatures of 40–45°C in Yugoslavian rivers. In temperate regions summer discharge temperatures are usually between 25 and 32°C (Langford, 1972; Spigarelli, 1975; Hannon, 1978;) but in more extreme climatic regions temperatures often reach 35–45°C as a result of high natural temperatures (Coutant, 1962;

Churchill and Wojtalik, 1969; Profitt, 1969; McNeely and Pearson, 1974; Kamath, *et al.*, 1975).

In Britain, high power demand during dry winter periods has resulted in parts of some rivers, notably the Trent, Witham, Great Ouse, and Lea, running at *summer* temperatures (i.e. 17–22°C) for several kilometres downstream of power station outfalls over many years (Alabaster, 1969; Langford, 1972; Severn–Trent Water Authority, 1978). The increased use of cooling towers in the last 20 years has however resulted in considerably reduced temperatures, in most rivers.

In the sea, high ambient temperatures can also lead to high discharge temperatures, especially where cooling-water is drawn from surface layers of near-shore waters. Thus in regions such as Florida and Puerto Rico, maxima of 40–42°C have been recorded at outfalls (Thorhaug, *et al.*, 1974; Thorhaug, 1974; Kolehmainen, *et al.*, 1975). In temperate zones maximum temperatures of up to 27–33°C are more usual. Naylor (1965a) however recorded maximum temperatures of 23°C in winter and 37°C in summer at a power station discharging into an enclosed dock in Britain.

(b) Areas of effect (mixing zones)

The area of detectable temperature increase or 'mixing zone' around any thermal discharge outfall depends on many factors in any situation and the definitions of 'mixing zone' vary (Mount, 1971; Jeter, 1977; Macqueen and Howell, 1978). Mixing depends to a great extent on the angle and depth of the discharge and on the size, and turbulence, of the water body. For example the warm effluent from Ironbridge power station was normally thoroughly mixed with river water 300–400 m downstream of the outfalls mainly because of turbulent riffles and the angle of discharge in the river (Langford, 1970, 1971a). Fully mixed temperatures rarely exceeded 27–28°C though a ΔT of up to 8°C could be detected 2 km downstream at times of low flow. In the River Trent similar mixing occurred in the faster reaches but not in the slower, deeper ones (Alabaster, 1962, 1969; Langford, 1971b). In deep rivers, warm effluents tend to stratify, soon after the point of discharge (Fig. 2.6), and the temperature of the river bed is not usually raised.

Lateral, or horizontal stratification also occurs in rivers where an effluent discharges at an acute angle to the flow (Fig. 2.6). It also occurs in large rivers where the downstream flow is large enough to restrict the warmer water to the edge of the channel and lateral mixing is slow (Nakatani, *et al.*, 1971). Generally, the mixing zone becomes larger as river flow decreases. In some smaller British rivers recirculation together with wind has resulted in heated areas of surface water 3–4°C above ambient occurring up to 0·5 km *upstream* of power station intakes (Langford and Aston, 1972).

In lakes, the area of the mixing zone also depends to a great extent on outfall siting and design, topography, weather, and many studies of plume configurations have been made (E.E.I., 1969; I.A.E.A., 1971, 1975a; Hannon, 1978).

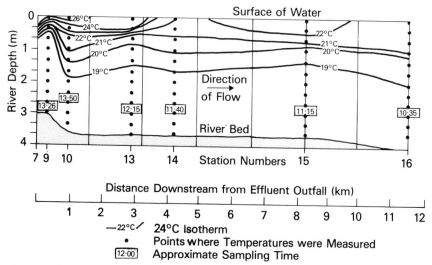

Fig. 2.6. Temperatures along the centre of River Thames at Earley, 20 August 1958 showing stratification of heated effluent water from a power station, redrawn from Alabaster (1969).

Surface mixing zones 5–6 km² in area have been recorded at some Great Lakes power stations (Effer and Bryce, 1975; Kyser, *et al.*, 1975) but these areas (i.e. above 1°C ΔT) are small compared with the total area of water. Biologically, however, their local significance may be considerable. In smaller lakes, such as Lake Trawsfynydd in North Wales, or the Konin Lakes in Poland, the low dilution and high water-use by a power station result in a very large proportion of lake being warmed, so that *intake* temperatures may be above ambient for much of the year. Trawsfynydd Lake, with an area of 550 ha and mean depth of 6 m, acts as the cooling pond for Trawsfynydd nuclear power station. The total lake volume is circulated through the power station cooling-water system every 48 hours. Intake temperatures range from about 8 to 17°C and outfall temperatures 17–32°C over the year (Whitehouse, 1971b). Heated lagoons and channels comprise over 10 per cent of the lake area and here some parts of the lake bed may be warmed, though marked vertical stratification occurs with increased distance from the outfall (Burnett, *et al.*, 1974). In the Konin Lakes, intake temperatures are always 2–5°C above ambient and discharge temperatures of 37–38°C have been recorded. In a Texas reservoir with an area of only 330 ha serving a 700 MW fossil-fuelled power station, summer surface temperatures near the outfalls reached 42·2°C (McNeely and Pearson, 1974). The lake bed was heated by several degrees in spite of marked stratification.

Stratification of thermal discharges in lakes may increase the temperature difference between epilimnion and hypolimnion in summer by 8–15°C. The extent and stability of the stratification may, however, depend on the outfall design and siting (see 2.3.5.a).

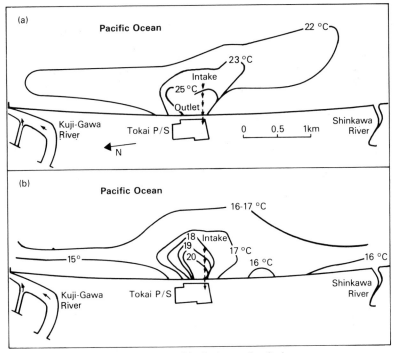

Tokai: Temperature Distribution on Sea Surface

(a) Date: 13 September 1968
Flow Rate of Condenser Cooling Water:
   13.3 m³ s⁻¹
Intake Temperature: 21.0 °C
Sea Water Temperature (Surface): 22 °C
Outlet Temperature: 28.8 °C
Wind Direction: NW to NNW

(b) Date: 29 November 1968
Flow Rate of Condenser Cooling Water:
   10.0 m³ s⁻¹
Intake Temperature: 18.8 °C
Sea Water Temperature (Surface): 16-17 °C
Outlet Temperature: 25.4°C
Wind Direction: NE to NNE

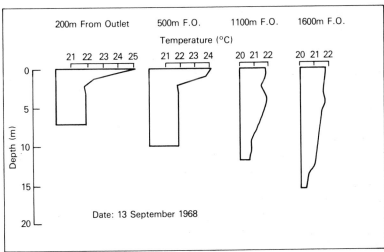

Tokai: Vertical Temperature Distribution in Sea Water

Fig. 2.7. Horizontal and vertical temperature distribution in the vicinity of a marine thermal discharge, Tokai, Japan, redrawn from Yoshioka, *et al.* (1971).

In estuarial and coastal waters similar effluent stratification may occur. Here again wind and weather conditions can markedly affect mixing (Fig. 2.7) (Yoshioka, *et al.*, 1971; Spurr and Scriven, 1975; Golden, 1975). Tidal movements also affect stratification and in some situations where there is not complete tidal flushing, warm water may oscillate within an estuary for long periods (DeTurville and Jarman, 1965). Where a thermal discharge crosses a flat shore the areas exposed at low tide are subjected to undiluted cooling water, but at high tide this is dispersed and diluted (Spencer, 1970). Offshore discharges may be distributed onshore (or, in an estuary, upstream) by incoming tides, while the ebb tides do the reverse (Swain and Newman, 1957). In tidal areas where fresh and saline water occur together (Raymont and Carrie, 1964; Carstens, 1975), the vertical distribution of a thermal plume may be affected by salinity gradients. For example saline thermal discharges in Southampton Water were found to be trapped between a surface layer of cold fresh-water and a bottom layer of cold saline-water (Pannell, *et al.*, 1962; Raymont and Carrie, 1964). Cooling was slower because of the lack of surface heat exchange.

In Norwegian sill-fjords, where natural barriers and freshwater in-flows trap pockets of saline-water and cause unusual salinity gradients, the rate of cooling of heated saline discharges is slow (Carstens, 1975). Where thermal discharges are released to partly enclosed bays or docks, plumes may mix or disperse slowly and high temperatures may occur on the bottom (Naylor, 1965a; Roessler, 1971).

(c) Diurnal and short-term fluctuations

When operating at full output, thermal power stations produce constant temperature rises (ΔT). However, considerable variations can occur in operating schedules at any one power station. The age of the machinery, fuel availability or costs, fluctuating demand, breakdowns, and management or government policies can dictate whether a power station operates constantly, partially, or only intermittently.

Usually older plant operates for about 10–16 h per day (two-shifting), depending upon demand. Because of such operation the diurnal temperature fluctuation in the River Severn, downstream of the Ironbridge 'A' outfalls, varied from 0·5 to 8°C. Ambient temperature fluctuations were 0–3·5°C (Langford, 1970). Fluctuating flows and weather conditions affected both upstream and downstream temperatures (Fig. 2.2). Fluctuating electricity demand was reflected also in the marked differences between weekdays and weekends. McNeely and Pearson (1974) showed that diurnal fluctuations of 5·5°C occurred at the surface and 1·5°C on the bottom, near the outfall of the power station on North Lake, Texas.

At some power stations in the U.S.A., ecological problems have occurred where power plant failure has caused outfall temperatures to *fall* by 10–15°C in a very short time. In winter such a sudden decrease can be lethal to acclimatized organisms as will be seen in chapter 5. In Britain, increases in oil prices, government policy on coal use, and decreased demand for electricity have meant that some oil-fired stations operate much less than was originally intended and thermal discharges may be intermittent or fluctuating. For example, the 2,000 MW Fawley power station was operating on very low output on some days in 1978 and 1979 resulting in a relatively small thermal discharge instead of a steady $65 \text{ m}^3 \text{ s}^{-1}$ at a ΔT of $8{-}12°C$. Reduced night-time output at Fawley also caused a fall in outfall temperature of approximately $6°C$. Such changes can produce wide diurnal variations in small areas of coastal waters where the natural diurnal range is $1{-}2°C$.

The stability of the temperature changes near a power station outfall may thus be strongly influenced by economic and political decisions in addition to hydrographic and climatic conditions.

(d) Effects of intake and outfall siting and design on thermal plumes

The area of the mixing zone in any water body is affected by the siting and design of the cooling water intakes and outfalls. Generally, the siting and location of each is such that, unless de-icing is required, the effluent should not encroach on the intake (Effer and Bryce, 1975), though as we have seen recirculation occurs in rivers (see 2.2.6). It is more or less universally agreed by engineers that surface dispersal is the most efficient method of losing heat from a water body (Edinger and Geyer, 1965; Krenkel and Parker, 1969a; Nakatani, et al., 1971; Spurr and Scriven, 1975). Surface dispersal may not always be the most advantageous *ecologically*, especially where the warmest water may drift on to a nearby shore, or very warm surface water may affect animals and plants living near the surface. In recent years, therefore, attempts have been made to discourage stratification of thermal discharges by rapid mixing (Gartrell, et al., 1971). At a Mississippi power station a side-jet surface outfall was replaced by a multi-port diffuser (pepper-pot) which ensured rapid mixing (Eiler and Delfino, 1975). Before the diffuser was installed, areas of the Illinois shoreline of the Mississippi were subjected to water temperatures up to $8.4°C$ above ambient 200 m downstream. After installation, no area more than 183 m downstream of the outfall was warmed by more than $1{-}2°C$ above ambient.

In stratified waters, intakes may be sited at depth taking in cooler, 'hypolimnetic' water and discharging effluent at much lower temperatures than if surface water has been used (Mangarella and Van Dusen, 1973). This type of system may also affect the chemistry and nutrient distribution within lakes or coastal waters (see chapter 3).

Temperatures may also be reduced by the release of hypolimnetic water from upstream reservoirs in impounded rivers, or by operation of hydro-electricity facilities upstream in peak summer periods (Parker and Krenkel, 1969; Jobson, et al., 1979). The combination of thermal, hydro-, and pumped-storage power plant operation to alleviate environmental effects is discussed at length by various authors (see Karadi, et al., 1971).

(e) Effects of enhanced cooling on thermal discharges

The effect of cooling towers or ponds is to reduce the amount of heat discharged to the receiving water, because most of this heat is lost through evaporation.

As examples a typical British 2,000 MW tower-cooled power station abstracts some $2 \cdot 0 \, m^3 \, s^{-1}$ (40 m.g.d.) and returns about $1 \cdot 4 \, m^3 \, s^{-1}$ (28 m.g.d.) at a temperature of 6–9°C above intake temperature. In contrast an older direct-cooled 200 MW station would abstract and return some $12 \cdot 5 \, m^3 \, s^{-1}$ (250 m.g.d.) at discharge temperatures of 6–10°C above ambient (Langford, 1970). In a 2,000 MW estuarine or coastal sited direct-cooled station the whole circulation of some $65 \, m^3 \, s^{-1}$ (1,300 m.g.d.) is returned to the receiving water at temperatures of 6–10°C above ambient. The effect of enhanced cooling is also to reduce the area of the mixing zone, and in British rivers cooling water discharges from closed-cycle 2,000 MW stations are often not readily detectable more than 1–200 m downstream of the outfalls (Langford, 1975; Langford and Daffern, 1975; Langford, et al., 1979).

(f) Infra-red thermal mapping

A technique used for the determination of surface plume dispersal during the past 10–15 years has been infra-red aerial photography or thermography whereby the heated water is shown on a photograph as an area of different colour. For example, in black and white photography it may appear lighter than the ambient water. Fig. 2.8 shows the dispersal pattern of plumes from a coastal station in Britain at various states of the tide (Spurr and Scriven, 1975). Although thermal mapping gives adequate snapshots of plume dispersal direct measurements are necessary to obtain absolute values (Moore and James, 1973; Marmer, et al., 1975). Thermal mapping only indicates surface dispersal, and vertical stratification cannot be assessed by this method.

2.3.6. Heat balance, heat budgets, and plume modelling

The 'energy-budget' or 'heat-budget' approach for predicting temperature changes was first used by Schmidt (1915) to estimate normal ocean evaporation and is widely used to predict the cooling-capacity of waters receiving thermal discharges (Edinger and Geyer, 1965; *Modern Power Station Practice*, 1971; Walker and Lawson, 1977).

The heat balance in a water body is established between the incoming energy from the sun and outgoing heat losses. Heat is dissipated by evaporation, long-wave radiation, and a combination of conduction and convection. Spurr and Scriven (1975) used the equation of Cummings and Richardson (1927) for expressing the heat balance of a natural lake:

$$I = B_n + E_n + C_n + W_n + S_n$$

where I is the natural heat input from the sun, which is the sum of the total

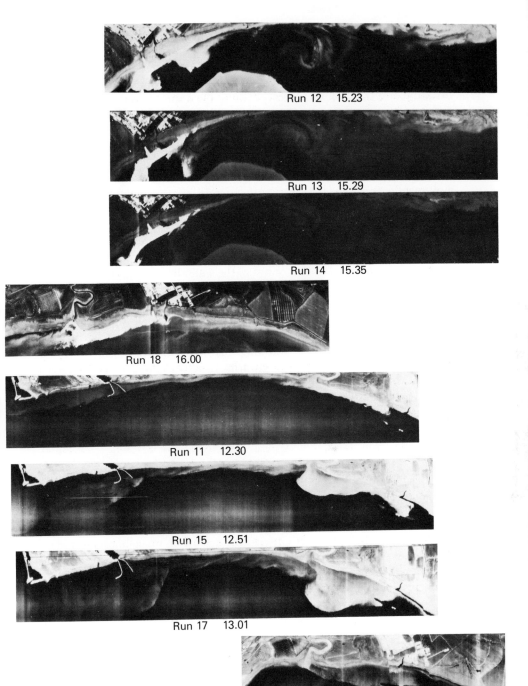

Run 12   15.23

Run 13   15.29

Run 14   15.35

Run 18   16.00

Run 11   12.30

Run 15   12.51

Run 17   13.01

Run 21   13.19

Fig. 2.8. Infra-red thermographs of cooling-water discharge from a coastal power station showing movement of plume as tide changes.

short-wave radiation reaching the water surface and the long-wave radiation from the atmosphere. The short-wave radiation is a function of the sun, the presence of clouds, and the reflectivity of the water surface. Long-wave radiation is a function of air-temperature, the atmospheric vapour content and the presence of clouds.

B_n is the heat loss by long-wave back radiation and depends on the temperature of the water surface.

E_n is the heat loss by evaporation, i.e. the latent heat of evaporation dissipated by the mass of evaporated water. It is a function of the air movement over the water surface, surface water temperature, the atmospheric temperature, and humidity.

C_n is the heat loss by conduction and convection, i.e. the sensible heat loss from the body of water to the atmosphere. It can be expressed as a function of evaporation.

W_n is the heat advected into or out of the water body and is the net energy gained or lost through flows into or out of the water body, for example, rainfall, surface run-off, inflowing streams, outflows, and seepage. The effect of these on the total energy budget depends on their magnitude and temperature in relation to water body temperature and volume.

S_n is the heat consumed warming the water body indicated by the temperature difference as the water warms.

When heat is supplied to the water body by thermal effluents, this natural heat balance is destroyed and a new balance has to be established between the incoming solar heat plus effluent heat and the increased rates of loss by evaporation, radiation, and sensible heat loss. The heat balance equation is then written as

$$I + H = B_a + E_a + C_a + W_a + S_a$$

where a indicates the artifically heated condition and H the heat rejected by the power station.

Given this equation, there are many methods of deriving the data for each term. Spurr and Scriven (1975) discuss the derivation of each term in order. Milanov (1973) shows the methods of deriving values for each term from climatic conditions.

Edinger and Geyer (1965) show the mechanisms by which heat is transferred across water surfaces and Bergstrom (1968) demonstrates the relationship between the heat dissipation rate and the increase in water surface temperature over natural ambience. Heat dissipation rates are higher for a given temperature rise in summer than in winter and dissipation by evaporation is higher at the same time.

Krenkel and Parker (1969a, b) consider that Bergstom's calculations support the contention that heated effluents should be discharged in their most concentrated form for quickest dissipation, assuming of course that ecological considerations are neglected.

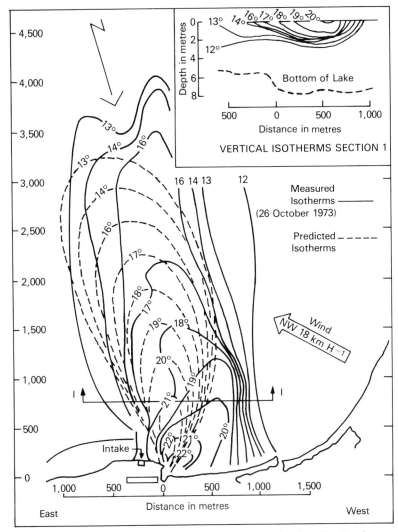

Fig. 2.9. Pickering Nuclear Generating Station. Measured and predicted surface temperatures due to thermal discharge, redrawn from Effer and Bryce (1975).

Milanov (1973) has used the heat-budget approach together with atmospheric and climatic data to estimate the areas of warmed water which would be more than 0·1 °C above ambient at different latitudes as a result of a 2,000 MW waste heat discharge with temperatures of 10 °C above ambient. Warm water plume areas decrease with latitude in January because of low air temperature and increase in April, July, and October. Except in relatively small bodies of water receiving discharges, the heat balance and over-all predicted temperature rise may be ecologically of little importance. The ecological significance of a discharge depends more on the areas affected by various temperature increments, and the mobility of the isotherms under varying conditions is the primary requirement for predictions (Fig. 2.9) (see chapter 5). Many studies have been carried out using data from existing power plants. Computation of predicted temperature changes is now standard practice for most proposed sites and many studies are continuing to produce improvements to mathematical models (Krenkel and Parker, 1969a; Brady, *et al.*, 1969; Young, *et al.*, 1971; Wills, 1972; Hindley and Miner, 1972; Naudascher and Zimmermann, 1972; Sweers, 1974; Paul and Lick, 1974; Burnett, *et al.*, 1974; Spurr and Scriven, 1975). These are only a selection of the available literature but all have bibliographies to which readers can refer.

2.4. SUMMARY

The physical changes caused by hydro-electricity schemes in rivers are, without doubt, fundamental and permanent, at least upstream of a dam. Water flow, sedimentation, degree of light-penetration, and temperature are all entirely changed from those of previous free-flowing river regimes. The changes permanently affect the whole of the relevant river reach whatever its size. Downstream effects on flow and temperature depend very much on the depth of turbine intakes, the balance of 'compensation' and 'generation' water, and the operating schedule of the power station. The effects of thermal generation and heated discharges on flow and temperature are less likely to involve the whole of a river, lake, or tidal water, unless these are particularly small. Even then such features as horizontal and vertical thermal stratification can restrict the actual zone of influence of the discharge. Further, heat is relatively quickly dissipated and high temperature zones (mixing zones) rarely extend over very large areas. Social as well as hydrographic conditions can affect heat inputs. It is important to note that the basic effect of thermal discharges from nuclear power stations is similar to that of other thermal stations, except that in some cases ΔT may be up to 5°C higher and volumes used may increase by 50 per cent per MW output.

Very generalized predictions of the effects of any thermal discharge or a hydro-electricity scheme on the temperature regime of any water body may not be adequate for ecological studies. Detailed studies are usually necessary for predictive work.

Chapter 3 Effects of electricity generation on water chemistry

3.1. INTRODUCTION

The chemical characteristics of natural upland streams are mainly a cumulative result of atmospheric precipitation, the solution of mineral matter from the rocks and soils from which they originate and over which they flow, and of atmospheric gases dissolved in the surface water (Cummins, 1972; Golterman, 1975). As streams flow towards the sea their chemistry alters considerably. Mineral concentrations increase and gaseous exchanges alter because of natural, physical, and chemical factors, the effects of the biota and the physical, chemical, and biological results of man's activities (Hynes, 1970). Lakes, usually situated at some point within a natural river system can cause further chemical alterations, usually in dissolved gases and minerals (Hutchinson, 1957). The chemistry of the sea is influenced by the cumulative effects of the chemistry of the inflowing rivers, precipitation and its own evolved chemical, physical, and biotic factors (Hood, 1963; Sverdrup, *et al.*, 1963). Of all these habitats, the sea tends to be the most stable environment chemically though minor local variations occur as a result of specific geographical, geological, or human influences (Yonge, 1949).

Small, freshwater spring outflows are often chemically stable, but streams, rivers, estuaries, and lakes show complex temporal chemical fluctuations both in the long and the short term. Many streams have characteristic annual and diurnal patterns in the concentration of dissolved minerals and gases, and although fluctuations may be less pronounced, diurnal variations in dissolved gases also occur in lakes and tidal waters.

Considerable chemical changes also occur as a result of high river discharge rates (Hynes, 1970).

Apart from these temporal variations the chemical properties of fresh waters vary with geology and geography. Even within small regions, the chemistry of the smaller water bodies can differ significantly (Hutchinson, 1957; Macan, 1963; Ruttner, 1963; Hynes, 1970; Cummins, 1972; Golterman, 1975).

The assumption must be therefore that although streams, rivers, and lakes may be categorized in various ways, almost every water body has some chemical characteristics by which it will differ slightly even from a near neighbour.

51

Macan (1963) noted that although there are ninety-two elements dissolved in natural waters, chemical analyses rarely cover more than a few, and rarely the more complex chemicals such as pesticides or vitamins which may be vital.

The increase in pollution control throughout the world has resulted in a massive increase in chemical analyses of natural waters and many of these data are not even compiled or summarized once collected. In Britain, data have been collected by water authorities (for sewage disposal, pollution control, and water supply), by many industries, and by some local authorities for various purposes, over 20–50 years at hundreds of sites on rivers and estuaries. At some power stations, river water used for cooling has been analysed at daily to monthly intervals for up to 50 years. Attempts are made by some institutions to collate and publish processed data (Trent River Authority, 1964–74). Given that such data are produced by most developed countries, immense amounts of detailed chemical information must be stored throughout the world at the present time.

The brief outline of water chemistry given here will follow a similar pattern to others (Macan, 1963; Hynes, 1970; Cummins, 1972) in that it will cover mainly those chemical constituents believed to be of most biological significance. Attempts to correlate the distribution of organisms with natural water chemistry have, as yet, met with limited success (Macan, 1963). The chemical constituents generally considered to be of most biological significance are oxygen, carbon-dioxide and several other gases, calcium, pH, total conductivity, dissolved solids, carbonates, metals and metal salts, and organic matter. Hydro-electric and thermal power generation may have effects on these. In addition, thermal generation involves the use of antifouling chemicals, and effluents may include these or their residues, together with ash, fuel, and sewage liquids (see chapter 1).

Further complications arise from synergisms, that is where chemicals which although toxic when present separately in water, become much more toxic when in combination. Their combined toxicity is therefore much greater than would be expected from the sum of component toxicities.

3.2. DISSOLVED GASES IN NATURAL WATERS

The naturally occurring gases most significant to this subject are probably oxygen, nitrogen, carbon dioxide, and ammonia. Dissolved gases in natural waters originate from the atmosphere, from substances or organisms in the water and in the substrate and from inflows. Soluble gases in contact with water generally dissolve until a state of equilibrium is reached in which the solution and emission of gas are balanced. Henry's Law expresses total solubility, that is *the concentration of a saturated solution of a gas is proportional to the pressure at which the gas is supplied*.

The solubility of any gas in a water body is dependent on many factors including partial pressure (that is mixture with other gases), temperature, concentration of dissolved solids, water-vapour content, saturation level of the water, and turbulence or wave action (Hutchinson, 1957). Solubility differs even when pressures of different gases are equal and different solubility constants apply for each gas in solution.

3.2.1. Oxygen

The oxygen concentration in any surface-water body depends on 'oxygen balance' or the 'oxygen economy' (Koppe, 1974). Thus, the consumption of dissolved oxygen by the chemical and biological constituents of the habitat is balanced against the uptake from the atmosphere and by input from the photosynthesis of aquatic plants.

The oxygen content of natural waters, its importance, and the factors which affect it, are discussed by several authors (Welch, 1952; Hutchinson, 1957; Klein, 1962; Macan, 1963; Sverdrup, et al., 1963; Hynes, 1970; Perkins, 1974) and only a brief summary of some relevant aspects is given here. 'The concentration of dissolved oxygen is probably the most significant single chemical factor determining the over-all selection of species of which any biocoenosis is composed' (Hawkes, 1962). Of all the chemical constituents, dissolved oxygen has probably been the most often measured in pollution studies and chemical surveillance programmes. The amount of oxygen in any volume of water may be expressed as volumes per litre (ml l^{-1}) or as mercury pressure which would produce that concentration.

It is also expressed either as a percentage of the expected saturation level at the particular temperature of the water body (per cent sat.) or as part per million (p.p.m.) (mg l^{-1}). The 'saturation' concentration of oxygen is inversely related to temperature, ranging from 14 mg l^{-1} at 0°C to 7·5 mg l^{-1} at 30°C (Macan, 1963).

To any organisms in the water body, the actual amount of oxygen available for respiration is the significant value and not the percentage saturation. However, as Macan (1963) states 'percentage saturation gives a valuable indication of what has been happening in a sample of water'.

(a) Running water

The oxygen economy changes as a stream develops. If the source is a groundwater spring, usually with little flora and a restricted fauna, oxygen concentrations are low. Water from a seepage or from a natural lake source may be more oxygenated (Hynes, 1970). If the source has a low oxygen content, saturation will increase as the stream increases in velocity and turbulence. At some point, not far from the source, where the water is clean and turbulent and has a developing flora the saturation will be on average 100 per cent. Water saturated to 100 per cent with oxygen contains 11·0 mg l^{-1} (or approximately 7·15 ml l^{-1}) at 10°C. As the stream becomes larger, deeper, and slower, oxygen saturation may exceed 100 per cent owing to plant photosynthesis in summer. In some rivers, particularly with dense algal or macrophyte populations, up to 200 per cent oxygen saturation may be recorded. Daily oxygen contributions by plants of up to $2 \cdot 27$ gm^{-2} day^{-1} on weed beds have been recorded in small British rivers (Owens and Edwards, 1961).

Under natural conditions most streams exhibit a marked diurnal oxygen rhythm, mainly because of the respiration and photosynthesis of the flora (Butcher, *et al.*, 1930; Hutchinson, 1957; Gameson and Truesdale, 1959; Owens and Edwards, 1964).

Generally, the more turbulent or agitated a stretch of water, the faster is the gaseous exchange across the surface. Fast-flowing upland streams with poor floras tend, if undisturbed, to show least diurnal variation. However, average values vary seasonally, being usually lower in winter owing to reduced photosynthesis. Also, prolonged ice cover may lower oxygen levels further in slow rivers by preventing free surface exchange.

Increasing depth and decreasing current velocities tend to reduce surface oxygen exchange and suppress diurnal rhythms mainly owing to effects on light penetration and reduced turbulence (Owens and Edwards, 1964).

Weirs and waterfalls make significant contributions to the oxygen balance by aerating de-oxygenated water or reducing supersaturation (Gameson, 1957).

(b) Lakes

In naturally stratified lakes, oxygen stratification is closely related to that of temperature (Hutchinson, 1957). Oxygen stratification develops owing to differences in the oxygen balance between the different strata. In the epilimnion, algal photosynthesis together with wind and wave action keep the oxygen concentration high. The epilimnion may show diurnal rhythms similar to those of streams.

In the hypolimnion, bacterial decay of organic matter on the bottom gradually uses up oxygen and with no exchange processes operating, the deoxygenated layer increases in depth and volume. Rapid oxygen depletion of the whole hypolimnion can occur and other factors such as bacterial activity in the water itself, together with the production of gaseous reducing agents, contribute to the oxygen depletion as it could not be accounted for from the extension of 'substrate decay' alone (Welch, 1952; Hutchinson, 1957).

The seasonal cycle of the oxygen regime also follows the thermal patterns, in that equinoxial overturns result in mixing and equal oxygen distribution in the lake (see chapter 2). Prolonged stratification can, in some eutrophic lakes, lead to deoxygenation of the whole water column. Prolonged ice cover can also result in total deoxygenation. Fish mortalities which occur in such lakes are known as 'summerkill' or 'winterkill' respectively.

(c) Tidal waters

Oxygen is about 20 per cent less soluble in sea-water than in fresh-water. The relationship between salinity and oxygen solubility is more or less linear.

Theoretically the oxygen concentration in sea-water can range from 0 to 14 mg 1^{-1} at 30 parts per thousand of chloride (Cl^-), though concentrations can vary depending upon many factors (Sverdrup, *et al.*, 1963). Stratification may occur, particularly in deep sheltered coastal waters such as fjords or in the deep oceans.

Supersaturation may also occur in surface waters and in shallow bays. For example, Adams (1969) recorded concentrations of 8·0–8·5 mg 1^{-1} in Californian coastal waters giving saturation values of 120–130 per cent and values up to 250 per cent saturation have been recorded in some waters (Perkins, 1974). In estuaries, oxygen stratification also follows salinity stratification so that deeper saline water may contain less oxygen than the surface layer of fresh-water (Dyer, 1973; Perkins, 1974).

(d) Pollution, oxygen demand, and depletion
The input of organic matter to rivers, lakes, and tidal waters is a major cause of oxygen depletion. Organic man-made wastes such as sewage, farm wastes, food-industry wastes, and inorganic reducing agents such as sulphides, sulphites, ferrous salts, and phenols, all demand oxygen for biological or chemical oxidation (Klein, 1962).

The degree of pollution in the receiving water and its effect on the oxygen regime are vital factors in assessing or determining the effects of a hydro-electric dam or a thermal discharge.

Mild pollution may cause only a small imbalance in the oxygen economy but where the pollution load is heavy and the oxygen demand high, a water body may become completely deoxygenated. In a river, the zone of minimum oxygen concentration may occur some way downstream of the discharge, so that an oxygen 'sag curve' develops (Hynes, 1960) (Fig. 3.1). The minimum

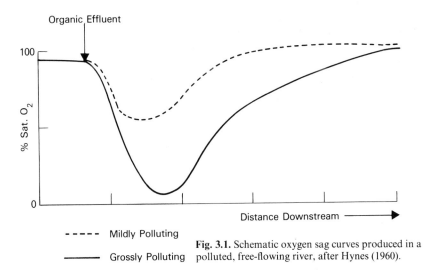

Fig. 3.1. Schematic oxygen sag curves produced in a polluted, free-flowing river, after Hynes (1960).

oxygen level depends on the type of organic waste, dilution, stream velocity, and presence of other pollutants, such as poisons, which can suppress biological activity.

The recovery of the normal oxygen levels (self-purification) depends upon many factors including turbulence, velocity, photosynthesis, surface exchange, and oxygen concentration in the water. Generally, the lower the concentration (i.e. the greater the imbalance between water and air) the faster the rate of re-aeration (Klein, 1962; Owens and Edwards, 1964). Provided that it is not completely devoid of oxygen (anoxic) and that there is an established flora, diurnal oxygen rhythms may still occur in a polluted river (Aston, 1973).

The build-up of nutrients in a river (eutrophication) is vital to the chemical development of an impoundment, and high nutrient loads lead usually to highly stratified lakes and reservoirs (National Academy of Science, 1969).

Most large rivers of the world are affected by some kind of organic pollution, except for those in the undeveloped tropical regions, but even these may have seasonal periods of low oxygen owing to natural organic loading from fallen leaves and decaying aquatic plants.

Hynes (1970) quotes a paper by Gessner (1961) who showed that in a period of high discharge, low dissolved oxygen concentration occurred in the Amazon because of the organic matter swept into the river and the suppression of photosynthesis in the turbid water.

3.2.2. **Other dissolved gases**

Two other important gases dissolved in natural waters are nitrogen and carbon dioxide. Hynes (1970) considers that nitrogen is not much affected by the biota and is largely inactive in water. In most waters, dissolved nitrogen gas is at around 100 per cent saturation. Other forms of nitrogen exist in both clean and polluted waters as a result of aerobic or anaerobic decomposition of nitrogenous organic matter (Klein, 1962).

Nitrogen in solution mainly originates from the atmosphere. In stratified lakes the hypolimnion can become supersaturated with nitrogen as a result of groundwater inputs or as lack of circulation, hydrostatic pressure and low rate of diffusion curtail gaseous exchange. This supersaturation can have serious biological consequences in impounded lakes (see chapter 4).

The sources of dissolved carbon dioxide (CO_2) in water are numerous, but are mainly exchange with the atmosphere inflowing ground water, decomposition of organic matter, and respiration of animals and plants. Carbon dioxide is also found in combination with calcium or magnesium as almost insoluble carbonates (e.g., $CaCO_3$, $MgCO_3$). In this state it is *fixed* or *bound*. *Half-bound* CO_2 is also found in bicarbonates of calcium or magnesium. In this form it is still available to plants for photosynthesis (Hutchinson, 1957).

Carbon dioxide is lost from water bodies mainly by plant respiration, production of calcium or magnesium carbonates by animals or plants (e.g., mollusc shells), turbulence and evaporation, which results in bicarbonates being transformed to precipitated monocarbonates.

In some waters gaseous CO_2 may be given off as bubbles from bacterial decomposition of organic matter. Kerr, *et al.* (1970), for example, showed that bubbles breaking at the surface of a small organically polluted stream contained 8–9 per cent CO_2 plus 6 per cent O along with other gases.

The other reducing gases are ammonia and methane mainly originating from aerobic organic decomposition and hydrogen sulphide originating from anaerobic decomposition.

3.3. EFFECTS OF HYDRO-ELECTRICITY INSTALLATIONS ON DISSOLVED GASES

3.3.1. Oxygen

As we have seen, a clean upland river has a balanced, unstratified, diurnally fluctuating oxygen economy with average saturation around 100 per cent. The impoundment of such a stream will usually have the predictable effect of creating the lacustrine situation with marked stratification developing soon after dam closure (Ackermann, *et al.*, 1973; Bond, *et al.*, 1978). Overturns will depend on regions and climate. If an impoundment is poor in nutrients (i.e. oligotrophic) hypolimnial deoxygenation may not be severe. If, on the other hand the nutrient concentrations are high, a deep, deoxygenated hypolimnion will develop during periods of stratification (Fig. 3.2). Thus, oxygen is unavailable for resident organisms at that depth. As most stream-dwellers are usually intolerant of low oxygen levels this loss has drastic effects. Around the shores and in the surface waters of an impoundment, oxygen is usually plentiful if the water is relatively unpolluted. In oligotrophic waters the effects of hydro-dams on water quality may not be so drastic. For example Geen (1974) suggested that a hydro-dam at Moran Canyon on the Fraser River in Canada would result in the formation of a large but oligotrophic reservoir. Thus, although Kidd (1953) estimated that 15.2×10^6 tonnes of inorganic sediment would deposit behind the dam annually, Andrew and Geen (1960) forecast that there would not be hypolimnetic development and no severe oxygen depletion. Further, Geen (1974) suggested that in summer both oxygen and nitrogen would be supersaturated to a depth of 10 m and that the supersaturated water could be drawn through the turbines and spilled from the surface spillways. When heavy organic pollution occurs upstream of impoundments the development of the deoxygenated zone may be more extensive than shown in Fig. 3.2 (Krenkel and Parker, 1969; Liepolt, 1972; Ruggles and Watt, 1975). An example is the St. John River in New Brunswick, Canada, where discharge of organic wastes from paper mills and food-processing plants enter the river upstream of the Grand Falls Dam. Even in the surface water, dissolved oxygen levels may be as low as $2\,mg\,l^{-1}$ in August,

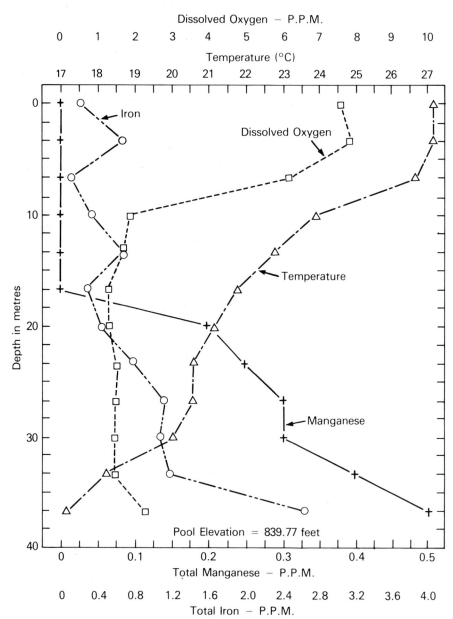

Fig. 3.2. Depth distribution of various water quality parameters in the Allatoona Reservoir, 28 August 1964, redrawn from Kenkel and Parker (1969a).

and the bottom is practically anaerobic (Ruggles and Watt, 1975). The reduced current speed also allows finely divided organic sediments to settle out (see chapter 2) and the reservoirs upstream of the dams have heavy deposits of wood pulp and vegetable washings on the bottom.

The effect of operating a hydro-electricity station on the oxygen regime upstream and downstream of a dam depends largely on the depth at which the upstream intakes are set. For example, where deep water intakes are used in the hypolimnion in summer, deoxygenated, cold, odorous water containing hydrogen sulphide may be discharged downstream after passing through the turbines (Ganapati and Screenivasan, 1968; McFie, 1973). High-level intakes in the epilimnion, will, in contrast, discharge oxygenated, warm water downstream. In most older hydro-schemes, low-level intakes are used and Fig. 3.3 shows the oxygen concentrations at the tail-race and 6 km downstream of such an installation, over 24-hour periods. The sudden upsurge of oxygen coincides with peak power output and it is evident that passage through the turbines (turbulence) has enhanced the re-aeration process, though the levels are still very much below what might have been expected in a natural clean stream.

Where dams and hydro-installations have surface overspill or outlets at various depths, oxygen depletion in the downstream reaches is lessened. Spillage over dams or passage of water through turbines from polluted impoundments can cause rapid re-aeration and a speeding up of the self-purification process (Starrett, 1972). The oxygen profile in the St. John River (Fig. 3.4) shows clearly that discharges from some impoundments enhanced re-aeration where very heavy pollution by pulp and municipal wastes had caused almost complete deoxygenation. The exception was the

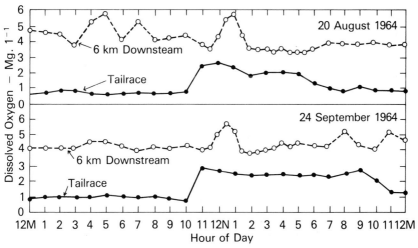

Fig. 3.3. Dissolved oxygen concentrations at the tailrace and 6 km downstream from the Allatoona Dam.

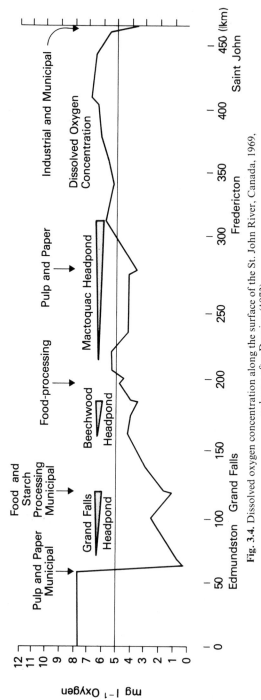

Fig. 3.4. Dissolved oxygen concentration along the surface of the St. John River, Canada, 1969, redrawn after Dominy (1973).

Mactaquac dam, where hypolimnetic discharge actually depressed oxygen concentration for some 40 km downstream of the outlets (Dominy, 1973; Ruggles and Watt, 1975).

The level of the turbine intakes can also affect the oxygen balance in the impoundment upstream of a dam (Orlob, 1969). A deep-water intake, drawing out hypolimnial water, can result in the lowering of the thermocline and an extension of the epilimnion (Orlob, 1969). Drawing water only from the epilimnion via intakes at different levels could also result in gradual diminution of the hypolimnion after which the newly exposed surface layers can become warmed and illuminated.

Krenkel and Parker (1969a) pose the question: 'Do we want high dissolved oxygen and temperature or low dissolved oxygen and temperature downstream?' For the best conditions downstream the probable answer is 'neither—but a compromise between the two which will enable the flora and fauna downstream at least to remain as little altered as possible'. This implies, therefore, the continuation as near as possible of pre-regulation conditions. The solution lies in balancing overspill (surface compensation) with turbine outlet water during the periods of stratification (Orlob, 1969). If a reservoir is not markedly stratified the problem is less acute. The major difficulties are encountered in eutrophic impoundments where the hypolimnion may occupy a major proportion of the reservoir depth. In some regions artificial aeration is used to oxygenate the hypolimnion during the stratification period, though this may be expensive (Speece, 1971; Davis and Collingwood, 1978).

3.3.2. Effects on other dissolved gases

Dissolved carbon dioxide concentration tends to increase as oxygen decreases although the full ecological significance of CO_2 is not always understood. Increased CO_2 concentrations do, however, increase the acidity of the water (Harding, 1966; Begg, 1970).

Nitrogen concentrations are of considerable biological significance below hydro-electricity dams with hypolimnial discharges, because supersaturation leads to the phenomenon known as gas-bubble disease of fishes. This is described and discussed in chapter 4. Concentrations in the Columbia River were found to be above 120 per cent saturation in June and over 100 per cent in April for much of the length. In February, levels were below 100 per cent (Trefethen, 1972). The cause of nitrogen supersaturation in the Columbia is believed by Trefethen to be spillage over the high-level spillways at dams, as he noted that high supersaturation did not occur in the turbine outlet area of the hydro-power station. Ruggles and Watt (1975) disagree. They attributed nitrogen supersaturation in the St. John River directly to power station operation. They believed the source to be air vented into the turbines to reduce negative pressure during low generating levels. In the free-flowing downstream reaches, excess nitrogen is given up to the atmosphere. The ratio of oxygen to nitrogen in air-supersaturated water is important as high oxygen/nitrogen ratios appear to cause greater fish mortalities than the reverse (see chapter 4).

Hydrogen sulphide (H_2S) formation in reservoirs may cause important operational problems at hydro-electric stations (Begg, 1970; McFie, 1973). Around Lake Kariba for example, H_2S is formed in rivers infested with the weed *Salvinia* sp. and passes into the impoundments. In Tasmanian rivers the noxious smell of H_2S can be detected for several miles downstream of some hydro-dams (McFie, 1973). Chlorination is used as a biocide but the H_2S renders the chlorine ineffective and turbines drawing on H_2S-laden water have overheating problems due to algal accumulation on the intakes.

Being a strong reducing agent, H_2S also uses up oxygen and increases hypolimnial deoxygenation.

3.4. EFFECTS OF THERMAL GENERATION ON DISSOLVED GASES

3.4.1. Oxygen

Most general papers describing the ecological effects of thermal discharges predict that oxygen concentration should be reduced because of the lower solubility at higher temperatures. Percentage saturation may of course increase. This is applied to unpolluted and polluted rivers, lakes, and tidal waters. Actual observations do not always substantiate this generalized prediction because of factors other than temperature (Fig. 3.5).

Theoretically, a static body of inert water at any temperature holds more oxygen than the same body at a higher temperature.

However, natural surface waters are not inert. They are usually physically, chemically, and biologically 'active' and the simplistic rule does not apply. As we have already seen, water bodies can be, and often are, supersaturated with oxygen (see 3.2.1).

Agitation and turbulence also have marked effects on the oxygen balance, and during passage through pumps and culverts in a power station, and during discharge from most outfalls, heated cooling water is subjected to considerable turbulence. Thus oxygen concentrations may actually increase, the magnitude of the increase depending upon the original concentration in the cooling water (Aston, 1973) (Fig. 3.5).

Decreases in dissolved oxygen concentrations were, however, found by Proffit (1969) below the heated discharge from the Indiana Power and Light Company plant on the White River (U.S.A.). Reductions from intake to outfall were from 12 mg l^{-1} to 9 mg l^{-1}; 15 mg l^{-1} to 10·5 mg l^{-1}; and 13 mg l^{-1} to 10 mg l^{-1} at various times. Maximum discharge temperature was 42°C. Parker and Krenkel (1969a) also illustrated reduced oxygen concentrations in a heated Polish stream but they did not show if the heated stream contained pollutants in addition to the heated discharge. Examples of 'nil-effect' or actual increases in dissolved oxygen caused by power station discharges are

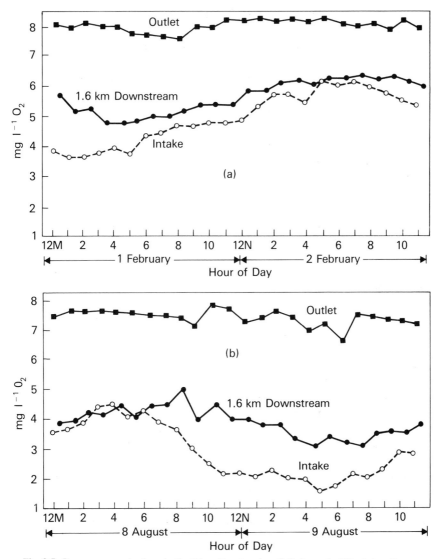

Fig. 3.5. Oxygen concentrations in the River Trent and outfall channel of Drakelow Power Station, redrawn after Aston (1973).

more numerous (Markowski, 1959; Adams, 1969; Clement, 1972; Aston, 1973).

In a Texas reservoir with a high proportion of water used for cooling a power station, oxygen concentrations were altered very little and stratification was less pronounced than in similar unheated reservoirs (McNeely and Pearson, 1974).

Re-aeration is generally pronounced where cooling-towers are involved,

Table 3.1. *Dissolved oxygen levels in river and circulating water at two British power stations*

		Meaford			Rugeley		
Date		Temp. °C	Oxygen mg l^{-1}	Date		Temp. °C	Oxygen mg l^{-1}
19 February 1964	R	7·0	5·30	22 July 1963			
	C	12·6	9·45				
27 February 1964	R	11·0	4·70	25 September 1963			
	C	19·7	10·05			No data	
12 March 1964	R	7·5	6·40	October 1963			
	C	15·5	7·70				
8 April 1964	R	11·0	6·40	17 October 1963			
	C	18·0	8·00				
16 April 1964	R	12·0	6·30	18 November 1963		9·0	5·30
	C	19·5	6·30			19·5	7·75
23 April 1964	R	12·0	6·30	28 November 1963		6·5	6·80
	C	18·5	7·40			16·5	7·85
30 April 1964	R	12·0	5·10	6 December 1963		6·0	8·35
	C	19·0	8·00			16·0	8·80
7 May 1964	R	14·0	4·55	12 December 1963		5·0	8·85
	C	21·0	7·10			16·5	9·40
				18 December 1963		4·0	8·65
						14·5	8·90

R = River water; C = Circulating water. (From Gammon, 1969).

and this is perhaps why analyses of British cooling-water discharges often show high oxygen levels. Data from two tower-cooled English power stations sited on the River Trent, are shown in Table 3.1. Rugeley Power Station is situated on a cleaner reach than Meaford but at both sites there were marked increases in dissolved oxygen concentrations between intake and outfall. Davies (1966) showed similar increases in dissolved oxygen at other British power stations. Aston (1973) in his studies on the most polluted reach of the Trent, showed that as much as 2·0–3·9 tonnes of oxygen were injected into the river each day by the Drakelow Power Stations. The discharge at the time was composed of cooling-water from a direct-cooled and a tower-cooled station. A 24-hour survey in August (Fig. 3.5a) showed a diurnal fluctuation at the intake, owing to photosynthesis. Minimum levels were very low at night as a result of the pollution load. Downstream of the power station the same pattern occurred but on a higher level, while oxygen concentrations in the outfall channel fluctuated less. Percentage saturations were from 17 to 49 per cent at the intake; and 84 to 100 per cent in the discharge channel. The average temperature rise was 9·3°C.

In a February survey, oxygen levels were higher at all sites because of increased river discharge. Oxygen concentrations were again highest in the outfall channel and in the river downstream (Fig. 3.5b). In the most polluted reaches of the Thames estuary, artificial aeration of a once-through power station discharge was attempted during 1962. An air injection system was installed on the outlet of Belvedere Power Station cooling-water system and one unit added about 1.2 mg 1^{-1} of oxygen to the discharge. When the power station was operating on all four units, the total injection of oxygen from the discharge, with the addition of artificial aeration was 5 tonnes day^{-1}. However, at low tides, free-fall over the weir and turbulence in the culvert added more than three times as much as the aerator (D.S.I.R., 1964).

Alabaster and Downing (1966) concluded that, in general, cooling-water discharges tended to increase dissolved oxygen in British rivers when the intake water had low percentage saturations, and to decrease oxygen levels when intake water was supersaturated. Thus, the discharges enhanced equilibration with the atmosphere and 'balanced' the oxygen economy.

Hypolimnial water may well be aerated by thermal power stations abstracting from depths of reservoirs, lakes, or seas, and discharging at the surface. It is possible in small lakes that such a system may break down stratification. Even though a discharge does initially aerate the water in the vicinity of the outfall, the over-all increase in the temperature of a small river may eventually lead to some oxygen depletion further downstream as a result of increased bacterial activity (D.S.I.R., 1964; Kosheleva and Tkachenko, 1973). Several authors (Krenkel and Parker, 1969a) have suggested also that temperature increases of 6–14°C may in fact reduce the capacity of rivers to assimilate organic wastes and inhibit self-purification.

3.4.2. Effects on other dissolved gases

Polluted waters are often rich in ammonia originating from the decay of proteinaceous wastes or from chemical factories. Sewage effluents may also be used directly to supplement cooling-water supplies (Humphris, 1977). Carbon dioxide concentrations vary with the source and quality of water (Klein, 1962). The concentrations of both gases are usually altered during the passage of cooling water through a power station, particularly when cooling towers are incorporated.

Ammonia is generally oxidized in cooling-towers to form nitrites and nitrates (Davies, 1966; Ross and Whitehouse, 1973; Stratton and Lee, 1975) and where ammonia-rich water is used for cooling, nitrate levels may be very high in the discharge (Humphris, 1977). Nitrification is a result of bacterial action in cooling towers and is severely restricted when bacteriocides (e.g. chlorine) are present (Stratton and Lee, 1975). Ammonia concentrations may decrease sharply in cooling-tower ponds immediately after chlorination, probably because of the formation of chloramine compounds (White, 1972; Brown and Aston, 1975; Stratton and Lee, 1975; see also Jolley, et al., 1978). The rate of nitrification and the toxicity of ammonia are directly related to temperature and pH (E.I.F.A.C., 1970). High levels of nitrate may have effects on plant growth (Stratton and Lee, 1975) and in drinking water may be harmful to young children causing the condition known as 'methaemoglobinaemia' (Klein, 1962). The oxidation of ammonia reduces the toxicity of cooling-tower effluents in polluted rivers though some of the additives used in the cooling system are in themselves toxic.

Spraying and droplet formation in cooling-towers together with the formation of thin films on tower packing, effectively scrubs carbon dioxide from the water during cooling. Deposition of calcium carbonate may occur on the wooden packing and in one cooling-tower in the U.K. timbers collapsed under the weight of an estimated 900 tonnes of such deposits (Davies, 1966; Ross and Whitehouse, 1973).

3.5. EFFECTS OF POWER GENERATION ON OTHER ASPECTS OF WATER CHEMISTRY

3.5.1. Hydro-generation

Thermal and gaseous stratification is often coincident with other chemical stratification in impoundments. The concentrations of iron (Fe^{++}) and manganese (Mn^{++}) may increase with depth, probably because of the low oxidation-reduction potential at the mud–water interface which cause the mobilization of the metal ions from deposits (Fig. 3.2). Where a river is polluted by heavy metal salts, any reservoir in the system will act as a sink in which these will accumulate, mainly through adsorption on to particles and bioaccumulation. In an Oklahoma reservoir, zinc, lead, and cadmium, brought in from upstream, were found to accumulate in the deepest zones and only small amounts removed in the out-flow (Pita and Hyne, 1975). Hypolimnial discharges contain higher concentrations of these elements than epilimnial discharges (Oglesby, et al., 1978). The development and seasonal death of phytoplankton in eutrophic reservoirs, together with river contributions, leads to accummulation of organic matter on the bottom. Generally, where stratification occurs, conductivity, alkalinity, and pH decrease with depth, while nutrients, i.e. nitrates, phosphates, and other mineral salts, increase (Begg, 1970; McFie, 1973; Egborge, 1979; see also Lowe-McConnell, 1966; and Ackermann, et al., 1973; Ward and Stanford, 1979).

The nutrient budget of a reservoir and the effects of discharge on the chemistry of the river downstream will depend largely on the depth from which the discharge originates (Martin and Arneson, 1978). Thus if the discharge is from the epilimnion, nutrient-poor water leaves the reservoir usually because algal blooms in the epilimnion have utilized the nutrients. In contrast, nutrient-*rich* water is discharged from the hypolimnion. Reservoirs and lakes are in fact considered by Hynes (1970) as fertility traps which accumulate organic matter, whereas a free-flowing river tends to transport materials downstream. Thus the impoundment of a river may reduce the amounts of nutrients reaching the estuary. Where pollution occurs downstream of an impoundment any reduction of flow will intensify ecological effects because of decreased dilution. Releases of deoxygenated hypolimnial water also reduce the capacity of the river downstream to assimilate waste and in consequence suppress self-purification processes.

Impoundments, including hydro-impoundments are blamed for reducing water quality still further in polluted Canadian rivers. Ruggles and Watt (1975) considered that these deleterious effects were underestimated in the planning of hydro-dams on the St. John River and that the reduction in the capacity of the river to assimilate wastes was not, they state, fully recognized. The change to lacustrine conditions has been mostly to blame. Water which normally used to take 3 days to flow 300 km from Edmonton, to a point downstream of the Mactaquac hydro-installation, may take now up to 3 months. This slowing down gives the biochemical oxygen demand (B.O.D.) more time to consume the oxygen and re-aeration is reduced because of increased depth, lack of turbulence, reduced current velocity and stratification. Although, as has already been mentioned, some re-aeration takes place immediately downstream of hydro-electricity turbine outfalls, this is evidently insufficient to counteract the otherwise deoxygenating tendencies of hydro-impoundments in the St. John (Fig. 3.4).

Another important effect is that wood pulp and other organic waste solids are deposited in the reduced current velocities upstream of the dams and this pulp accumulates as a loose ooze on the bottom. This increases deoxygenation in the hypolimnion and when scoured out as the flood-gates are opened each spring also pollutes the downstream reaches. Of the impoundments on the St. John, the largest, the Mactaquac, acts as a recovery zone unlike the others, mainly owing to lower organic loadings. All the impoundments are rich in nutrients with total nitrogen values of 870–970 mg m^3 compared with an average of 185 mg m^3 in a nearby unpolluted reservoir. Although adverse effects appear to have occurred as a result of damming the St. John River, Starrett (1972) considered that in the Illinois River (U.S.A.), even though the movement of water is restricted, spillage over and through dam installations speeds the recovery from pollution. In the absence of dams, the normally low stream gradient and the nature of the pollutants would retard re-oxygenation and spread the ecological impact of the pollution further downstream.

Diversions and water transfers associated with hydro-electricity and pumped storage schemes can have considerable chemical consequences, for example by transfer of water from an alkaline stream into an acid stream or vice-versa. As has been stressed, streams and rivers have individual chemical characteristics and these may be of immense importance to homing fish and to the survival of other organisms (see chapter 4). Therefore mixing of two water types may have considerable ecological implications (Berry, 1955; M.A.F.F. and A.R.A., 1972; Geen, 1974). Pollution may be worsened or alleviated by diversions and transfers depending upon the respective water qualities concerned (Starrett, 1972). Changes in nutrient loads caused by impoundment can also affect the chemistry of the respective estuary or coastal region.

Further, if river flows are severely restricted, the infiltrating tongue of saline water in most estuaries may reach further upstream. Thomann (1972) has demonstrated this effectively for the Delaware River in North America. Geen (1975) forecast, on the other hand, that the dam at Moran Canyon on the Fraser River would, by increasing the freshwater flow from December to April, also increase the area of the freshwater plume in the Strait of Georgia and because of the peculiarities of stratification of salt and fresh-water in the Strait, increase the productive area.

3.5.2. Thermal generation

Becker and Thatcher (1973), in a comprehensive review on the toxicity of chemicals used in power installations, stressed that there was no universal chemical composition of a power plant discharge. The composition of a discharge is unique to each situation and depends on such factors as intake water quality, additives used for neutralizing or passivating the water (i.e. removing dissolved gases from boiler feedwater), preservatives for wooden structures, corrosion inhibitors, descaling compounds, and fouling inhibitors. The composition of any discharge is further complicated by the reaction of many chemicals with each other and with naturally occurring substances. Becker and Thatcher gave a list of 130 chemicals which they say are associated with power station operation, though generally most are present in minute quantities if at all. A detailed bibliography of literature on the chemical effects of cooling water is given by Opresko and Hannon (1979) and a summary is included in a biological account by McGuire (1977) and Chu and Olem (1980).

The chemicals can be divided into two main types namely: those associated with the cooling-water circuit, for example, additives or concentrations of natural substances in the inlet water; and those associated with non-thermal discharges, which may be combined with cooling-water discharge or may be discharged separately, for example, ash effluents, sewage, and boiler-cleaning wastes.

(a) Chemicals in cooling water

(i) *Originating from intake water*
There are usually few major chemical changes in the cooling-water as it passes through a direct-cooled power station. There are, however, considerable changes in water quality in cooling-tower or pond circuits, some of which have already been discussed (see 3.4). The evaporation of water in tower circuits results in concentrations of most non-gaseous substances found in the inlet water, with concentration factors ranging from 1·5 to 2·5 depending upon the power station (Ray, N. J. 1965; Davies, 1966; Gammon, 1969; Stratton and Lee, 1975; Humphris, 1977; Macqueen, 1978).

The pH of water in cooling-tower circuits is often higher than ambient, owing to the stripping of CO_2 and the solution of sodium bicarbonate (Davies,

1966). Also, deposits of calcium phosphate occur where inlet water is rich in nutrients. This can cause operational problems by blocking condenser tubes at some sites (Ross and Whitehouse, 1973; Flook, 1978). Increases also occur in total hardness, conductivity, chloride-dissolved solids (Gammon, 1969), and in concentrations of trace elements. If the inlet water is heavily polluted by metal salts, high concentrations of metals in solution may occur in a cooling-tower system (Romeril and Davis, 1976).

(ii) *Additives to cooling circuits*
Cooling-water discharges, whether from indirect-cooling or direct-cooling systems may contain additives.

The major additives are biocides, used to control growths of bacterial and fungal slimes in condenser tubes and in pipework, and also to prevent fouling by aquatic organisms in intake culverts. The chemical most commonly used for this control is chlorine, injected into the cooling-water systems as a gas in solution or in the form of hypochlorite. Other biocidic substances which may be used are chlorophenols, quaternary amines, and metallic-organic compounds, but these have much less widespread use than chlorine. A number of major publications have been produced recently dealing with the practical, chemical, and biological aspects of chlorination and antifouling, namely by White (1972), Jensen (1977), and Jolley, *et al.* (1978), and for further details the reader should refer to these.

Becker and Thatcher (1973) and Cole (1977) outlined the normal chlorination procedure in power stations in the U.S.A. Doses averaging $0 \cdot 1$–$0 \cdot 5$ mg l^{-1} are injected into the system for 30 min, three times daily. This is also the standard practice in most British inland power stations (Brown and Aston, 1975), though Ross (1959) quotes concentrations of $0 \cdot 1$–$0 \cdot 75$ mg l^{-1} depending on water quality. Where mussel fouling is a serious problem, in some marine or estuarine situations, doses of up to 4 or 5 mg l^{-1} of chlorine may be applied for periods of up to 4 h. In extreme cases Becker and Thatcher (1973) state that doses of 12 mg l^{-1} have been applied. Beauchamp (1969) recommended that doses of up to $0 \cdot 5$ mg l^{-1} applied continuously would keep marine intakes clear and give equally effective antifouling protection within the plant. The chemistry of chlorine in fresh- and salt-water is complex and though chemical reactions in the former are reasonably well known, those in sea-water are not yet fully understood (see Jolley, *et al.*, 1978).

The basic reactions between chlorine and water are:

$$Cl_2 + H_2O \rightleftharpoons HOCl + H^+ + Cl^-$$

The hypochlorous acid then dissociates further:

$$HOCl \rightleftharpoons H^+ + OCl^-$$

Generally the amounts of OCl^- and $HOCl$ are controlled by temperature and pH in fresh-water but the system is more complex in sea-water because of reactions with bromine compounds and the subsequent reactions with organic compounds (Morgan and Carpenter, 1978).

When chlorine is added to any natural water, initial reactions occur with organic matter, dissolved gases, and inorganic salts. The amount of chlorine utilized in these reactions is known as the 'chlorine demand' and only after this demand is satisfied is there a free-chlorine residual. The chlorine demand of any water therefore depends largely on organic material concentrations so that sea-water and polluted or eutrophic waters have much greater demand than clean fresh-water.*

Generally, the dose rate at a power station intake depends on this demand, supposedly balanced so that the dose required at the inlet is sufficient to allow biocidal activity within the system, without allowing environmentally damaging residuals at the outfall. Needless to say this is not always successful owing to mechanical difficulties and the fluctuations in composition of the intake water. Also, once the chlorine demand is satisfied at the intake, the residual will not decrease until contact with the receiving water body at the outfall. Chlorine residuals found in power station effluents vary considerably (Morgan and Carpenter, 1978). A study at the Carbo plant on the Clinch River (Dickson, et al. 1974) showed that there was a marked peak in total residual chlorine some 30–40 min after dosing. Total residuals at the intakes ranged from 0·64 to 1·31 mg l^{-1} and free chlorine ranged from 0 to 0·35 mg l^{-1} in the discharge from the cooling towers. Ammonia levels in the cooling waters at the intake ranged from 0·05 to 0·25 mg l^{-1}. Figure 3.6 shows the reduction in chlorine residuals after discharge. Brown and Aston (1975) found similar patterns some minutes after dosing in the cooling-water circuits at Ratcliffe-on-Soar Power Station on the River Trent (U.K.). Dosing ran for 10–20 min every 8 h for each operating generating unit. Peak total residuals recorded were 0·6–0·8 mg l^{-1} in condenser outlets and 0·3–0·5 mg l^{-1} in cooling-tower ponds. Ammonia levels in the intake water from September to November 1974 ranged from 0·4 to 2·3 mg l^{-1}. In the cooling circuits the range was 0·07–0·56 mg l^{-1}, a considerable reduction over the inlet water. Usually, however, dosage of 0·5 mg l^{-1} at the inlet results in residuals of 0·1–0·2 mg l^{-1} at outfalls of British power stations (Coughlan and Whitehouse, 1977). Once discharged, chlorine residual levels drop sharply because of chemical reactions and dilution (see Jolley, et al., 1978), though in some cases toxic levels may persist for several hours (Lee, 1979).

Mixtures of chromates, zinc phosphate, and silicates are used for corrosion control in cooling-water systems. Copper compounds are also used to inhibit bacterial degradation of wooden structures in cooling towers (Capper, 1974). Copper is also used as an alloy constituent in condensers and may be discharged via cooling-water as a result of corrosion (Becker and Thatcher,

---

* In all subsequent references to effects of 'chlorine' it is accepted that measurements are usually of total residual oxidants (TRO) particularly in seawater, i.e. effects of the chlorination process rather than of the element itself.

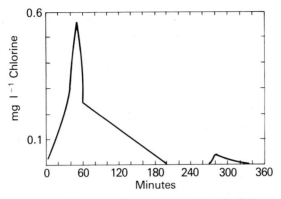

(a) 9 metres Downstream of the Outfall.
0 Time = 13.00 hours, 12 September 1972.

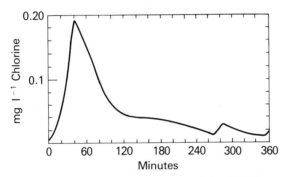

(b) 35 metres Downstream of the Outfall.
0 Time = 19.00 hours, 13 September 1972.

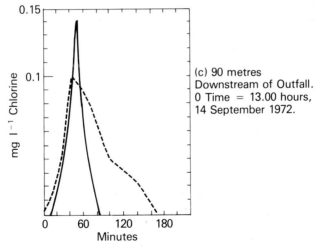

(c) 90 metres
Downstream of Outfall.
0 Time = 13.00 hours,
14 September 1972.

Fig. 3.6. Residual chlorine concentrations at different points downstream of the cooling-water discharge from the Carbo Power Plant, redrawn after Dickson, *et al.* (1974).

1973; Stratton and Lee, 1975). Organic phosphates, amino-ethylene phosphates and polyesters are also used to prevent corrosion and calcium scale formation. Acids or alkalis may be used to control the pH of boiler feed and cooling-water. Hydrazine may also be used to remove dissolved oxygen from the feed water. Small concentrations of these substances may be discharged in various effluents. Copper, zinc, and chromium found as impurities in anticorrosion agents have all been found to increase in concentrations as a result of concentration factors in cooling-towers, often exceeding limits set by pollution-control authorities (Capper, 1974; Dickson, et al., 1974; Stratton and Lee, 1975; Romeril and Davis, 1976).

(b) Non-thermal discharges

(i) Ash effluents
The effects of any ash-lagoon discharge will depend on the efficiency of pre-discharge treatment and subsequently on dilution in the receiving water, but basically ash effluents have three ecologically damaging components, namely suspended solids, heavy metal ions, and pH (Watt and Thorne, 1965; Brackett, 1967; Brown J., et al., 1976; Dreesen, et al., 1977; N.T.I.S., 1977; Coutant, et al., 1978). In most U.K. ash effluents a high pH is normal but in other studies low pH has been found as a result of high sulphate levels (Cherry, et al., 1979). The levels of each component vary markedly with the composition of the coal burned and as well as with the efficiency of treatment (U.S. Dept. of Interior, 1953; Watt and Thorne, 1965). Suspended solid concentrations in normal lagoon outlets in Britain vary from 25 to 100 mg l^{-1}. Trace metal concentrations are also highly variable, often over short periods of time, and pH may vary from 8·0 to 13·0. Often ash effluents are mixed with acid boiler-cleaning wastes to buffer the pH changes though this increases metal concentrations in the discharge. Ash effluents may be discharged via the cooling-water which offers high dilutions (Dreesen, et al., 1977; also see chapter 1). The possible re-use of ash effluents to augment cooling-water supplies may involve increased concentrations of minerals in discharges.

(ii) Gas-washing, boiler-cleaning, sewage, and water-treatment wastes
Gas-washing and boiler-cleaning wastes generally contain acids and metal salts. Sewage is mainly domestic with small amounts of laboratory wastes, while water-treatment wastes are generally concentrations of natural substances found in local potable or river water, together with acids or alkalis. The volumes of these are generally small, and unless they are discharged to small streams or pools, their chemical effects are unlikely to be of great significance (Ross, 1959). In many places, however, legislation may control the composition of each effluent, irrespective of its discharge point.

(c) Radio-active effluents
Radio-toxicity is probably more a physical than chemical phenomenon

though most of the radio-toxic elements are isotopes of naturally occurring elements. All elements having a greater atomic number than 80 have some radio-active isotopes, and all isotopes of elements with atomic number greater than 83 are radio-active (Eisenbud, 1973). The concentrations of radio-active substances found in aquatic releases from nuclear power stations vary markedly with sites, design and operation of the plant and discharge constraints (Polikarpov, 1966; I.A.E.A., 1966, 1969, 1971, 1974a, b, c, 1975b; Eisenbud, 1973).

The most common radio-active elements monitored in aquatic discharges from British power stations are ^{106}Ru, ^{134}Cs, ^{90}Sr, $^{239/240}$Pu, ^{241}Am, ^{137}Cs, ^{144}Ce, ^{95}Zr/^{95}Nb, ^{65}Zn, ^{3}H, ^{60}Co (Mitchell, 1977a,b). Radio-active isotopes are generally much more toxic than the stable isotopes of the same element, though the amounts of the stable forms found in the environment are generally several orders of magnitude greater than those of the radio-active substances (Eisenbud, 1973). The major problem is, however, that as with the stable heavy metal ions, radio-active metals accumulate in sediments and organisms. Concentration factors of 10^3–10^6, depending on exposure and the organism, have been recorded (see chapter 7). Also the toxicity decreases only slowly as the radio-activity properties may decay over thousands of years (Eisenbud, 1973). Generally the concentrations of radio-active material released into natural waters from generating stations are much lower than those from nuclear processing plants or from weapon factories. Some radio-active elements are also produced from non-nuclear thermal power generation, mainly from concentrations occurring naturally in coal, oil, or additives (Eisenbud, 1973).

(d) Aerial emissions and water quality

It is very difficult to define the extent to which the pH of rivers and lakes in North America and Scandinavia has declined specifically as a result of airborne sulphur and nitrogen oxides from power stations. Indeed there is not general agreement among Scandinavian authors that airborne acidification is solely or partly responsible for any decline. Rosenqvist (1977, 1978) has suggested that any pH changes in lakes and rivers are because of soil processes and changes in agriculture and forestry. Several others have provided evidence to suggest that acid precipitation is, however, a significant factor (Jensen and Snekvik, 1972; Brosset, 1973; Hultberg, 1975; Leivestad, et al., 1976; Siep and Tollan, 1978). pH changes of 1–2 units have been reported in some lakes over 30–40 years (Siep and Tollan, 1978), but the data are not fully conclusive. The mean pH of precipitation has declined in several regions and Siep and Tollan report an average decline of 0·5–1·0 units at three Norwegian sites, with a rapid change between about 1963 and 1968. Where poorly buffered soils border acid streams and lakes, acidified rain or snow can depress pH of drainage water such that fish mortalities occur. Changes of 0·5–2·0 units have been recorded after snow-melts and hydrogen ion concentrations have been shown to increase by 2–13 times in Ontario streams because of melting acidified snow (Jeffries, *et al.*, 1979). For further details of chemical changes and the possible reasons the reader is referred to Gorham (1976), Rosenqvist (1977, 1978), Siep and Tollan (1978), and Howells and Holden (1978).

(e) Synergisms

The composition of any power station effluent may be, as we have seen, quite complex. Although few actual toxicity tests have been done on non-thermal discharges (see chapter 7), it is known that many of the metal salts found in these effluents produce synergistic effects when mixed (Jones, J.R., 1964; Brown, V., 1968). Thus, even small concentrations of metal salts in complex mixtures may be toxic to some organisms. Toxicity of most poisons is also affected by other factors such as low dissolved oxygen, salinity, pH, and temperature (Jones, J. R., 1964; Cairns, *et al.*, 1976, 1978). Exposure to a toxicant can in turn increase sensitivity to other stresses such as increased temperature (Cairns and Messenger, 1974; Cairns, *et al.*, 1978). Synergistic effects of mixtures of radio-toxins may also occur (Rice and Baptist, 1974). The toxicity of certain chemicals may be reduced by increases in the hardness of waters or even by the presence of other salts. For example, glucose or sodium chloride added to toxic solutions of mercuric chloride were found to reduce its toxicity to fish, possibly as a result of an osmotic effect (Jones, J. R., 1964).

3.6. SUMMARY

The potential chemical alterations in natural waters as a result of electricity generating installations are numerous, and may be far reaching. The major problems arise from impoundment, particularly of polluted rivers, where gaseous and chemical stratification can affect both upstream and downstream reaches, altering the water quality and the capacity for self-purification. The chemical effects of electricity generation and reservoirs both upstream and downstream of a dam depend very much on the depth of the turbine intakes.

The observed effects of thermal discharges on oxygen regimes of natural waters are not always as damaging as generally predicted, even though decomposition of organic matter may be speeded up at higher temperatures. Thermal generation involves the use of chemical additives in the cooling system. Residues in the discharge may be ecologically significant. The concentrations of all the potentially harmful substances in ash effluents, radio-active wastes, and other wastes vary enormously from site to site and, probably over fairly short periods owing to fluctuating operating schedules at power stations. The effects of airborne SO_2 and NO_2 from power stations and other industries on the pH of water are not clear, though some acidification of lakes and rivers is indicated in specific regions.

Chapter 4 The biological effects of hydro-electricity generation

4.1. INTRODUCTION

The relationships between the biological, chemical, and physical components of aquatic habitats are complex and the more the complexities of these relationships are revealed, the more difficult it is to divide them into clearly separate categories for discussion.

It should be clear by now that the major physical change *upstream* of a hydro-electricity dam or any other impoundment, is the formation of a lacustrine habitat from a riverine habitat. The main alterations *downstream* are to the flow and chemistry, and the magnitude of any change depends on the design and operation of the scheme. The biological consequences of such major physical and chemical changes may be far-reaching and there are likely to be effects at all trophic levels in the ecosystem. Lagler (1969) stated that 'although reservoirs are generally planned to solve one or more primary problems such as a need for hydro-electric power, their construction generates innumerable secondary problems, many of which have proved to be serious'. Of all the far-reaching effects, the changes in the aquatic habitat itself are only a small part. Other problems such as the social and economic upheavals involved in resettlement of flooded towns, changed patterns of agriculture, forestry, and archaeological salvage are outside the scope of this book (see Lowe-McConnell, 1966; Lagler, 1969; I.C.S.U., 1972; Ackermann, *et al.*, 1973).

Unlike thermal power stations almost all hydro-electricity stations are sited inland on rivers or lakes and this chapter deals, therefore, almost exclusively with freshwater flora and fauna. Experience of actual ecological changes resulting from tidal hydro-electricity generation is limited.

Both operational and environmental biological problems occur at hydro-developments in most parts of the world (Lowe-McConnell, 1966; Ackermann, *et al.*, 1973) and often the flora and fauna involved are specific to that region (Straskraba and Straskrabova, 1969; Hall, 1971; Leentvaar, 1966, 1974; Mordukhai-Boltovskoi, 1979; Ward and Stanford, 1979).

In this section it is not possible to deal with every situation; rather an attempt is made to describe the general scope of biological consequences using well-studied situations as illustrations. More details are obtainable from the references included in this chapter.

There are a number of recent books dealing with the ecology of the freshwater flora and fauna (Macan and Worthingon, 1951; Hynes, 1960, 1970; Macan, 1963, 1970; Bardach, 1964; Oglesby, et al., 1972; Whitton, 1975b) and it would be superfluous to attempt more than a brief outline of some relevant aspects here. For a more detailed coverage the reader is referred to these excellent accounts and their bibliographies.

The main point to be borne in mind here is that organisms are physiologically, structurally, or behaviourally adapted for survival in a specific habitat (see references). Thus the creation or destruction of such habitats can either lead to multiplication or elimination of relevant species often with chain-reactions on other dependent or competing species. These changes are usually irreversible so long as the physical and chemical changes persist.

4.2 BACTERIA, FUNGI, AND PROTOZOA

4.2.1. In natural waters

Bacteria, fungi, and protozoa are important primary links in any aquatic ecosystem (Cairns, 1977), breaking down dead organic matter, releasing gases and nutrients and at the same time providing a valuable direct food source for many invertebrates. The numbers of bacteria and protozoa in fresh waters are highly variable from perhaps a few hundred to some millions per millilitre, depending on the habitat. In most natural waters the bacterial flora is probably derived from the surrounding soil flora (National Academy of Science, 1969; Hynes, 1970; Fjerdingstad, 1975; Jones, J. G., 1975).

In clean, oxygenated water, aerobic bacteria break down organic compounds, without stressing the oxygen regime. In polluted water, with excess organic material, bacterial and chemical oxygen demand exceeds supply and the water becomes deoxygenated. Anaerobic bacteria then replace the aerobes and their decomposing activities lead to septic conditions, the formation of hydrogen sulphide (H_2S), methane, and other odorous compounds. In polluted running water large growths of 'sewage fungus', a community consisting of fungi, filamentous bacteria, and colonial protozoa, may develop (Hynes, 1960; Hawkes, 1962). In naturally stratified lakes bacterial stratification can also occur, with the largest numbers being found in the hypolimnion (Fliermans, et al., 1975). Species composition may also vary with depth during stratification and mixing occurs with the overturns (Cairns, et al., 1973; Cairns, Kaesler, et al., 1976). Pollution also causes changes in the community structure of protozoa (Cairns, et al., 1972).

4.2.2. Effects of hydro-electricity schemes

It is reasonable to suppose that the development of a reservoir will, if stratification occurs, change the aerobic bacterial population of the previous river bed, to an anaerobic system in the hypolimnion. The subsequent production of H_2S and CO_2 may lead to acidic conditions which causes corrosion of turbine blades and pipework. McFie (1973), Coutant (1963), and Hammerton (1972) reported that discharges of hypolimnial water from stratified reservoirs contained large quantities of bacteria. As a general pattern bacterial numbers increase soon after impoundment as the flooded organic matter decays and nutrients increase. Subsequently, as the reservoir stabilizes the numbers decline. Impoundments may enhance self-purification processes (Imhoff, 1965) provided that the pollution load is not so heavy that oxygen is permanently depleted. Biochemical oxygen demand is often reduced though algal material may be discharged from epilimnial outlets (Straskra-bova, 1975). In grossly polluted reservoirs such as those on the St. John River (see 3.3 and 3.4) the absence of aerobic bacteria and low microbial activity is probably responsible for the reduced capacity of these impoundments to assimilate wastes (Ruggles and Watt, 1975). Where reservoirs are used for water supply in addition to hydro-electricity generation, contamination of the water by faecal bacteria is important. However, as these do not survive for long periods in water, retention in impoundments can result in significant reductions in coliform concentrations to the benefit of the supply (Houghton, 1966; Mitchell, R., 1972; Grabow, *et al.*, 1975). Attached growths of 'sewage-fungus' often occur in free-flowing reaches below turbines where organic matter and nutrients are discharged (McFie, 1973).

4.3. ALGAE

4.3.1. **In natural waters**

The major freshwater habitats, namely springs, streams, rivers, ponds, and lakes, contain characteristic species, though other species may occur in most habitat types. In the sea, algal communities may be more or less related to geographical regions and to major currents, for example the Gulf Stream or Humbolt Current. In any habitat the algal community may be further subdivided into categories, based on their micro-habitat (Hynes, 1970; Round, 1973; Whitton, 1975a). *Epilithic* algae live attached to stones, and many are physically adapted to withstand currents of considerable velocity. *Epithytic* algae live attached to other plants such as thalloid or filamentous algae or aquatic macrophytes, in both running and standing waters. Epiphytes tend to increase in abundance as water currents become slower and nutrient levels increase. Species composition and abundance also vary with the host.

Epipelic algae are those which are free-living on sediments and attach themselves only lightly and temporarily. Most species are unicellular and highly motile and are able to return to the surface of sediments if they become buried. In larger flowing rivers, such as the Danube, algae may be entirely limited to the littoral zone because of the inhospitable mobile gravel in the centre of the channel (Ertl and Tamajka, 1973).

Other algae are adapted for very specialized habitats but these are generally of less importance to this discussion (Round, 1973).

The floating community of algae found in lakes and slowly flowing rivers constitutes the *phytoplankton*. As a general rule, faster-flowing rivers have no permanent phytoplankton. However, if a river originates as a lake outflow some of the lake phytoplankton spills over, though few cells persist for any distance downstream (Round, 1973; Whitton, 1975a). They are usually deposited on macrophytes or ingested by the filter-feeding invertebrates which characterize lake outflows (Hynes, 1970). Other floating cells may occur in faster rivers, as a result of the scouring or 'sloughing off' of epilithic, epiphytic, and epipelic algae. Only in slower water, however, will this 'potamoplankton' be self-sustaining. Nutrients and light are generally the major factors determining the standing crop of algae in a waterbody. Algae in lakes occupy the same types of micro-habitat as in larger slower rivers and many species may be common to both.

In many lakes the littoral algal flora (largely epipelic) may contain more species than the phytoplankton and can be of considerable importance to the over-all productivity (Round, 1973). In most natural stratified lakes algae are limited to the epilimnion and the littoral areas. In these littoral regions there may be some species zonation with depth. Where pollution occurs the algal flora alters, usually tending toward changes in species composition and decreased species diversity (Butcher, 1947; Patrick, 1969; Cairns, *et al.*, 1972). Increases in filamentous forms such as *Cladophora* occur and their subsequent decay may exacerbate pollution problems (Hynes, 1960).

In tidal waters the most obvious algae on rocky shores are the large thalloid forms which, together with their epiphytes, extend from the upper edge of the intertidal zone down to the sub-littoral. Less conspicuous but often abundant are the microscopic epipelic algae. Sandy and muddy shores are mainly colonized by these species. The phytoplankton, which is often abundant, has distinct community groups, for example 'coastal' (neritic), 'open water', and 'oceanic' (Round, 1973).

4.3.2. Effects of hydro-electricity generation

(a) **Upstream of dams**

Lacustrinization, as might be expected, results in the elimination of the riverine benthic algae (epipelic and epilithic) mainly as a result of reduced light penetration, reduced gas exchange, and increased sediment deposition. Algal production is limited, as in natural lakes, to the photic zones—the epilimnion, the littoral, and sub-littoral zones, depending upon turbidity. The community of the littoral zone is also very dependent on the extent and duration of drawdown (see 2.2.2). A permanent phytoplankton develops in most reservoirs where the retention of water is of sufficient duration. Stratification of the living phytoplankton tends to be closely related to thermal and oxygen stratification, the density generally decreasing with depth. Water supply

reservoirs have considerable problems with blooms of algae, mainly Cyanophyta (blue-green algae), and these have been studied in great detail (Round,
1973).

Pre-impoundment studies on rivers are somewhat scarce though post-
impoundment work is plentiful. Brook and Rzoska (1954) showed that the
phytoplankton in the free-flowing Nile was limited to a few diatom species,
but as the Nile changed from a riverine to a lacustrine habitat behind the Jebel
Aulia Dam, there was a 100-fold increase in the standing crop of phytoplankton (Fig. 4.1). Blue-green algae were dominant and massive blooms occurred
near the dam, which increased the turbidity of the water considerably. Similar
blooms and increases in primary productivity were recorded for other Nile
areas (Hammerton, 1972; Rzoska, 1976). Usually the phytoplankton was
restricted to the upper 3–4 m of the epilimnion. The basic pattern of nutrient
release and phytoplankton bloom is characteristic of almost any new
impoundment in any part of the world (see Lowe-McConnell, 1966; National
Academy of Science, 1969; Ackermann, *et al.*, 1973; Mordukhai-Boltovskoi,
1979). Hypolimnial development of a new reservoir may be exaggerated in the

Fig. 4.1. Effect of the Rosieres Reservoir on
phytoplankton development in the Blue Nile,
redrawn after Hammerton (1972).
The dam was completed in June 1966.

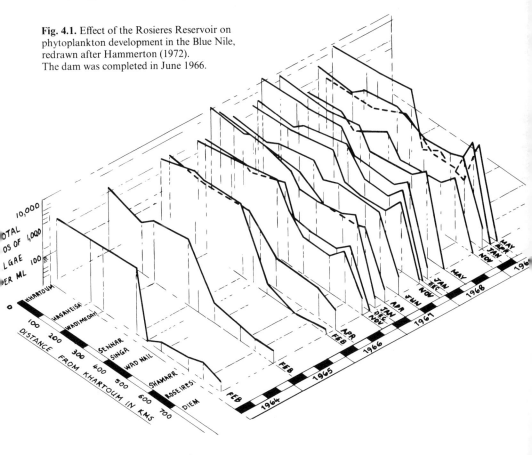

first few years by the decay of flooded plant material. The subsequent release of nutrients may also stimulate blue-green algal blooms but eventually most reservoirs develop an equilibrium between nutrient input and uptake (Levadnaya and Kuz'mina, 1972; Soltero and Wright, 1975). Studies in regulated Canadian lakes before and after impoundment showed that blue-green algae *Dactylococcopis* spp. and *Oscillatoria limnetica* replaced the former algal community of *Dinobryon* spp. *Asterionella formosa*, *Tabellaria* sp., *Rhizosolenia* sp., and *Melosira* spp. (Duthie and Ostrofsky, 1975).

In oligotrophic lakes the impact of regulation on lake phytoplankton may be small (Rodhe, 1964; Sinclair, 1965). Where drawdown is rapid and low water levels persist for longer periods, littoral algae will not develop. Species are not able to grow quickly enough to establish themselves. If drawdown is relatively slow, littoral algae may however be able to grow quickly enough to follow the changing water levels (Quennerstedt, 1958).

(b) Downstream of dams

The growth of benthic algae in rivers immediately downstream of hydro-dams depends to a great extent on the depth of the turbine and outlets in the lake and on generation schedules. Nutrient-rich hypolimnial discharges may stimulate production of some algae e.g. the filamentous genera *Cladophora, Oedogonium,* and *Spirogyra*. If the flow patterns are constant *and* turbidity and erosion are reduced below a dam, standing crops of algae may actually increase (Blum, 1957; Neel, 1963; Stober, 1964; Ward, 1974). However, if scour and substrate movement are increased because of sudden surges during electricity generation, standing crops of algae may be much lower than before impoundment (Douglas, 1958; Kobayasi, 1961a,b). Temperature changes downstream caused by discharges from the hypolimnion may also affect the algal populations (Spence and Hynes, 1971). Ward (1974) also showed that a cold-water summer discharge from the hypolimnion of a Colorado reservoir encouraged the cold-water chrysophyte, *Hydrurus* sp. in the stream below the dam. Also, warmer winter temperatures from the discharge for the first 9 km below the dam promoted winter growth of *Cladophora* and *Ulothrix*. Seasonal changes in the dry weight of the epilithon below the release were, however, closely correlated with the flow stability during the 10 days prior to sampling.

Below the surface spillways of the Sennar Dam on the Nile large numbers of phytoplankton cells were still identifiable up to 350 km downstream (Talling and Rzoska, 1967). Hammerton (1972), on the other hand, found few plankton cells below the Rosieres Dam where the turbine outlets were in the hypolimnion (Fig. 4.1). Phytoplankton cells from surface overspill can influence the composition of the invertebrate fauna downstream of a reservoir by encouraging filter feeders. Where diversion or flow regulation reduces downstream discharge so that areas of river bed are exposed, total algal production may be seriously curtailed. Also, where a series of reservoirs are produced by river impoundment the discharge of plankton and nutrients from upstream reservoirs may well affect the productivity and community composition of downstream reservoirs (Cowell, 1970). The nutrient supply to estuaries can be markedly altered by hydro-schemes, though the effects will depend on the design. Releases of nutrient-rich waters and decreased turbidity may enhance primary production (Fraser, 1972). On the other hand, in the Nile, reduced flows have actually reduced total loads to coastal waters with consequently low productivity of algae and food for fishes and wildfowl (Hammerton, 1972; Rzoska, 1976).

4.4. MACROPHYTES

4.4.1. In natural waters

Aquatic macrophytes are divisible into three ecological groups, namely 'rooted' forms, having roots into the substratum, 'free-floating' forms, and 'attached' forms which are fixed by a 'holdfast' arrangement to solid objects such as rocks (Hynes, 1970).

The most common macrophytes in fast-flowing mountain streams are mosses and liverworts. Mosses are specially suited to such habitats, as their very small but strong rhizoids attach to hard surfaces. Liverworts are thalloid and flattened and lie close to the substrate. Further downstream, as nutrient levels increase and soft sediments are deposited, the true rooted plants become abundant, including water lilies (*Nymphaea* spp., *Nuphar* spp.), water buttercups (*Ranunculus* spp.), and pondweeds such as *Elodea* spp., *Potomogeton* spp., and *Callitriche* spp.

Few angiosperms are adapted to fast-running water in temperate regions though in tropical rivers there are representatives such as the *Podostemaceae* which grow in riffles and on waterfalls (Hynes, 1970; Leentvaar, 1966, 1974).

Floating plants generally occur in lakes, in canals, or in silted backwaters of slow-flowing rivers. Examples are the duckweeds (*Lemna* spp.) of temperate waters and the water hyacinth (*Eichhornia* sp.), the water ferns (*Salvinia* spp.), and water lettuce (*Pistia* sp.) in tropical waters. Current velocity, substratum, light-penetration, nutrient levels, pH, and pollution all affect the distribution and ecology of macrophytes (Whitton, 1975b).

Macrophytes in rivers also affect flows by causing restrictions and siltation, often to such an extent that weed-cutting is necessary for efficient land drainage. Plants can also alter water chemistry by taking up nutrients and carbon dioxide from the water and producing large quantities of oxygen (Owens and Edwards, 1964). Temperature changes may also occur in lake surface waters with dense weed mats (Dale and Gillespie, 1976).

4.4.2. Effects of hydro-electric schemes

(a) Upstream of dams

Where a fast-flowing river is impounded the reduction of light penetration, sedimentation, and the chemical changes in deep water would be expected to eliminate all those macrophytes specially adapted for the original riverine habitat (Peltier and Welch, 1969, 1970; Boyd, 1971; Geen, 1974, 1975; Whitton, 1975b). Leentvaar (1966, 1974) found that although the fast-water *Podostemaceae* were found on rocks in the Suriname River Rapids (South America) only floating duckweeds and water hyacinth were found after inundation formed the Brokopondo Lake.

The development of a 'rooted' macro-flora after impoundment depends on water chemistry, light penetration, shore erosion, and the extent of draw-down. Where drawdown is extensive and prolonged any rooted aquatic plants growing in the littoral zone are eliminated, either by heating and desiccation in summer, or freezing and desiccation in winter, depending upon the use of the reservoir and the region. This also applies to natural lakes regulated to augment river flows or provide storage for hydro-schemes. Before regulation of Lake Blasjon (Sweden), *Ranunculus* spp., *Myriophyllum alterniflorum*, *Potamogeton gramineus*, *Sparganium* spp., *Eleocharis acicularis*, and *Subularia aquaticus*, were common in the submerged vegetation, while *Equisetum fluviatile* was common in protected coves. A few years after the lake was regulated all of these species were eliminated, and of the larger plants only sparse strands of the algae *Nitella opaca* remained. Freezing of the exposed areas in winter was believed to be the major cause of the lack of vegetation (Quennerstedt, 1958). Reduced light penetration and stratification prevented the development of a deep-water flora. Nilsson (1978) could, however, find no evidence of major floristic changes in an impounded Swedish river, though the distribution of some species was altered.

Erosion occurs partly because of the lack of vegetation in the drawdown zone, and Geen (1974) described a 'drawdown' zone in the regulated Buttle Lake, Canada, as little more than a sterile gravel bar, caused by wave action and erosion. Erosion may, however, be controlled by the introduction and maintenance of suitable vegetation in the drawdown zone (Little and Jones, 1979).

Light penetration determines the extent to which the flora of the sub-littoral (sub-drawdown) zone penetrates. In oligotrophic reservoirs some macro-vegetation may, therefore, develop (e.g. *Littorella* spp.) and maintain itself throughout fluctuations in water level, provided that water clarity enables light to penetrate during the 'full' periods (Hynes, 1961; Hunt and Jones, 1972). In more eutrophic and hence more turbid reservoirs, algal blooms inhibit light penetration and will prevent any macro-flora from developing in the deep water during the 'full' periods.

Where there are extensive shallow areas with soft substrates, for example in some of the Tennessee River impoundments, large growths of rooted macrophytes occur during the long periods of summer drawdown, usually to such an extent that they interfere with recreation, fishing, and navigation (Peltier and Welch, 1969, 1970; Elliot, 1973).

(i) *Nuisance plants*

Probably the greatest problems in tropical reservoirs are created by the explosive growth of floating or rooted macrophytes which occur soon after filling. Impoundments may be invaded by either indigenous lake species or introduced species. The most extensive studies have probably been made on the African lakes (Lowe-McConnell, 1966; Lagler, 1969; Boyd, 1971; Hammerton, 1972; Mitchell, 1973), though problems have also been reported in many areas including South America, the U.S.A, and New Zealand (Little, 1966; Mitchell, 1969, 1973; Elliot, 1973; Leentvaar, 1974). The dense mats of floating plants impede navigation and fishing by choking propellers and nets and prevent turbine operation by blocking screens and intakes. Further, the chemical and biological changes in the water column caused by the decay of dead plants, the reduced light penetration, reduced gaseous exchange at the water surface, and the formation of hydrogen sulphide may be drastic (McFie, 1973). Some plants, notably sedges, grasses, and the water hyacinth carry the invertebrate vectors of both human and cattle diseases which are of tremendous importance to the local populations (Brown and Deom, 1973). Whether an impoundment is used for hydro-electricity, or for other purposes, the development of these nuisance plants is a common factor in the tropics. The species of macrophytes receiving most attention in African impoundments have been in the water ferns *Salvinia molenta* and *S. auriculata*, the water lettuce *Pistia stratiotes*, and the water hyacinth *Eichhornia crassipes*. Other species may also cause problems in localized bays or flooded areas (Mitchell, 1969, 1973; Leentvaar, 1974). In temperate impoundments floating macrophytes rarely cause major problems, though *Lemna* spp. are sometimes abundant.

In Lake Kariba *Salvinia* developed so rapidly that only 3 years after first being sighted the weed covered over 650 km^2 (Harding, 1966; Mitchell, 1969). *Salvinia* mats form in sheltered areas and may be broken up by winds and waves. Clearance of flooded forest areas has, over the years, reduced mat-formation areas considerably. Other vascular plants, including grasses and sedges, are often established on *Salvinia* mats and the roots of these bind the mats together, forming the much more stable and less destructible 'sudd' (Mitchell, D., 1969). Sudd, once established, is an important habitat for the establishment and maintenance of insect disease vectors. *Pistia* is also common in the central African impoundments but tends to be excluded by *Salvinia* with which it competes. Both *Salvinia* and *Eichhornia* are South American native weeds, the latter being first recorded in Africa (Sudan) in 1957. It reached nuisance proportions in a very short period, being particularly suited to the Nile impoundments (Hammerton, 1972). *Eichhornia* also significantly increased the evaporation of water from reservoirs (Little, 1966; Hammerton, 1972) and it is also host to the snails *Bulinus* spp. and *Biomphalaria* sp. which transmit schistosomiasis.

(ii) *Control of nuisance plants*

Many attempts have been made to destroy nuisance plants. Mechanical, chemical, and biological control methods have been used but as yet chemical herbicides have been most effective and economic, at least on the floating or emergent vegetation (Little, 1966; Boyd, 1971). Control of submerged weeds is, however, much more difficult. In the Tennesssee Valley chemical and physical control of water milfoil (*Myriophyllum*) has been successful in the past but eradication is almost impossible (Elliot, 1973). Aquatic herbicides have been used extensively for controlling *Salvinia*, *Pistia*, and *Eichhornia*, though restrictions on the use of the more potent substances, such as 2-4.D, in the past decade have increased control problems (Little, 1966). Mechanical cutting and removal over large areas is impractical, and biological controls using snails, insects, herbivorous fish, or manatees will always involve an imbalance between the herbivore and the plant populations. The grass carp *Cteropharyngodon idella*, has been used successfully to control weeds in some Russian, North American, and European waters, but the full implications of such introduction to new areas are not fully understood (Bouquet, 1977). The turnover of nutrients by nuisance plants may cause further problems of eutrophication in reservoirs. The control of submerged weeds by lake fertilization, which increases turbidity and cuts light penetration as a result of phytoplankton blooms, is apparently successful in some regions (Little, 1966). Drawdown often strands and kills many floating plants and in some reservoirs it is used as a control method. Also, where large overspill channels are incorporated into dams, seasonal spates break up and wash away the troublesome weed mats, much as in a normal river system (Mitchell, 1973). There is little doubt, however, that invasions of nuisance plants cause major operational, fishery, navigational, and medical problems in many regions, offsetting the benefits of water storage, flood-control, and hydro-electricity.

Control of plants also adds to the cost of operating reservoirs.

(b) Downstream of dams

River regulation may reduce the available wetted area for macrophytes below some dams, thus reducing the total crop. Nutrient-rich releases from the hypolimnion may, however, actually result in increased production of aquatic angiosperms below some dams (Hilsenoff, 1971). Ward (1974) recorded aquatic angiosperms, such as *Potamogeton filiformis*, *P. crispus*, and *Ranunculus aquatilis* below the Cheeseman Dam in Colorado, though few plants were present in the first kilometre below the dam. The conclusion was that the low temperature immediately downstream of the hypolimnial discharge during summer did not stimulate growth.

Generally, the development of a macrophyte flora depends on the generation schedules at the power station and on the magnitude of the discharge fluctuations. The more stable the flow regime, the more successful the establishment of macrophytes.

4.5. ZOOPLANKTON

4.5.1. In natural waters

The zooplankton of free-flowing rivers is generally derived from other sources, usually lakes or pools in the catchment, though in lowland reaches some self-maintaining zooplankton may be found in backwaters or bays (Hynes, 1970; Whitton, 1975b). Protozoa, Rotifera, and small Crustacea are usually most abundant components of the zooplankton in fresh and tidal waters. Fish and mollusc larvae are abundant seasonally and, in fresh waters, larvae of insects as *Chironomidae* may also be abundant (Hutchinson, 1957; Sverdrup, *et al.*, 1963; Macan, 1970; Perkins, 1974; Whitton, 1975b).

In stratified natural lakes the zooplankton tends to be restricted to the epilimnion as does the phytoplankton, though in many lakes planktonic organisms migrate vertically in a diurnal rhythm. Oligotrophic lakes generally contain a sparser plankton than the more eutrophic lakes, and species compositions differ (Hutchinson, 1957). Zooplankton discharged from a natural lake into a stream may be distributed well downstream, though most organisms are removed by the normal lake outflow filter-feeders (see 4.5.2).

4.5.2. Effects of hydro-electricity schemes

(a) Upstream of dams

The rate of development of the zooplankton in a new reservoir formed from a fast upland river will depend partly on the proximity of other waters with an established zooplankton, which will provide an inoculum. In the Nile system the natural lakes such as Victoria and Albert are excellent sources of inocula as they contribute vast quantities to the river. In the free-flowing reaches small numbers of Copepoda can exist, though in the impoundments Cladocera are usually the more abundant (Brook and Rzoska, 1954). Like the phytoplankton, the zooplankton increases in abundance up to 200 times as the water approaches dams (Hammerton, 1972), and this type of effect is common to most new impoundments (Lowe-McConnell, 1966; Ackermann, *et al.*, 1973; Mordukhai-Boltovskoi, 1976). Hynes (1970) notes that the retention time of water in an impoundment is important for the establishment of zooplankton, as even active species are unable to maintain themselves against flows of over several millimetres per second.

Where existing lakes are used as a basis for regulation a zooplankton fauna will already exist, and unless regulation is very severe, changes may be slight. However, some changes can occur. For example, Sinclair (1965) showed that although the total abundance of zooplankton was unaltered after the regulation of Lower Campbell and Buttle Lakes in British Columbia, the cladoceran, *Holopedium* sp., was prominent following regulation, though it was not recorded prior to impoundment.

In the unstratified Lake Blasjon in Norway, Grimas (1964) showed that most cladoceran species were found near the lake bottom. The only species found to decline following regulation and drawdown was *Diaphanosoma brachyurum* which was normally most abundant in the upper portions of the littoral zone. *Eurycercus lamellatus* and other species in the lower littoral zones were not markedly affected.

The major production of Cladocera occurred earlier in the year in Blasjon than in non-regulated lakes. In flooded lakes used for hydro-schemes and in other water storage reservoirs the production of Cladocera has been found to be intensified during the early years of regulation, though this is a temporary effect (Grimas, 1961; Duthie and Ostrofsky, 1975) and is related to the rapid nutrient release and consequent phytoplankton production (see 4.3.2). In the natural lakes regulated for Canadian hydro-schemes increases in particulate feeders are reported, with *Diaptomus* spp, *Daphnia*, and *Bosmina* dominant (Duthie and Ostrofsky, 1975).

(b) Downstream of dams

In stratified lakes and reservoirs the depth distribution of zooplankton often tends to follow that of the phytoplankton (Hutchinson, 1957). The abundance of organisms discharged from a dam outlet in any season will, therefore, depend upon the depth and position of the turbine outlets. The Rosieres Dam on the Nile, for example, discharges very little zooplankton from its deep hypolimnial outlets (Hammerton, 1972).

The discharge of zooplankton from reservoirs has the same effect on the downstream fauna as that of the plankton from natural lakes, in that filter-feeding invertebrates become abundant in the benthos, feeding on the drifting plankton (Hynes, 1970; Hammerton, 1972; Armitage, 1977a). Planktonic animals are usually eliminated within a short distance downstream of the outflow by the filter-feeders, though in the Missouri planktonic organisms can persist for many kilometres below hydro-dams (Neel, 1963). In a study of a flow regulation reservoir in England, where water was discharged simultaneously from the surface and the bottom, planktonic crustacea were found to represent 29 per cent by number and 3 per cent by weight of the organisms drifting downstream in the reach below the dam during July (Armitage, 1977a).

It is possible that some planktonic crustacea in natural lakes may actively avoid outflows, though where the retention time is short in a reservoir and water movement more than 1 cm s^{-1}, this seems unlikely (Hynes, 1970). Cowell (1970) showed that the abundance and composition of the plankton fauna of a reservoir on the Missouri River was affected by the discharge of species from an upstream reservoir.

4.6. BENTHIC MACRO-INVERTEBRATES

4.6.1. In natural waters

The benthic macro-invertebrate fauna is probably the most studied part of aquatic ecosystems, and books to which reference has already been made deal extensively with the subject in great detail (Welch, 1952; Hynes, 1960, 1970; Macan, 1963, 1970; Whitton, 1975b; Newell, 1970; Perkins, 1974). All of these books contain extensive bibliographies for further reading. The term 'macro-invertebrates' here includes all those invertebrate animals visible to the naked eye, but not always identifiable to species without a microscope. This is a practical, though hardly quantifiable, working definition. In fresh waters the best-studied and most obvious groups are insects (mayflies, stoneflies, true-flies, caddis-flies, etc.), molluscs (snails and bivalves), crustaceans (shrimps and crayfishes), flatworms, hydras, sponges, and oligochaete worms. In tidal waters insects are scarce, and crustaceans (crabs, shrimps, prawns, barnacles, etc.) are much more abundant. The fauna of near-shore and inshore waters also includes other groups, such as tunicates (sea-squirts), anemones, and jelly fishes which are rarely, if ever, found in fresh waters.

The diversity of adaptive mechanisms among these invertebrates, which fit them for their specific micro-habitat or niche, is vast. For example, in fast-flowing upland streams with hard substrates animals may be streamlined, flattened, have strong hooks or suckers, or swim strongly; all features designed to prevent them being washed downstream. In silted and poorly oxygenated habitats organisms may have gill covers to prevent clogging respiratory surfaces, pigments to enhance oxygen absorption, or adaptations for direct air-breathing (Hynes, 1960, 1970). On rocky sea-shores many species have strong attachment organs to prevent displacement by waves while others live in sheltered crevices. Most littoral zone organisms are adapted physically or physiologically to withstand exposure to desiccation and temperature fluctuation during tidal cycles (Newell, 1970).

Apart from these basic adaptations, feeding, respiratory, and breeding processes are also often highly modified for a particular niche.

4.6.2. Effects of hydro-electricity schemes

(a) Upstream of dams

The flooding of a river bed upstream of a dam eliminates most of the obligatory 'fast-water' species of the original river bed because of silt deposition and changes in dissolved gases. In a number of studies of impoundments, the profundal fauna of the new reservoirs was found in most cases to be characterized by fly-larvae of the family Chironomidae and by oligochaete worms (Nursall, 1952; O'Connell and Campbell, 1953; Aggus, 1971; Isom, 1971; Hammerton, 1972; Radford, 1972; Geen, 1974; Armitage, 1977b; Kaster and Jacobi, 1978). These invertebrates are particularly tolerant of silted and deoxygenated conditions and are either burrowing or tubicolous. They also have characteristically high haemoglobin content in their body fluids which enhances oxygen absorbtion in low oxygen tensions (Macan, 1963), and are thus particularly abundant in polluted waters where oxygen depletion has occurred (Hynes, 1960; Hawkes, 1962). The elimination of predatory riverine invertebrates in reservoirs probably also allows their successful initial colonization. Given that such a basic profundal and sub-littoral fauna occurs in most stratified reservoirs, there exist in addition the faunas of open shores and vegetation and the species-composition of these invertebrate communities depends on wave-movements, substrates, amounts of organic material present, extent and type of vegetation, and the drawdown regime. The rate of development of this 'superimposed' fauna and its species composition also depends on the survival of some of the more adaptable species from the river (Nursall, 1952) and the immigration of rapid colonizers from nearby lakes and pools. For example, water bugs (*Corixa* spp., *Microvelia* spp.) and water boatmen (*Notonecta* spp.) which are active fliers, are often among the first species to appear in new reservoirs and pools.

In the Mozhaisk Reservoir, on the River Moskva in the U.S.S.R., only 56 of the 150 fast-water (rheophilic) species survived 9 months after flooding with only the mud-fauna (Oligochaeta and Chironomidae) flourishing (Rjoska, 1966). Detrital-feeding mysid shrimps and the zebra mussel *Dreissenia* sp. were also found to flourish in new reservoirs in the Volga Basin (Borodich and Havlena, 1973).

In the Nile the 'hard-bottom' riverine clams (*Corbicula* spp.) disappeared in the reservoir upstream of the Jebel Aulia Dam owing to siltation and stratification. Above the Rosieres Dam the large freshwater bivalve *Etheria elliptica* also disappeared for the same reasons. In the impoundments caused by the huge dams on the Tennessee River (U.S.A.) commercially fished mussel (clams) declined sharply after the dams were closed, though other species invaded. The number of species was reduced from about 100 to 52 (Isom, 1971). In the newly flooded parts of the Kananaskis River silt-dwelling *Pisidium* spp (pea-shells) became dominant in the benthos after flooding (Fillion, 1967). Where reservoirs do not stratify the profundal fauna tends to contain more species (Grimas, 1961). Species-diversity generally decreases with depth in stratified impounded waters. In many reservoirs, where complete hypolimnial deoxygenation occurs for long periods in summer, even the tolerant chironomid and oligochaeta fauna may disappear (Ferraris and Wilm, 1977).

In some eutrophic reservoirs artificial destratification, brought about by either pumping air into the hypolimnion or by circulation of epilimnial water to the hypolimnion, has resulted in a re-invasion of the deeper substrates by oligochaetes and chironomids as a result of increased oxygen concentrations. Further, in such lakes the sub-littoral species tend to penetrate to deeper waters (Ferraris and Wilm, 1977). The extent of development of a weed-dwelling fauna is related to the effects of impoundment and drawdown on the vegetation (see 4.4.2).

(i) *Effects of drawdown*
The abundance and composition of the sub-littoral faunas of reservoirs depends considerably upon the rate, extent, and duration of drawdown. Drawdown varies from 1 m to more than 40 m, and the duration from several hours in a pumped-storage reservoir to several months in water supply reservoirs.

In hydro-electricity reservoirs drawdown depends very much on the water capacity in relation to power demand. Maximum shore exposure may occur in summer or in winter depending upon the region and the electricity demand pattern (see 2.2.2.b). Generally, drawdown occurs too quickly for most invertebrates to follow the receding water level. Most studies show that drawdown zones contain sparse invertebrate faunas, though during full periods some recolonization may take place (Grimas, 1961; Lowe-McConnell, 1966; Isom, 1971; Efford, 1975; Kaster and Jacobi, 1978; see also Mordukhai-Boltovskoi, 1979). In most cases the maximum diversity of species and maximum biomass occur in the zone immediately below the lowest water level reached during drawdown (Geen, 1974, 1975). Drawdown is also the major factor affecting the macro-invertebrate fauna of natural lakes regulated for hydro-schemes.

In a number of such lakes species have disappeared or declined once drawdown began. For example, in Lake Blasjön the shrimp *Gammarus lacustris* declined in abundance, and the later *Asellus aquaticus* was eliminated soon after regulation began (Grimas, 1961). In some Canadian lakes the amphipod *Hyalella azteca* also disappeared after regulation (Geen, 1974), and in a Norwegian reservoir the crustacean *Lepidurus arcticus*, which was an important food item, declined, to the detriment of the brown trout population (Borgström, 1973). Drawdown was also responsible for the elimination of shore-dwelling stoneflies and mayflies in the Canadian lakes (Geen, 1975). Molluscs (snails, limpets, and clams) are particularly prone to stranding during drawdown and in most cases may be completely eliminated in the drawdown zone (Grimas, 1961; Sinclair, 1965; Isom, 1971; Kaster and Jacobi, 1978).

Kaster and Jacobi (1978) noted that although some clams (*Lasmigona complanata*) were capable of moving quickly enough to follow the receding water in a storage reservoir, they burrowed instead and died within about 5 days as the substrate froze. In contrast the oligochaete worms *Stylaria* sp. and *Pristina* spp., apparently active swimmers, kept pace with the receding water. Recolonization of the drawdown zone took about 3 months. Some months after re-filling, the inundated zones contained higher numbers and a greater biomass of benthos than before drawdown, mainly Chironomidae and Oligochaeta.

Plate 4.1. The drawdown zone of a reservoir, showing exposed substrata in different areas. (Photographs supplied by Dr P.D. Armitage)

Recolonization of the drawdown zone may be more rapid than expected where the duration of 'low-water' is less than a month, because of several mechanisms for resisting drought shown by some invertebrates (Hynes, 1958). For example, large oligochaetes burrow into exposed substrata and may survive for several months (Kaster and Jacobi, 1978). Also the cocoons of several species can survive in damp areas, hatching on re-inundation. Grimas (1961) also found that 80 per cent of the benthic fauna of the Blasjôn survived after 3 months of freezing in mid-winter drawdown. Finally, benthic Chironomidae may re-colonize rapidly because they breed profusely in newly flooded areas and the larvae hatch out from eggs in a few days. Faster-moving animals such as amphipods may also move quickly in-shore as the water rises.

In a pumped-storage reservoir, where the fluctuation cycle may be only several hours, it is likely that many invertebrate species will survive and prosper provided there is a substrate suitable either for burrowing or for providing damp areas and shelter during drawdown.

In a reverse of the usual situation the normal autumn drawdown of Lake Francis Case on the Missouri River was reduced by regulation from about 12 m to about 7 m. The fauna of the newly 'wetted' zone, which was previously dominated by a chironomid/oligochaete community, was soon characterized by an increasing abundance of the burrowing mayfly *Hexagenia limbata* (Benson and Hudson, 1975). This species has also been found to be common in other stabilized reservoirs (Isom, 1971).

The floating vegetation in tropical and sub-tropical reservoirs provides a habitat for some invertebrates which feed either directly on the plants or on their epiphytes (Lowe-McConnell, 1966; Petr, 1968; Hammerton, 1972; Junk, 1977). Different species of plants have different characteristic communities (Petr, 1968; Junk, 1977).

To summarize the changes upstream of a dam, therefore, it is evident that the basic benthic fauna of either a flooded river or regulated lake is dominated, at least in the profundal zone, by Oligochaeta and Chironomidae. The composition of the shallow or 'superimposed' community depends on species remaining from the previous habitat (river or lake), and the proximity of habitats which will provide colonizers. Drawdown definitely decimates the shallow fauna but recolonization may be more rapid than expected. There are suggestions that once a reservoir 'matures' species which seemed to have disappeared soon after impoundment may become re-established (Hynes, 1961; Isom, 1971; Hunt and Jones, 1972), though more evidence needs to be obtained. In reservoirs on the Volga River (U.S.S.R.) mysid shrimps have been actually introduced to provide extra food for resident fishes (Borodich and Havlena, 1973; see also Mordukhai-Boltovskoi, 1979).

(b) Downstream of dams

The amount of plankton and drifting organisms in the outflow water of a reservoir or lake determines to a great extent the composition of the fauna downstream. As a result of this food resource the invertebrate fauna of rivers below natural lakes and surface-spill reservoirs are often dominated by filter or net-feeding species, such as the net-spinning caddis flies (Hydropsychidae) and the black flies (*Simulium* spp.). The total invertebrate biomass may be much greater than before impoundment because of the enhanced food supply (Chutter, 1963; Hynes, 1970; Spence and Hynes, 1971; Walburg, *et al.*, 1971; Hammerton, 1972).

Below hydro-electricity turbine outlets there may be, on the other hand, a reduction in the fauna caused by low temperatures and poor water quality of a hypolimnial discharge, by scour at high discharge, or by flow restriction during non-operating periods (Radford and Hartland Rowe, 1971; Covich, *et al.*, 1978). Lemkuhl (1972), for example, found that none of the twenty mayfly species found in the river upstream could survive immediately downstream of the Gardiner Dam on the Saskatchewan River. Only seven species were present up to 110 km downstream. Temperature changes were considered to be the main causes (see 2.3.2.b). Some species did not have the period of cold required for completion of their life-history in winter, while low summer temperatures prevented other species emerging. Temperature changes were also reported as the reasons for the faunal differences below hypolimnial releases from the Belwood Lake, Ontario (Spence and Hynes, 1971), and the Cheeseman Dam, Colorado (Ward, 1974), though neither were hydro-electricity dams. Chironomidae are common to most lake and reservoir outflow areas and often exist where few other species are found (Spence and Hynes, 1971; Isom, 1971; Walburg, 1971; Lemkuhl, 1972; Ward, 1974; Covich, *et al.*, 1978).

In the Kennebec River, where hydro-electricity generation caused major fluctuations in the flow regimes (Trotzky and Gregory, 1974), the absence of many species below the dams was considered to be a result of the slow current velocities during non-operating periods, though scour could also have been a major factor during generation. The biomass of insects was between 2·5 and 32 times greater in riffles above the dams than in those below. Fast-water species were absent in the areas of lowest current velocities downstream of dams. Covich, *et al.* (1978) also found lowest densities of Chironomidae in areas of greatest flow fluctuations below a dam on an Oklahoma river.

The drying-out of large areas of river bed during non-operating periods in badly regulated rivers reduces the available habitat for invertebrates and fish (Baxter, 1961; Trotzky and Gregory, 1974; Davies, B., 1975), though the recolonization of some areas may occur through downstream drift, upstream migration, migration upwards from within the substrate, and aerial invasions (Williams and Hynes, 1976).

Downstream drift may, however, be severely restricted by a dam and colonization by upstream migration of some species may be highly significant in such cases (Spence and Hynes, 1971). Where multiple level outlets are used to maintain temperature and water quality certain invertebrates may flourish.

For example, the hard-bottom bivalve molluscs eliminated from U.S.A. rivers by lacustrinization flourish in outlets from hydro-electricity turbines (i.e. tailwaters) (Walburg, *et al.*, 1971; Isom, 1971; Hammerton, 1972, 1976). Also, filter-feeders flourish, as do many obligatory fast-water species. Where flows are drastically reduced slow-water species may be able to survive better in some reaches (Trotzky and Gregory, 1974).

Thus to summarize, hydro-electricity generation will cause a decline in the diversity and abundance of species downstream if regulation is severe and the discharge hypolimnial. In less severely regulated rivers, or where multiple-level outlets are used, the abundance of the fauna may actually increase, though usually diversity is reduced.

(i) *Effects on invertebrate drift*
In running waters, benthic organisms are commonly found drifting in the water column. These organisms are of considerable importance as fish food and to the colonization of downstream habitats (Horton, 1961; Waters, 1961, 1969, 1972; Bailey, 1966; Hynes, 1970; Elliott, 1973; Gibson and Galbraith, 1975). Generally, there is a diurnal pattern of invertebrate drift with maxima around dawn and dusk (Waters, 1962; Elliott, 1965).

Drift abundance also tends to be closely related to flow (Waters, 1962; Elliott, 1967; Minshall and Winger, 1968; Anderson and Lemkuhl, 1968; Kroger, 1974; Brooker and Hemsworth, 1978; see Ward and Stanford, 1979), and the fluctuations caused by hydro-electricity generation would be expected therefore to affect drift of benthic invertebrates living below turbine outlets. The sudden release of water and increased flows during power station operation will scour resident benthic organisms in the same manner as artificial freshets, or natural spates (Radford and Hartland Rowe, 1971).

In the tailwater fauna of the Gavins Point Hydro-electric Station on the Missouri River most of the drift fauna originated in the reservoir, though some drifting Chironomidae were from the tailwater benthos (Walburg, *et al.*, 1971). In the River Wye in the U.K., 48-h discharge from an impoundment caused up to a sevenfold increase in the numbers of drifting benthic organisms (Brooker and Hemsworth, 1978). Some organisms, however, retained their normal diurnal drift periodicity in both the Wye and in the tailwaters of the Gavins Point Power Station.

(c) **Nuisance invertebrates**
There are many records of invertebrates reaching nuisance proportions in impounded waters, either causing human health problems (Lowe-McConnell, 1966; Lagler, 1969; Hammerton, 1972; Brown and Deom, 1973) or competing with species which are commercially viable (Isom, 1971; Walburg, *et al.*, 1971).

In the Nile, *Simulium damnosum*, the blackfly species, which transmits human onchocerciasis, flourishes in the fast water below the turbine outlets of the Rosieres Dam. Also, *Tanytarsus* spp. (midges) became a nuisance in the Wadi Halfa area near the Aswan Dam, causing allergies and a form of asthma (Hammerton, 1972). The midges bred in the shallow, newly flooded margins where reeds and weed beds developed. The snail vectors of schistosomiasis, and insect vectors of malaria, yellow fever, sleeping sickness, and filariasis are all able to survive and flourish either in the water or on the vegetation of new impoundments in tropical and sub-tropical areas, and though some of these organisms have been at least partly controlled so far with insecticides and molluscicides, restrictions on use of the most powerful chemicals may well allow the diseases to take hold again (Lagler, 1969; Brown and Deom, 1973).

Competition with the commercial bivalve species of the Tennessee River by Unionidae (swan mussels) and the Asiatic clam, *Corbicula manilensis*, has caused problems and the latter has also become a nuisance in water supply pipes and water intakes (Isom, 1971).

4.7. FISHES

4.7.1. **In natural waters**

Nikolsky (1963) summarized the basic requirements and the variety of ecological relationships of many freshwater and marine fishes. The fishes of fresh waters can be divided into three major ecological groups which are relevant here, namely:

Rheophilic species, i.e. those species which live, feed, and reproduce entirely in running water and depend on flowing water for survival. They may be physically modified for maintaining position in fast-water, for example by the development of a friction plate, such as bullheads (*Cottus* spp.) or suckers *Garra* spp. or alternatively they may have evolved strong musculature and a slim shape to swim well against a current, like salmonids.

Migratory or diadromous species, i.e. those which maintain a two-phase life-history involving extensive migrations between the sea and fresh waters. The seaward migrations may be to reproduce as in the eel (*Anguilla* spp.), or to feed as in salmon (*Salmo* spp.), sturgeon (*Acipenser* spp.), shad (*Alosa* spp.), and lampreys (*Petromyzon* spp.). Some species migrate within fresh waters usually between spawning streams and the lakes in which they feed. These are known as potamodromous. Most migratory species rely on the natural characteristics of swift-running water for successful reproduction and egg-development, that is clean stones or gravel with intergravel spaces through which the water flows, supplying and removing metabolic products.

Limnophilic species, i.e. those which live best in slow flowing or standing water, for example bream (*Abramis* spp.), carps (*Cyprinus* spp.), and whitefishes (Coregonidae).

A number of species may be found in both running and standing waters, including trout (*Salmo* spp.), roach (*Rutilis* spp.), and other Cyprinidae, perch (*Perca* spp.), and pike (*Esox* spp.).

Most of the limnophilic species spawn and feed among weed-beds or on stones, though some species, for example whitefishes (Coregonidae), require clean stony shores with considerable water movement for successful spawning and egg-development.

Nikolsky (1963) and Hynes (1970) showed clearly how the body shapes of fishes relate to the current velocities in which they are generally distributed. The slow-water species, for example bream and carp, appear to have much more rounded and less streamlined bodies than the strongly swimming, fast-water species like trout and salmon, though some species such as the slim bleak (*Alburnus* sp.) found in slow waters do not fit easily into the 'slow-water' shapes. The European system of river zonation and classification is based on the normal current-velocity requirements of resident species (Hawkes, 1975). The tolerance of each species to water-quality conditions varies considerably but generally the rheophilic and salmonoid species are less tolerant to low oxygen levels and pollution than the limnophilic species (Jones, 1964).

4.7.2. Effects of hydro-electricity schemes

Impoundment of natural lakes and rivers may have both deleterious and beneficial effects on fish and fisheries. There are, predictably, changes in the species compositions of impounded rivers, but these changes may lead to greater total fish production and at least to greater availability of recreational areas, including those for fishing (Lowe-McConnell, 1966; Hall, 1971; see also Ackermann, *et al.*, 1973; Mordukhai-Boltovskoi, 1979). For the purposes of this section, 'fishes' are divided arbitrarily into resident and migratory species, irrespective of habitat, though it is difficult to adhere strictly to this division. There is little doubt, however, that the special problems of the migratory fishes have caused most controversy in the development of hydro-electricity schemes (Pyefinch, 1966; see also Lowe-McConnell, 1973; Efford, 1975; Mordukhai-Boltovskoi, 1979).

(a) Resident (non-diadromous) species

These include rheophilic and limnophilic species, which do not migrate between fresh and saline waters. They may, however, migrate within river or lake catchments for feeding, breeding, or dispersal and such potamodromous fishes are known from many parts of the world (Nikolsky, 1963).

(i) *Upstream of dams*

Species composition. The effects of lacustrinization as they apply to other aquatic organisms also apply to fish, in that resident riverine (rheophilic) species decline or disappear to be replaced by lacustrine (limnophilic) species as the reservoir fills and matures (Lowe-McConnell, 1966, 1973; Petr, 1968; Hall, 1971; Holcik, 1977; Blake, 1977a,b; Mordukhai-Boltovskoi, 1979).

In stratified reservoirs the changes in the fish community after impoundment tend to be much more drastic than in non-stratified reservoirs (Pyefinch,

1966; Aass, 1973). The general patterns of change are similar to all parts of the world, though the species differ from region to region. Such changes are reported from many major river systems including the Volga (Rzoska, 1966; Mordukhai-Boltovskoi, 1979), the Tennessee, Missouri (Ruhr, 1957; Hall, 1971), the Danube, Nile, Columbia (Oglesby, *et al.*, 1972; Rjoska, 1976), the Zambesi, Niger, Amazon (Lowe-McConnell, 1966, 1973; Lagler, 1969; Blake, 1977b) and many others (see also Ackermann, *et al.*, 1973; Ward and Stanford, 1979). Tributary systems have also suffered invasions of limnophilic species further and further upstream (Ruhr, 1957; Erman, 1973; Mordukhai-Boltovskoi, 1979). Only a few examples of the many published are used for illustration, though reference to the others can be found via the bibliography. Lowe-McConnell (1973) also gives an excellent summary of the effects of impoundment on fisheries in different parts of the world.

In a Missouri reservoir, studied for the first 7 years after impoundment it was found that of the 68 river species, 6 did not appear in the impoundment and 13 species declined sharply, probably due to unfavourable spawning conditions and competition from limnophilic species. Several species, including gizzard shad (*Dorosoma cepedianum*), bluegills (*Lepomis* spp.), carp (*Cyprinus carpio*), and channel catfish (*Ictalurus punctatus*), became much more abundant. The species which flourished in the impoundment were mainly those originally found in slow weedy backwaters of the river (Patriarche and Campbell, 1958). Similarly, in Lake Kariba above the Kariba Dam, species common immediately after impoundment were those previously found in slow weedy streams (Jackson, 1966). Only 9 of the original 29 riverine species produced significant populations in the reservoir after inundation. Post-impoundment declines may be followed by gradual increases in abundance as a reservoir stabilizes, though some families or species may adapt to lacustrine conditions faster than others (Blake, 1977a) (see also Mordukhai-Boltovskoi, 1979). In the Volga River, sturgeons, other migratory fish, and rheophilic species which inhabited the river channels and tributaries declined with the massive expansion of the hydro-electricity schemes. The limnophilic species of the flood plain and lakes and the pelagic planktivorous species have invaded and become dominant in the reservoirs. Bream, roach, and pike now constitute the major part of the exploited stocks. As in other regions, the riverine fishes maintain small populations in the tailraces (see Mordukhai-Boltovskoi, 1979).

In some North American impoundments the number of species present has also increased after impoundment, often to the over-all benefit of the productivity, though the most 'prized' fish for angling are often eliminated (Hall, 1971). In the Kananaskis hydro-reservoirs in Canada the longnose sucker (*Catostomus catostomus*) disappeared, allegedly as a result of the loss of river spawning areas (Radford, 1972), though Nelson (1965) found similar declines in the species populations in non-impounded rivers. Other factors, therefore, may have been responsible for the decline, and this case emphasizes the need for good controls or parallel studies of a species in different places. Predictions for other Canadian rivers being impounded for hydro-electricity schemes suggest that riverine species such as grayling (*Thymallus thymallus*) will be replaced by lake trout (*Salvelinus namaycush*) and whitefishes (*Coregonus* spp.) (Geen, 1974).

Spawning habits may also influence the relative success of anadromous species in impounded rivers. In the St. John River the anadromous alewife (*Alosa pseudoharengus*) and the blue-back herring (*A. aestivalis*) have increased, while anadromous species, for example salmon (*Salmo salar*) and shad (*Alosa alosa*), have declined. The reason is that the former spawn in still water, the latter in fast clean water (Dominy, 1973). The pattern is basically very clear. The prediction of future species-composition in an impounded river can be based on a knowledge of the relevant local backwater, pond, and lake species in the vicinity. However, in some lakes artificially introduced species may become dominant, especially if the indigenous populations are cleared before stocking or if selective cropping is carried out (Jenkins, 1970; Hall, 1971; Lowe-McConnell, 1966, 1973). In all regions, however, the extension of lacustrine conditions to higher altitudes extends the geographical range of the limnophilous species.

Depth distribution. In many stratified reservoirs fish are restricted to the epilimnion for much of the summer by availability of oxygen and food (Lowe-McConnell, 1966; Coke, 1968; Fast, 1971; Hall, 1971) (Fig. 4.2). In

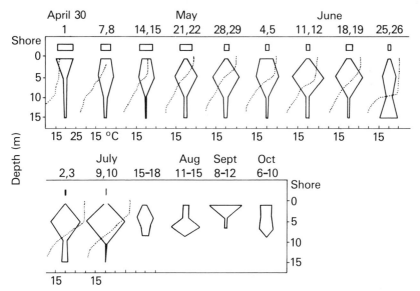

Fig. 4.2. Relative vertical distribution patterns of young gizzard shad expressed as percentage of the catch at each depth on the date indicate in area 2 of Beaver Reservoir, Arkansas (U.S.A.) 30 April–10 October 1969, redrawn after Netsch, *et al.* (1971).

highly eutrophic reservoirs, high temperatures and deoxygenation in the surface layers may cause heavy mortalities (Fast, 1973).

Young fish are often concentrated in the epilimnion (Netsch, *et al.*, 1971; Storck, *et al.*, 1978), so much so that large-scale mortalities of larvae and juveniles can occur in power stations using epilimnial turbine intakes (Walburg, 1971; Lewis, *et al.*, 1968). Walburg (1971) estimated that 24-hour losses of yearling fish through the surface-spill discharge from the Gavins Point Hydro-electric Station (U.S.A.) totalled over 11.5×10^6 individuals of several species. In contrast, losses from the Lake Francis Case hydro-electric station, upstream from Gavins Point were almost nil, because the discharge was from a depth of 40 m, well into the hypolimnion. Most fishes discharged from Gavins Point were less than 25 mm long and were probably unable to withstand the water currents created during flushing. Summer survival of some species was found to be inversely related to the July–August flushing rate. In contrast, in a flood-control reservoir on a tributary of the Columbia River, young Chinook salmon were found to be more abundant in the hypolimnion than the epilimnion and many emigrated via the hypolimnial outlet during flushing, usually dying as a result of pressure changes. The hypolimnion was generally oxygenated and cool for much of the summer (Korn and Smith, 1971).

In some smaller reservoirs the depth distribution and consequent potential standing crop of some fish species has been extended by artificial destratification (Fast, 1971, 1973).

Growth. The growth of fish in a new reservoir generally depends on their adaptability to changing food resources and on competition from other fish. The picture is often complicated by deliberate introductions either of invertebrates (Borodich and Havlena, 1973) or of prey fish species (Hall, 1971). There are a number of post-impoundment studies of growth (Hall, 1971; Nelson and Walburg, 1977; Pivnicka and Svatora, 1977; Spanovskaya and Grygorash, 1977), but very few pre- and post-impoundment comparisons of the same species in the same area, usually because the most abundant species before impoundment are not the same as those afterwards (Elrod and Hassler, 1971).

As a general rule, planktivorous and phytophagous fishes grow well in the early years of reservoir development when nutrient release and plankton production are at their peak (Patriarche and Campbell, 1958; Lowe-McConnell, 1966, 1973; Hall, 1971; Dudley, 1974, 1979). Young fish which feed on plankton show good growth rates but these are also restricted later on if there is not sufficient food for the larger fish (Hall, 1971; Pivnicka and Svatora, 1977). The growth of the freshwater drum (*Aplodinotus grunniens*) varied in relation to age and food availability in the Lewis and Clark impoundment (see Hall, 1971). Young fish (0–1 year olds) grew well in the early years after the reservoir was filled because of increases in primary productivity and zooplankton. Growth rates of older fish were poor at this time. Later, as the reservoir stabilized and plankton declined, growth-rates of juveniles slowed down. However, growth of older fish increased sharply as a result of the development of a large population of the mayfly (*Hexagenia limbata*), on which the fish fed.

In very large reservoirs the growth of fish may vary in different sections owing to the presence or absence of suitable food. For example, Varley, *et al.* (1971) found at least ten statistically separable rainbow trout populations in the Flaming Gorge Reservoir, based on differences in growth.

Production and yield. There is little doubt that the lakes created by impounding major tropical and sub-tropical rivers have provided exploitable food resources and are valuable producers of cheap protein for the developing human populations (Lowe-McConnell, 1973).

However, the initial increases in production needs to be maintained for the reservoir to be of long-lasting benefit, and there is ample evidence that in many impoundments fish production declines after the first few years unless species are introduced to make full use of the lacustrine environment. The main cause of the decline is usually that the 'nutrient–phytoplankton–zooplankton' explosion is soon spent and the lake stabilizes. Overfishing may also be an important factor. For example, Harding (1966) showed that the catch per unit effort on the north bank of Kariba decreased from about 17·65 kg/100 m of gill net in 1959–60 to 6·4 kg/100 m in 1961, mainly because of overfishing. Catches increased later as fishing became controlled, and in 1963 estimates put the total commercial yield as 3,500 tonnes, up from 500 tonnes in 1960.

In a Kentucky reservoir, angler catches declined from 1·48 to 0·81 fish per man hour between the first and fourth year after filling. Actual effort in man hours increased incredibly from 28 per day before impoundment to 1,050 per day after impoundment. Also, the species caught altered as the reservoir matured (Turner, 1971).

In the Marvann Reservoir in Norway the brown trout yield declined almost to zero once regulation began, mainly because of the lack of food and the emigration of fish through the outlets (Borgström, 1973).

Fishery problems are not, however, peculiar to hydro-electricity reservoirs. Management usually requires stocking programmes, introduction of new species, and some form of catch limitation (Hall, 1971), and many waters are managed on a supply and demand basis (Jenkins, 1970; Templeton, 1971). In most cases the total yield of fish from a reservoir stabilizes at a much higher level than from the river prior to impoundment (Lowe-McConnell, 1973).

The major factors influencing production and yield are dissolved solids, age of reservoir, storage ratio, shore stabilization, area, mean depth, and to a lesser extent, outlet design, stratification, and management (Jenkins, 1970).

Effects of drawdown on feeding and reproduction. We have already noted some of the drastic effects of drawdown on the littoral faunas of hydro-electricity and water supply reservoirs. Many of the invertebrate species are important in the diet of resident fishes in a natural lake. Also weed beds on lake or river margins provide spawning cover for many limnophilic species (Nikolsky, 1963; Jackson, 1966; Hall, 1971).

There are many lake-spawning fishes in all parts of the world. Some of these spawn on gravel banks and shoals in littoral zones. Spawning in many limnophilous fishes is triggered by rising water levels in wet seasons (Nikolsky, 1963).

The reduction in littoral food supply is reported to have been a major factor in the disappearance of the Dolly Varden (*Salvelinus malma*), rainbow trout, and cut-throat trout from the Kananaskis River impoundments (Geen, 1974). Graham and Jones (1962) and Haram and Jones (1971) suggested that although brown trout (*Salmo trutta*) and the Welsh coregonid, the gwyniad (*Coregonus clupeoides pennanti*), survived in Llyn Tegid after regulation, both species appeared to have partly changed their diets. Littoral species of insects and *Gammarus*, which disappeared for some time after regulation, were replaced in the diets by sub-littoral species and planktonic crustacea. Changes in diet also occurred in bullheads (*Gobio gobio*), minnows (*Phoxinus phoxinus*), and trout (*Salmo trutta*) in a North of England reservoir after impoundment (Crisp, *et al.*, 1978). Trout fed on inundated terrestrial material for 3 years after inundation, while the other two species adapted to zooplankton and chironomid diets. Borgström (1973) considered that the almost complete destruction of the crustacean (*Lepidurus arcticus*), in the Marvann Reservoir in Norway, was a major reason for the falling yield and declining condition factors of brown trout. After 2 years of increased water level fluctuation the total yield of fish fell from an estimated 350 kg in 1969 to practically zero in 1971 and 1972. Increased erosion and turbidity were considered to be the main cause of the destruction of the *Lepidurus* population. In other Scandinavian reservoirs drawdown reduced the littoral food supply of brown trout but char, which are plankton feeders, were not affected (Aass, 1957; Nilsson, N., 1965).

We have already noted the introduction of mysid shrimps to the Volga reservoirs to provide food for resident fish (see 4.6.2.a). These mobile shrimps would not be left stranded during drawdown and would survive and prosper. In reservoirs which contain Cyprinidae (carp, roach, and bream) and northern pike (*Esox lucius*), which use macrophytes as spawning cover, virtual destruction of weed beds during drawdown may eliminate some species. In some Eastern European reservoirs, however, roach and white bream (*Blicca bjoernka*) will spawn on rocky shores in the absence of weed, though the eggs of these species would be killed by exposure during drawdown. The 'littoral-gravel' spawning fishes such as lake trout, large-mouth bass, and coregonids also suffer large egg mortalities where summer drawdown exposes late spring eggs to desiccation and winter drawdown exposes autumn-laid eggs to freezing and subsequent death (Hall, 1971; Geen, 1974, 1975).

In the Nasser Reservoir, behind the Aswan Dam, extensive and rapid drawdowns left tens of thousands of littoral-zone mating stations of *Tilapia* spp. exposed to desiccation. Isolated, shallow pools in the drying-out littoral zone contain many fry which are exposed to predatory birds and high water temperatures (Ryder and Henderson, 1975). The fishes fitted to survive in reservoirs, therefore, are those which either spawn during filled periods, spawn at depth or in open water, or in feeder streams. Van Velson (1974) showed that rainbow trout (*Salmo gairdneri*) spawned in streams feeding a Nebraska reservoir and sustained their population size. Severe drawdown before the spawning runs may, however, make the streams inaccessible (Van Velson, 1974). Fish with buoyant eggs such as the fresh-water drum (*Aplodinotus grunniens*) can maintain populations even in fluctuating water levels (Lowe-McConnell, 1973). Aass (1971) described a most unusual mass migration of Arctic char (*Salvelinus alpinus* L.) in two Norwegian hydro-electric reservoirs during the period of drawdown. In natural lakes the winter distribution of char differed little from the autumn distribution. However, in hydro-reservoirs, such as the high boreal lakes Tunhovdfjord and Paalsbufjord, the onset of drawdown set off extensive migrations. The post-spawning and early winter movements were mainly toward the nearest shore for feeding. Later in winter, as lake levels lowered, char accumulated downstream in large numbers within currents near the inlets to the lake outflow tunnels. Fish also accumulated in the narrow channels which developed as water levels were lowered. The nearest fish entered the fast-flowing areas first, and as winter progressed fish from increasing distances joined the group. Eventually, more than 50 per cent of the whole lake population migrated to the faster currents. Some fish may have participated in these migrations in several subsequent years. The reasons for migration were not clear. There was little food supply in the currents, and as Aass stated 'the downstream movement makes little sense as active feeding migration'. He concluded eventually that drawdown caused the fish to move to deeper water and the downstream movement was a passive displacement. An important factor was that the accumulation of fishes was readily exploited by anglers. In summer, fish became redistributed over the reservoir as water levels rose.

The conclusion is from the evidence, so far, that for most species feeding and spawning on a near-shore substrate, drawdown can only be detrimental to population size.

Drawdown and management. Drawdown may not always be entirely injurious to fisheries as a whole. In fact 'deliberate drawdown' is sometimes used as a management technique to reduce stocks of undesirable species, or to thin out unwanted vegetation (Jackson, 1966). To reduce carp populations (*Cyprinus carpio*), for example in U.S.A. game fishing reservoirs, 'artificial' drawdown after spawning is used to expose and destroy most of the eggs normally laid in weedy shallows (Shields, 1958). In many temperate and tropical fisheries deliberate drawdown is a major conservation measure to reduce competition for the more desirable fishes. Drawdown also allows lake bottom deposits to dry out and become aerated. Re-inundation then stimulates the nutrient release, much as in new reservoirs (Jester, 1971). Drawdown zones may also be major feeding grounds for migratory wildfowl which dig up stranded invertebrates. In Hungary, pigs feed on mussels, water snails, and aquatic weed roots on mud-flats in drawdown zones.

(ii) *Downstream of dams*

Where a regulated river has adequate good-quality compensation water, or where regulation has resulted in an actual increase over the 'natural' minimum flow (Pyefinch, 1966; Havey, 1974), resident river fish populations below a hydro-scheme may be maintained or even increase.

Tailrace fisheries at hydro-generating stations are a significant part of river sport fishing in the U.S.A. utilizing both migratory and resident species (O.R.R.C., 1962; see Hall, 1971). Below the Gavins Point Dam 29 species were recorded in the tailwaters of the hydro-electricity station and only 20 species in the reservoir. Seventeen species were common to both (Walburg, *et al.*, 1971) and most species appeared to grow better in the tailwater than in the reservoir. Few were caught in the fastest water and most occurred about 0·5 km downstream of the outlets. It was also concluded that some species were spawning in this habitat. Food included spilled-over reservoir organisms such as plankton, chironomids, algae, and terrestrial insects. Studies on a Volga River reservoir showed that benthos-feeding fish grew better in the reservoir than in the tailwaters, while predators showed the reverse trend (Mordukhai-Boltovskoi, 1979).

In larger rivers, like the Volga and Tennessee, hydro-generation probably uses only part of the flow at any one dam, and fluctuations in current and discharge are not too violent. In smaller rivers this is not usually the case. Where flows are reduced to practically zero during non-generating periods the downstream fish will suffer, either by being stranded or from lack of food (Kraft, 1972; Trotzky and Gregory, 1974; Erman and Leidy, 1975). Spawning areas may also be dried out by diversions or drastic regulation (Berry, 1955).

Rainbow trout stranded in pools in an intermittent stream were found to survive short periods of low flows and low oxygen but were more prone to predation by birds or other vertebrates (Erman and Leidy, 1975). Kraft (1972) showed clearly that the number of 1-year-old trout in riffle areas was reduced by 62 per cent when 90 per cent of a stream flow was diverted for 3 months. Smaller fish did not decrease in numbers, however, even under these conditions. Where fish spawn in such reaches competition for spawning space is intensified and reduced current velocities will affect the interstitial flow through the gravel in which eggs are laid. High temperatures in summer may also be lethal (Fraser, 1972).

Many fish species in tropical rivers depend upon the seasonal flooding of riverside vegetation for spawning (Fraser, 1972; Hammerton, 1972). River regulation for hydro-generation and multi-purpose use has generally reduced these floods in many regions. In other rivers natural floods have caused mortalities of salmon and trout eggs as a result of silting (Withler, 1952; Needham and Jones, 1959; Wickett, 1959) and regulation which reduces such floods may reduce these mortalities, provided compensation flow is adequate (Berry, 1955).

Below hypolimnial outlets the combined effects of temperature and chemical changes may stress fish populations so that diversity is reduced, often because a few species are abundant (Edwards, 1978).

(b) Migratory (diadromous) species

Migratory fishes have always been faced with natural obstacles on their journeys in the form of waterfalls, landslides, or perhaps even beaver dams. Mill dams and fish weirs have been the subject of fishery law for many hundreds of years (Netboy, 1968; Gregory, 1974) suggesting that their adverse effects on fish were well understood. Many migrating fishes can usually overcome such obstacles provided they do not exceed 1·4 m in height and water flow is adequate for progress. Atlantic salmon (*Salmo salar*), for example, can leap heights up to 3·3 m (Jones, 1959) while 'even the sluggish carp (*Cyprinus carpio*) has been observed surmounting low head dams' (Hynes, 1970). Eels may move upstream by wriggling over the edges of waterfalls and cascades or through wetted water margins.

The life-cycle of most diadromous fish involves the movement upstream during one phase of the life-cycle and downstream in another phase. In an anadromous fish, such as the salmon which spawns in fresh-water, adults go up, the juveniles (smolts) down. The catadromous eel shows the reverse pattern (Sinha and Jones, 1974). Thus hydro-electricity schemes present obstacles to both upstream and downstream movements.

(i) *Migration distances and dams*

The larger dams built for hydro-electric power generation and river regulation are insurmountable by any fish without some bypassing facility and it was realized very early on that for migratory fish stocks to survive, such bypasses were necessary. Economics did not initially allow this at many sites, with devastating consequences for the fish (Larkin, 1958; Netboy, 1968; Schweibert, 1977).

Many species migrate a long distance to spawn. For example the sturgeons (*Huso huso*, *Acipenser galdenstadtii*, and *A. stellatus*) migrate up to 2,000 km in the Danube. The migration distance was however reduced to 750 km by the insurmountable Iron Gate Dam on the Yugoslavian/Rumanian border. In the Volga, white salmon (*Stenodus leciethys*) and the lamprey (*Caspiomyzon* sp.) may migrate up to 2,800 and 2,400 km respectively (Hynes, 1970; Liepolt, 1972), and even non-diadromous fish are reported as travelling up to 600 km during feeding and breeding migrations (Lowe-McConnell, 1973). In Sweden migratory fishes became almost extinct at one time in rivers which were so completely impounded that almost every available metre of head was utilized for electricity generation (Pyefinch, 1966).

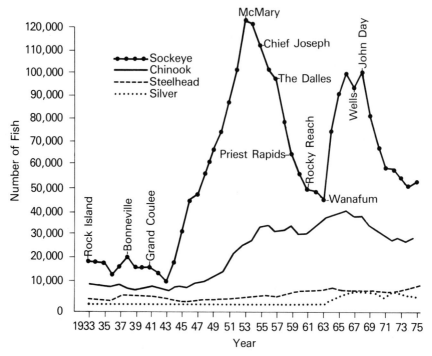

Fig. 4.3. Columbia River Rock Island counts of salmon based on 5-year moving averages, redrawn after Leman in Schweibert (1977).

Of the many rivers in the world affected by dams, the Columbia in North America and the Volga in the U.S.S.R. are perhaps the most studied (Pruter and Alverson, 1972; Trefethen, 1972; Schweibert, 1977; Mordukhai-Boltovskoi, 1979). Irrigation dams and diversions built on the Columbia tributaries in the latter part of the nineteenth century effectively blocked off upper spawning areas for salmon, mainly in the Yakima, Okagon, and Snake Rivers (Trefethen, 1972). The first hydro-electricity developments in the river systems were on waterfalls at Spokane in 1885 and Willamette in 1889. Trefethen (1972) considered that although fish passes were incorporated into the larger dams at hydro-electricity installations, many passes were ineffective and fish did not pass upstream to spawn (Fig. 4.3). In 1933 the Rock Island Dam was built on the mainstream Columbia River. By 1964 ten large dams controlled the main river flow and by 1972 over seventy dams regulated the discharge from the tributaries and main river. Most of these were used at least partly for hydro-electricity generation. It is remarkable that even as late as 1933 the Grand Coulee Dam was built without any provision for fish passage, blocking off 1,600 km of available spawning river. In contrast, every dam built in Scotland since the 1860s was obliged by law to have a fish pass (see chapter 9), though here the topographical problems were not as huge as in other countries. In Canada there were 365 hydro-electricity projects, affecting 200 rivers by 1972. Installations at 80 of these dams could each produce over 100 MW (Efford, 1975). This is probably the largest development of hydro-power in any region of the world except perhaps the U.S.S.R. In both Canada and the U.S.S.R. considerable expansion of hydro-electricity schemes is in progress which may have drastic effects on the natural migratory fisheries (Yourinov, *et al.*, 1974; Efford, 1975).

(ii) *Problems for downstream migrants*
Young fish such as salmon smolts can descend over small obstacles safely because of the cushioning effect of water. In hydro-electricity reservoirs the only downstream exits are usually fish passes (if provided) and the turbine intakes (Fig. 4.4) unless surface overspill is incorporated, whereby fish can pass directly over the dam. However, the distance from dam crest to base may be so great as to cause injury to falling fish. Fish passing through turbines may also be seriously injured or killed (Berry, 1955; Eicher, 1970). Estimates of the mortality of young fish during passage through turbines vary. Pyefinch (1966) quotes 10–20 per cent physical injuries in salmon smolts passing through turbines with hydrostatic heads of over 55 m in Scotland.

Trefethen (1972) quotes 10–20 per cent directly killed in the Columbia with another 15 per cent being lost through predation after injury. Schoenemann,

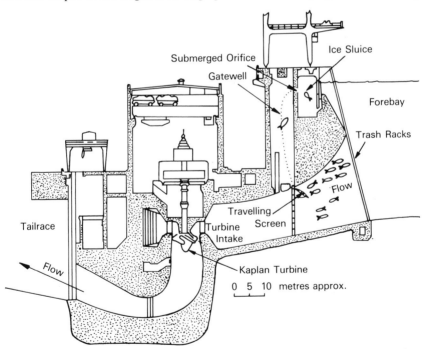

Fig. 4.4. Section through typical hydro-electricity turbine unit showing a travelling screen for diverting fish into the sluiceway, redrawn from Trefethen (1972).

et al. (1961) give a figure of 11 per cent mortality at the McNary Dam on the Columbia. Liscom (1971) quotes 9–13 per cent at McNary, 33 per cent at Ice Harbour and 38–50 per cent in three lower dams. On this basis, cumulative losses of fish passing through turbines in the whole Columbia system could be highly significant with estimates ranging from 5–100 per cent.

Dominy (1973) estimated mortalities of 9–55 per cent of salmon and steelhead trout smolts at dams on the St. John River. Assuming an average 25 per cent, the cumulative loss of fish passing through turbines could be 58 per cent between the spawning reaches and the sea. All these estimates, however, apply only to those fish passing through turbines. Many fish go downstream via fish passes and these are not always counted.

Raymond (1968, 1969) showed that juvenile Chinook salmon (*Onchorhynchus tshawytscha*) took on average 22 days to migrate downstream through an impounded reach of the Columbia instead of 13–15 days before impoundment. The lack of downstream currents which the fish utilize for migration was suggested as the reason for the delay. At low flow periods the fish could take up to 31 days to travel between the two dams (Bentley and Raymond, 1976). Such delays also expose the juvenile fish to predation by birds and piscivorous fish. Mills (1964) showed that pike and goosanders preyed heavily on salmon smolts upstream of Scottish hydro-dams. Geen (1974) speculated that losses above a proposed dam on the Fraser would be 25–85 per cent, as a result of 'residualism' in reservoirs.

Pyefinch (1966) suggested the delays are because of slow water currents and possible disorientation, and he speculated that as a result there may be some loss of the 'migratory urge' in young salmonids, though this is difficult to verify. The complexity of density currents in a reservoir may also disorientate migrants (Wünderlich, 1971).

Accurate estimates of delay-effects and turbine-losses are, however, difficult to make and the figures should be viewed with some reservations.

Many resident reservoir fishes survive passage over dams and through turbines. For example, most fish species inhabiting the tailwaters of Tennessee River hydro-stations were derived from the reservoir upstream (Walburg, *et al.*, 1971), though large numbers were killed as they passed through turbines. Water quality and dissolved gas concentrations may also affect migrants, and the effects of these are noted in the next section.

(iii) *Problems for upstream migrants*

Much of the concern expressed by objectors to dams has been over the passage of upstream migrating adult fishes and, in essence, this would seem to be a fairly straightforward problem. There are, however, a number of complicating factors beside that of a physical barrier, notably, the suppression of natural flow and flow variability either preventing fish moving upstream or removing their stimulus to do so; the quality of water discharged from hydro-stations over fish waiting to ascend which may deter or even kill them, the lack of velocity or the complex density currents in reservoirs possibly leading to disorientation in naturally rheophilic fish.

River discharge. River regulation either as a consequence of dams, diversions or hydro-generation may, as we have seen, either stabilize flow regimes or cause diurnal or short-term extremes of drought and flood (see 2.2.1).

It has generally been accepted that river discharge is a major factor affecting upstream migration of salmonids in many rivers and that any alteration may be harmful (Banks, 1969; Alabaster, 1970). Spates or freshets are believed to stimulate fish to move from an estuary into the freshwater reaches, and an adequate flow is necessary to maintain the upward migration. However, only rivers with highly variable discharge (spate rivers) fit into this pattern. Rivers with much more stable discharge patterns are not so dependent on spates for fish movement (Hellawell, et al., 1974). Here, other factors such as temperature or season may be important.

Baxter (1961) suggested that artificial freshets of 70–100 per cent of the average dry-weather discharge were necessary to stimulate upstream migration in spate rivers. Where downstream discharge is completely eliminated fish will not be able to negotiate dried-out channels but, provided there is sufficient depth of water in a channel for upstream movement to be possible, quite drastic reductions in discharge may not be as detrimental as predicted (Geen, 1974). In controlled rivers, which dry out when power is not being generated, the danger is that fish stimulated to move upstream during generation may be stranded once generation ceases. Ultimately, however, this is avoided only by proper management of compensation flows and generating patterns.

Temperature and water quality. The relationships between salmonid migrations and water and air temperatures have been discussed at length by Banks (1969) and Hellawell, et al. (1974). The relationship seems uncertain and as Banks says 'the evidence of the influence of temperature on upstream migration is both conflicting and inconclusive'. Most of the data came, as might be expected, from field experiments and observations where temperature was measured while other parameters, for example turbidity, light, etc., were not. Although Hellawell, et al. (1974) agree with the general conclusions of Banks about temperature it is of some interest to note that both Pyefinch (1966) and Menzies (1939) observed that migratory movements were reduced at water temperatures below about 4·5–5·0°C. Stewart (1968) suggested that movement occurred when air temperature was lower than water temperature, but evidence for this is limited and the presence of any mechanism for detecting air temperatures by fish is doubtful (Hellawell, et al., 1974). Major and Mignell (1966) deduced that water temperatures above 21°C blocked entry of upstream migrants from the Columbia to the Okagon River, but Banks (1969) quotes a number of authors who have observed quite unhindered migrations between bodies of water which differed in temperature. If low temperatures impede migration, then hypolimnial releases could be significant. As yet the relationships between temperature and migration are not clear and perhaps the design of releases to produce a temperature pattern as near to the original pre-impoundment pattern as possible is the 'safest' though not necessarily the most economic approach (Krenkel and Parker, 1969a; Orlob, 1969).

Water quality alone has been responsible for the elimination of many migratory fishes from many rivers of the world, though the proportional effects of pollution and of hydro-electricity developments on the decline of migratory fish in rivers affected by both is not understood. Trefethen (1972) suggested that in the Columbia, surface temperatures of 27°C in impoundments, *together* with low oxygen levels in the hypolimnion, were a potential danger to migrant salmon. In some of the rivers widely quoted as being grossly affected by both pollution and dams, notably the Delaware, the Tobique and St. John, the Illinois, and the Danube, it is obvious that where fish passage was facilitated, pollution was the prime agent eliminating fish stocks. In the Thames (U.K.) locks and weirs are still major obstacles to future fish migration now that many pollution problems have been alleviated.

Fish living in water with low dissolved oxygen concentrations are less tolerant to other stresses than fish in saturated water (Jones, 1964). The consequences of combined stresses or synergisms in polluted, impounded rivers are therefore likely to be significant (Andrew and Geen, 1960; Banks, 1969; Krenkel and Parker, 1969a; Starrett, 1972; Dominy, 1973; Geen, 1974, 1975; Efford, 1975). In most cases it has not been possible to establish a sole cause for anadromous fish decline, and usually low oxygen concentrations, overfishing, pollution, siltation, impoundment, or competition from introduced or favoured species have been specified either singly or in some combination (Thomann, 1972; Starrett, 1972; Liepolt, 1972; Ruggles and Watt, 1975). There is a danger that sudden discharges of deoxygenated hypolimnial water from turbines may cause direct mortalities of fish below dams, though records are scarce (Hall, 1971). Oglesby, *et al.* (1978) recorded a massive fish kill in a hatchery below a Georgia (U.S.A.) hydro-electricity station where the causal agents were probably a combination of low oxygen, sulphides, and toxic metals.

Gas bubble disease. This phenomenon has resulted in mortalities of fishes in many impounded rivers, notably in the U.S.A. and Canada (Beningen and Ebel, 1970; Trefethen, 1972; Rucker, 1972; Boyer, 1973; Dominy, 1973; Geen, 1974, 1975; Fickeisen and Schneider, 1974; Ruggles and Watt, 1975; Schweibert, 1977). The condition occurs because of the differential solubility of nitrogen and other gases in water at different pressures (Boyer, 1973). In heated waters supersaturation with oxygen produced by excessive photosynthesis of temperature changes may also cause the condition (Otto, 1976).

Fish in deep impoundments or in the tailraces of turbines are often exposed to hypolimnial water supersaturated with nitrogen. Body fluids become

anoxia

supersaturated as a result. Gaseous nitrogen is eventually released from the body fluids causing embolisms in blood vessels and tissues. Ultimately, blood flow is restricted and death or injury may ensue as a result of anoxia. The histo-pathological effects were described by Pauley and Nakatani (1967). Symptoms of gas bubble disease are notably bubbles in the mucous coat, gills, fins, or roof of the mouth, or bulging eyes, and have been noted in many species of fish downstream from hydro-electricity installations and thermal power stations (Westguard, 1964; Adair and Hains, 1974; Ruggles and Watt, 1975; Otto, 1976).

In water supersaturated with air, concentrations of between 106 and 120 per cent (1·06–1·2 atm) were found to cause decreased swimming activity in juvenile Chinook salmon though the fish recovered on transfer to 100 per cent saturated water (Schiewe, 1974). Mortalities did not occur below 110 per cent (1·10 atm) in experiments. Threshold levels for the onset of mortality were 120 per cent saturation (1·2 atm) after 77 hours' exposure (Nebeker, et al., 1976). However, in the Columbia, supersaturated conditions may persist for many miles downstream of dams (Hughes, 1973). Therefore even migratory fish may be exposed for periods of up to several days, though Boyer (1973) showed that fish swimming at depths of over 3·5 m would not suffer from gas bubble disease. Gray and Haynes (1977) tracked adult Chinook salmon in the Columbia and found that most travelled at depths of 3–6 m. At these depths the external pressure on the fish equals the internal dissolved nitrogen level at 137 per cent of the normal surface saturation. Problems arise mainly where fish are in shallow areas at dams and in ladders. The ratio of nitrogen to oxygen is an important factor, in that the higher the proportion of nitrogen in water the more rapidly death occurs (Rucker, 1976). Also, a rise in water temperature has been found to significantly decrease time of survival in some salmonid species (Nebeker, et al., 1979). Smaller fish are more resistant than larger fish. Different species have different tolerances (Ebel, 1969) and in some Columbia rivers stocks there may even be an inherited resistance to gas bubble disease (Cramer and McIntyre, 1975). The lesions caused by gas bubbles are also prone to fungus or bacterial invasions (Coutant and Genoway, 1968).

In Britain and other regions where hydro-electricity reservoirs are mainly oligotrophic, no mortalities caused by gas bubble disease have been recorded. The solution lies generally in balancing surface and hypolimnial water discharges or using artificial methods of destratifying reservoirs or modifying outlet structures to enhance gas exchange (Bernhardt, 1967; Fast, 1973; Fast, et al., 1975; Overholtz, et al., 1977; Schweibert, 1977). The large size of many mainstream river impoundments used for hydro-generation may, however, make artificial aeration impractical. The possibilities of discharging thermal effluents from nuclear power stations into hypolimnial regions, thus breaking down stratification, may be a partial solution for the future.

Disorientation and delay. Normally, rheophilic fish use water currents to orientate, and the tendency is for individuals to head into currents (Harden-

Jones, 1968; Arnold, 1974). It is likely, therefore, that river currents assist upstream migrants to find the spawning grounds. The current velocities and directions in hydro-electricity reservoirs are often complex (Krenkel and Parker, 1969a; Straskraba and Straskrabova, 1969; Wünderlich, 1971), especially during generation, and it is possible to speculate that migrating fish may be disorientated by both lack of velocity and complexity of water movements. Evidence for delay in smolt migration has already been given, but as yet there is little published evidence of drastic delays in upstream migrants passing through reservoirs. Residualism of upstream migrants either below dams or in lakes could also lead to exhaustion of metabolic reserves, as it has been suggested that stores of fat and protein in sockeye salmon are only just sufficient to allow for completion of spawning in the upper reaches of the Fraser River (Idler and Clemens, 1959). Also, if fish delayed in lakes feed, they may become infected with parasites through eating the secondary host, for example copepods or other fish (Geen, 1975). Evidence to support the 'metabolic reserves' or 'increased parasitism' hypotheses is as yet, scarce, though it is well known that limnophilous parasites increase in impoundments after several years (Hoffman and Bauer, 1971).

(c) Alleviation of effects of hydro-schemes

(i) Fish passage
Several methods have been used for facilitating the passage of upstream and downstream migrants around hydro-power dams. These include fish ladders (Fig. 4.5), lifts (Fig. 4.6), other fishways, and road transport. Many methods of deflecting fish towards fishways or away from danger zones are also in use or have been tried, including electric screens, air bubble screens, acoustic screens, louvres and travelling screens and water jets. These are discussed in chapter 6.

The problems of design, construction, and operation of fish passes have been reviewed extensively (Larkin, 1958; Andrew and Geen, 1960; Stuart, 1962; Clay, 1961; Aitken, et al., 1966; Mills, 1966; Pyefinch, 1966; Banks, 1969; Eicher, 1970). The effectiveness of fish passes has been questioned several times in these publications, though data from before and after installation are not usually comprehensive enough to allow real assessments.

Trefethen (1972) considered that the decline in the commercial salmon catch in the Columbia between 1911 and 1970 was caused entirely by dams, irrespective of passes (Fig. 4.3). On the other hand, Pruter (1972) in detailed analyses of commercial catches of several species of migratory fish in the Columbia found fluctuating patterns, and suggested that physical changes in the river were only partly to blame. Legislation, size limits, conservation practice, and public preference were also significant factors. In the estuary, catches of most species increased, with the exception of the Pacific sardine (*Sardinops sagax*) which suffered from overfishing. Pyefinch (1966) showed a

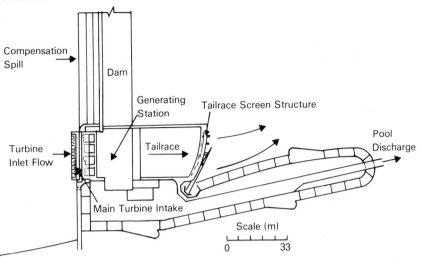

Pitlochry Fish Pass — Plan

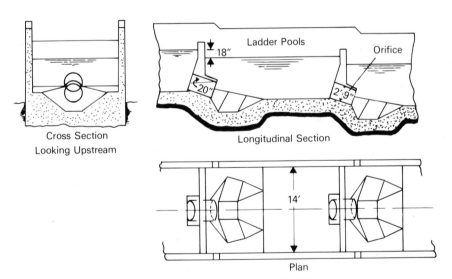

Cross Section
Looking Upstream

Longitudinal Section

Plan

Pitlochry Fish Pass — Details of Pool Orifice

Fig. 4.5. Conventional fish ladder for bypassing fish around dams, redrawn after Aitken, *et al.*
(1966).

decline in the numbers of salmon running after two Scottish dams were
finished in 1956, but by 1963–4 there was an increase, probably as a result of
artificial planting of fry. It was predicted that eel stocks in the Kariba lake
would disappear owing to the dam preventing upstream migration of
juveniles, but Balon (1975) found that stocks had not declined and concluded

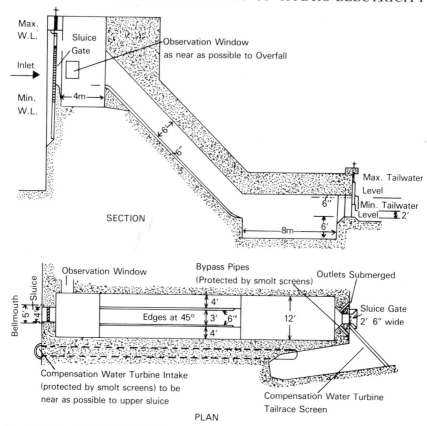

Fig. 4.6. Borland hydraulic fish-lift for bypassing fish around dams, redrawn after Aitken, *et al.* (1966).

that juveniles surmount the dam (not specifying how) in their second year. At one hydro-electricity station on the Volga 25×10^3 sturgeon and 10^6 lake herrings pass through the fish lock each year (Lowe-McConnell, 1973).

There is little doubt that once a fish enters a fish ladder or lift it will pass through (Banks, 1969). In 'endless' fishways Chinook and sockeye salmon have been found to ascend through 300 m (vertical distance), in 500–1,900 min (Collins, *et al.*, 1962), though fish were slower in a 1:8 (12·5 per cent) than in a 1:16 (6·3 per cent) gradient. One fish climbed over 2,000 m in a continuous trial over 5 days, though abrasions and fungal infections began to appear towards the end of the experiment.

The fish ladder (Fig. 4.5), usually a series of small 'steps', requires active movement by the fish. The fish lift, on the other hand (Fig. 4.6), is often operated hydraulically, carrying fish through the major portion of the vertical distance before releasing them (Clay, 1961; Stuart, 1962). Fishways of other designs may be used depending on gradients and discharge rates (Eicher, 1970).

Recent research in North America has shown that migrating fish will pass through simple inclined pipes, provided gradients, flow, and configurations are acceptable (Slatick, 1970). For high dams, spiral configurations may be an effective method of bypassing fish, and may be installed retrospectively and cheaply at dams which have no fish pass. Denil fishways, using vanes to reduce velocities in sloping channels, are also used with considerable success in the U.S.A. (Eicher, 1970).

Downstream migrants may also negotiate fish passes, though the problems of excluding salmonid smolts from turbine intakes at dams has been alleviated at some installations by allowing the fish to descend directly through the turbines during generation. By making minor variations in the start-up procedures and in turbine-blade alignment the small fish are allowed to pass through with little damage (Pyefinch, 1966; Eicher, 1970, Robbins and Mathur, 1976), though the extent of damage depends also on the design of the turbines. The attraction of both upstream and downstream migrants to the entrance of a fish pass still presents many difficulties, especially below hydro-power dams and diversions where water flows, temperature, and quality may fluctuate widely (Banks, 1969).

Generally, upstream-migrating fish are attracted to the higher flows and stronger currents coming from turbine outlets, rather than to the smaller volumes discharged from a fish pass. On the other hand, fish-pass discharges are usually of warmer oxygenated epilimnial water while turbine discharges may be hypolimnial and cold.

Records of the reluctance of fish to enter fish passes are few (Banks, 1969). In the River Shannon (Ireland) fish stunned by electric screens below turbine outlets kept returning towards them and would not enter a fish pass which had much lower flows.

Banks (1969) stated that 'the entrance to a fish pass or lift installation should be placed at the furthest upstream point which the fish can reach, though it is clear from the literature that the methods of attracting fish to fishways defy proper quantitative delimitation'. The success of passes seems to vary considerably (Andrew and Geen, 1960; Banks, 1969), though Pyefinch (1966) concludes that at least in Scotland success rates are high.

(ii) *Fish transport*

In a number of places, notably North America and Scotland, both upstream and downstream migrants have been collected at dams and transported by road to a suitable release point (Larkin, 1958; Mills, 1966; Mills and Shackley, 1971; Dominy, 1973; Efford, 1975; Mains, 1977).

Two examples serve to illustrate the principle. In the Conon River system in Scotland downstream migrating salmon smolts were trapped and transported 20 miles downstream of the dam for release. Of these, 99·8 per cent survived the journey and a significant percentage of marked fish were caught as adults in the river system in subsequent years. It was concluded that the method was biologically viable but the economics were not clearly defined (Mills and Shackley, 1971).

In the St. John River in Canada upstream and downstream migrants are transported around the Mactaquac impoundment to avoid both the dam and the most polluted reach of the river. Upstream migrants are collected in a trap, lifted into tankers, and released above the impoundment. Shads, salmonids, and other migrants are handled in a continuous operation during the major migration seasons. There is little doubt that this has been a major step in the conservation of the river's fish stocks (Dominy, 1973; Ruggles and Watt, 1975).

(iii) *Fish rearing and stocking schemes*

In many of the river basins affected by impoundment the potential losses of migrating smolts and adults are compensated for by introductions of hatchery-reared fry or smolts (Pyefinch, 1966; Schweibert, 1977). In the main-stream Columbia River, for example, about 147×10^6 salmon and steelhead smolts were released in 1973–4. Maintenance of facilities for hatching and rearing cost approximately $\$6 \times 10^6$ (U.S.A.) in 1974 and represented a capital investment of about $\$180 \times 10^6$ (Cleaver, 1977). Hydro-schemes may also incorporate hatcheries or artificial raceways where salmonids are encouraged to spawn naturally. In the U.S.S.R. sturgeon are bred and reared extensively to replace Volga stocks which have declined since impoundment (Lowe-McConnell, 1973; Mordukhai-Boltovskoi, 1979). Stocking is probably largely responsible for the maintenance of some migrating fish runs in several countries where hydro-electricity development has been intensive.

(d) Effects of diversions and water transfers

Problems which may arise from diverting water from streams and transferring water from one drainage basin to another (M.A.F.F. and A.R.A. Report, 1972; Geen, 1974) include:

> Reduced flows in a diverted stream not being adequate to attract and maintain migratory fish. Thus homing fish may be attracted by higher flows in the receiving stream.

> Mixing of waters of different chemical composition which may confuse fish homing by olfactory clues.

> Transfer of parasites or predators.

There is little evidence to substantiate the concern so far, provided that adequate compensation water is allowed in diverted streams (Pyefinch, 1966), but this is not always available in sufficient quantities, especially in older schemes.

Geen (1974) suggested that pike (*Esox lucius*) introduced to the Fraser River via diversions and transfer would act as a vector of the parasite *Triaenophorus crassus* and infected salmonids would show poor flesh quality. Arthur, *et al.* (1976) also suggested that the transfer of two parasites from one lake to another in a hydro-electricity scheme could have dramatic effects of the fish in the receiving lake. However, they found more parasite species present in the receiving lake than in the source lake. Their arguments were speculative and not convincing, though obviously little is yet known about such transfers.

New reservoirs often contain fish heavily parasitized with *Ligula intestinalis* which could be damaging if transferred, as the parasite tends to render certain species sterile (Hoffman and Bauer, 1971). As might be expected, fish parasites with intermediate zooplankton hosts tend to increase in abundance in impoundments. Parasite densities may, however, decrease soon after impoundment but increase in the following 3–5 years (Hoffman and Bauer, 1971). In the U.S.S.R., where reservoir fishes are a major source of food, parasite infestations are major problems and have been studied in great detail prior to and after impoundment (see Hoffman and Bauer, 1971; Mordukai-Boltovskhoi, 1979).

(e) Effects of hydro-schemes on other vertebrates
Regulation and hydro-electricity generation can also affect other vertebrates, particularly those which live or graze on the delta marshes of rivers. Both Blench (1972) and Geen (1974) note that the damming of the Peace Athabasca system reduced the spring flooding of the delta and the resulting vegetational changes are expected to cause declines in muskrat and bison populations. In addition, waterfowl may also suffer owing to loss of marsh nesting places. It is, however, difficult to assess the potential damage as good data from similarly affected areas do not exist. The use of drawdown zones by waterfowl and pigs has been noted earlier in this chapter.

(f) Effects of pumped-storage and tidal barrages
The effects of pumped storage schemes on lake and reservoir fisheries is not well documented. Research on the Conowingo Reservoir in the Muddy Run pumped-storage scheme in Virginia has shown that fish eggs and larvae have been transferred to the upper reservoir and have developed there, even after passing through the massive pumps. Also, 6·5 times as many fish passed upwards as downwards in the system (Snyder, 1975; Robbins and Mathur, 1976).

Estes (1971) found that large-mouth bass (*Micropterus salmoides*) and blue gills (*Lepomis macrochirus*) grew well in a pumped storage reservoir in spite of water-level fluctuations, though spawning was delayed. The elimination of the relict char (*Salvelinus perisii*) from Llyn Peris is the inevitable result of the construction of the Dinorwic pumped storage scheme in North Wales (C.E.G.B., 1976). The loss is mainly as a result of draining down of the lake during construction; the reduced water depth and higher temperatures during operation would not allow the species to survive. The species still exists, however, in an adjacent lake. The cutting-off of Llyn Peris from the natural drainage regime of the Seiont/Nant Peris river system for operating purposes could also have caused a block for migrating salmon. This has been alleviated by the drilling of a tunnel, bypassing the lake, allowing migrants to reach and leave the upper spawning grounds (Rogers, 1977).

In an earlier pumped-storage scheme in the same area a successful put-and-take trout fishery has been established though the maintenance of a permanent, self-sustaining fishery is doubtful where lake storage capacity is low in relation to operating requirements. This does not, however, apply to combined run-of-river, pumped storage schemes where the lower reservoir storage capacity is high (Schoumacher, 1976).

Little is yet known about the effects of tidal barrages on fish, though where tidal variation is reduced upstream, more waters may be available for the resident fish population. The dams are, however, likely to form barriers to migrants and the construction of fish passes and their siting will have to be carefully planned (Shaw, 1975; Severn Estuary Study Group, 1977; Gordon and Longhurst, 1979). Boyar (1961) forecast that, at the proposed Passama-quaddy tidal-power project in the New Brunswick, water currents would be too high for migrating herring to enter the bay. As yet, however, considerable research remains to be done on the ecological effects of both tidal and pumped storage schemes.

4.8. SUMMARY

Hydro-electricity schemes have had considerable effects on inland fish and fisheries because of impoundment, drawdown, physical blockage, flow regulation, and alterations in water quality. The problems for migratory fishes were recognized in the early years of hydro-developments but economic and physical limitations prevented solutions to these problems at many places. Fish can use most types of fish pass, but need to be attracted to the entrance by some means as they can be distracted by high flows from turbine tailraces.

Both pollution and impoundment have been major factors in the decline of fish populations in many rivers and it is not always possible to apportion blame. Provided that facilities for fish passage are incorporated into dams, there seems little evidence of serious effects on upstream migration. Where dams without passes have blocked spawning reaches, and where many kilometres of spawning gravels have been flooded upstream of a dam, there is certain to be some loss of stocks. These are often maintained or augmented by artificially reared fish. Downstream migrants suffer mortalities during passage over dams or through turbines, though these may be alleviated by modification of spillways and turbine operation, where possible.

Drawdown and discharge fluctuations are the main causes of changes in the flora and fauna of impounded and regulated rivers, though the chemical effects of hypolimnial releases and altered temperature regimes may also have considerable effects.

Whatever the use of a dam there are certain to be measurable and marked changes in the physical, chemical, and biological character of the relevant river. Whether these are considered significant, adverse, or beneficial, depends on the point of view and of the situation. Data on the effects of operating pumped storage and tidal power schemes are, as yet, scarce.

Chapter 5 The biological effects of thermal discharges

5.1. INTRODUCTION

As we have already seen, cooling-water discharges from thermal power stations may contain contaminants in addition to waste heat (see chapters 2 and 3). Thus, although a great deal of concern and research has been concentrated on the effects of waste heat on ecosystems, it is difficult in practice to dissociate temperature effects from those of other factors in field studies. It is possible that many early conclusions about 'thermal effects' and, consequently, some predictions were based on inadequate knowledge of the nature and composition of effluents and of the complexity of their discharge patterns.

This chapter deals mainly with the effects of heat on ecosystems in the vicinity of thermal discharges, bearing in mind that 'ecologically it is deplorable to deal with temperature in isolation as most ecological factors nearly always operate in conjunction with others to produce their effects. It is to be expected, therefore, that a discussion on temperature as a limiting factor will be involved and often inconclusive' (Macan, 1963). Temperature is one of the major enviromental factors which have preoccupied biologists for many years and most of the relevant data on the temperature relationships of aquatic organisms have been reviewed extensively. Some of the more recent reviews are given by Macan, 1963; Jones, J. R., 1964; Naylor, 1965b; Rose, 1967; Hawkes, 1969; Krenkel and Parker, 1969b; Hynes, 1970; Kinne, 1970, 1975; Eppley, 1972; Precht, et al., 1973; Perkins, 1974. General reviews and bibliographies which were specifically prepared for thermal discharge studies include those by Wurtz and Renn, 1965; Coutant, 1967, 1968a, 1969, 1970a,b, 1971a,b; Kennedy and Mihursky, 1967; Jensen, et al., 1969; Raney and Menzel, 1969; U.S.A.E.C., 1971; Coutant and Goodyear, 1972; Langford, 1972, 1974; Coutant and Pfuderer, 1973, 1974; Bush, et al., 1974; Coutant and Talmage, 1975, 1976, 1977; Howells, 1976; Langford and Howells, 1976; Water Research Centre, 1976; Hannon, 1978; Möller, 1978a,b; Becker, et al., 1979; Talmage and Coutant, 1980.

In addition, many data are included in Proceedings of National and International Symposia, including Krenkel and Parker, 1969b; C.E.R.L., 1971, 1975; I.A.E.A., 1971, 1972, 1975a; Gallagher, 1974; Gibbons and Sharitz, 1974; W.E.C., 1974; Esch and McFarlane, 1976; E.D.F., 1977; Lee and Sengupta, 1977; Thorp and Gibbons, 1978. Most publications contain extensive bibliographies.

5.2. TEMPERATURE RELATIONSHIPS OF AQUATIC ORGANISMS

5.2.1. In natural systems

Natural waters have a temperature range of about −3°C to almost 100°C, limited by the physical properties of the water itself (see chapter 2). As far as is known no single species of plant or animal can live in the whole range. The normal world-wide range of water temperature in which most living organisms are found is about −2 to +37°C though there are few places where this range is exhibited by a single body of water (see chapter 2). Organisms can be categorized into two major groups based on their temperature tolerance ranges, namely:

Stenotherms. Those species which tolerate only narrow temperature ranges.
Eurytherms. Those species which can tolerate wide fluctuations (Fig. 5.1).

Most aquatic organisms have little physiological control over their body temperature (i.e. *poikilothermic*), and their metabolism is greatly influenced by the temperature of the water in which they live. Some species may, however, exercise limited control by physiological or behavioural mechanisms (Fry, 1967; Kinne, 1970). Wholly aquatic, homoiothermic animals are relatively rare, e.g. whales or dolphins.

Outside the normal temperature ranges, thermal springs and streams are inhabited by bacteria withstanding temperatures of over 80°C, algae over 70°C, and insects up to 40–45°C (Hawkes, 1969; Brock, 1975). At the other extreme many organisms are active in the sea at −2°C (Crisp, 1964; Kinne, 1970; Precht, *et al.*, 1973). Intertidal animals and plants withstand very wide

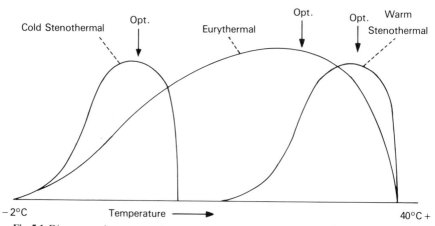

Fig. 5.1. Diagrammatic representation of the temperature tolerance ranges of aquatic organisms, redrawn after Hawkes (1969).

temperature fluctuations during tidal exposures (Strickland, 1969; Newell, 1970).

Temperature can influence geographical and local distribution, growth and metabolism, food and feeding habits, reproduction and life-histories, movements, migrations, and behaviour. Temperature may also influence the tolerance of an organism to other stresses such as parasites, disease, or pollution (Jones, J. R., 1964; Cairns, *et al.*, 1976).

Fry (1967) summarized the effects of temperature on organisms into three categories:

Lethal effects: those of high *or* low temperatures which will kill an organism in a 'finite' time within what would otherwise be its normal life-span, i.e. cause premature death.

Controlling effects: which influence processes such as growth, metabolism, activity, feeding, and reproduction.

Directive effects: for example temperature selection when presented with choice. This may also be related to activity and/or behaviour.

Apart from these, there are two other major effects relevant to aquatic organisms and thermal discharges, namely:

Indirect effects via water chemistry, for example changes in dissolved gases or synergistic effects.

Indirect effects via other organisms, for example the elimination of a food organism, or the enhancement of a predator or parasite.

These biotic effects can apply to natural systems undergoing warm or cold periods lasting up to several years, as well as to these systems affected by heat inputs from man-made sources such as power stations (Crisp, 1964; Naylor, 1965a,b; Hawkes, 1969; Kinne, 1970; Newell, 1970).

Experimental data to illustrate both direct and indirect effect of temperature and plentiful (see references), but direct field observations are seldom really conclusive as we will see, because of modifying enviromental factors and the complex interrelationships of organisms with each other.

5.2.2. Lethal temperature assessment

The concept of lethal temperatures appears simple, that is: 'organisms may be killed when heated above, or cooled below their usual tolerance ranges' (Table 5.1). In fact, lethality is a much more complex concept and the temperature which is lethal to any one organism at any give time may be dependent on several factors, including:

Rate of temperature change: Up or down the scale.

Duration of exposure: Generally the longer the exposure to unnatural temperatures, the more likely the mortality.

Acclimatization: Previous temperature history.

The size of the organism or life history state: Some species are more resistant in the egg, juvenile, or adult phases.

Season: Related to acclimatization.

Table 5.1. *Upper temperature limits for different groups of organisms*

Group	Temperature ($^\circ$C)
Animals	
Fish and other aquatic vertebrates	28
Insects	45–50
Ostracods (crustaceans)	49–50
Protozoa	50
Plants	
Vascular plants	45
Mosses	50
Eucaryotic algae	56
Fungi	60
Procaryotic micro-organisms	
Blue-green algae	70–73
Photosynthetic bacteria	70–73
Non-photosynthetic bacteria	> 99

Values are only approximate limits (from Brock, 1975).

Behaviour: Ability to avoid adverse conditions.

Physiological state: Starvation, osmotic conditions, effects of other stresses, for example toxins, changes in salinity, and oxygen levels.

Internal or behavioural control: Ability to maintain body temperatures or buffer effects of external influences to a limited extent.

Most of these modifying influences are well illustrated in the literature as will be seen later in this chapter.

The assessment and expression of lethal temperatures has occupied considerable research effort and several reviews contain a section on methods and terminology (Fry, 1967; Hawkes, 1969; Kinne, 1970; Newell, 1970). Three basic techniques have been used, namely: (1) Heating organisms at a standard rate, for example 1°C per min until death occurs. In rapid heating methods, however, assessment of the actual point of death is difficult (Fry, 1967). (2) Sudden transfer of organisms from acclimatization temperature to test temperature and recording time to death. (3) More recently, the technique has been to acclimatize test animals, transfer them to test temperatures for set periods, and then return them to the acclimatization temperature (Sprague, 1963; Ginn, *et al.*, 1976). The temperature after which they do not recover in a set time is regarded as the lethal temperature. Each method produces different kinds of data and each is applicable to specific circumstances, but the last two methods measure *dose-effects* rather than simple heating effects, and are perhaps more applicable to power station problems.

Probably the most widely used terms are those derived from temperatures causing 50 per cent mortalities during defined time periods (e.g. 24 h LT 50s,) which have been used as first-order approximations for predicting lethality in many species. A 50 per cent mortality criterion is, however, considered too high for safety by some ecologists and LT10s (temperatures lethal to 10 per cent of the population) may be more acceptable, though assessment of the significance of such low mortalities against controls may be difficult, Fry, *et al.* (1946) produced temperature tolerance polygons for several species, based on acclimatization and lethal temperatures (Fig. 5.2). This method clearly illustrates the upper and lower lethal temperature ranges. Zones of tolerance and zones of resistance, i.e. temperature ranges in which animals could tolerate for longer periods are indicated. In all these systems, death of all or of a proportion of the test individuals is the criterion.

Other methods have attempted to judge potentially lethal temperatures prior to the onset of death of the test animals. The critical thermal maximum (C.T.M.), or 'time-dependent-equilibrium-loss dose' (E.L.D) (Coutant, 1970b; Sylvester, 1975a; Becker and Genoway, 1979) is defined as the temperature above or below which the organism loses its locomotory function and cannot take avoiding action. This is perhaps useful for highly mobile organisms such as fish but much less so for slow moving and sessile organisms where time of death is difficult to judge. Studies of preferred and avoided temperatures are probably more relevant to the assessment of the effects of thermal discharges on fish than the establishment of lethal temperatures (Alabaster, 1962, 1969; Cherry, *et al.*, 1975; Richards, *et al.*, 1977).

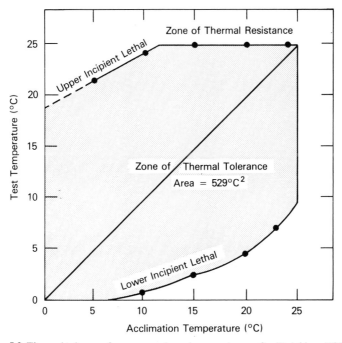

Fig. 5.2. Thermal tolerance for young spring salmon, redrawn after Hutchison (1976).

5.3. PREDICTED EFFECTS OF THERMAL DISCHARGES

As was shown in chapter 2, there is no way of generalizing about the effects of thermal discharges either spatially or temporally on the temperature regime in any water body. The prediction of ecological effects in the field, from experiments using either steady temperatures or regularly fluctuating temperatures is therefore of limited use.

Assuming that cooling-water discharges have significant effects on the temperature regime of habitats, there are, however, many potential biological consequences at all trophic levels, ranging from effects of decreased water viscosity on sedimentation, to effects of decreased ice-cover on plants, plankton, fish-parasites, and birds. Research has included studies of organisms from bacteria to aquatic reptiles, as will be seen in this chapter.

Changes in the local and geographical distribution of species may result directly from lethal conditions or indirectly through enhanced reproductive success or decreased competition.

It is possible that antifouling agents such as chlorine, metal ions originating from corrosion, anticorrosion agents, radioactive isotopes, acids, or alkalis from boiler-cleaning and from water-treatment plants could all pose hazards to organisms if discharged in cooling-water, especially if synergisms occur (see chapter 3). The effects of such contaminants are discussed in chapter 7.

5.4. SITE STUDIES OF THERMAL DISCHARGES

For the most part, the studies reviewed here are those directly concerned with operating power stations, though data from natural systems and experiments are used to illustrate certain aspects of biological change. At many power station sites, 'baseline' studies have been carried out, against which post-operational changes could be assessed. These are not in themselves of general application to other sites, except in the techniques or approach used, which may be of interest. Studies have concentrated on certain groups of organisms, usually because of ease of identification or sampling or possibly because of their direct importance to man. This means that algae, macro-invertebrates, and fish have generally received more attention in studies around power stations than bacteria, fungi, or higher vertebrates.

5.4.1. **Bacteria and fungi**

Bacteria have probably the widest temperature tolerance range of any group of organisms (Farrel and Rose, 1967; Oppenheimer, 1970; Precht, *et al.*, 1973). Hot-water species (thermophilic) have been recorded growing at temperatures of 70–80°C, cold-water species (psychrophilic) at 0°C. Most species, however, have optimal temperatures around 20–40°C (mesophilic).

Bott (1975), in a study of a natural stream with an annual temperature range of 1–23°C and a diurnal range of up to 10°C, found that periphytic bacteria, both unicellular and filamentous, decreased their generation times as temperatures increased, though optimal temperatures for most groups were 8–20° higher than the highest temperatures recorded. Several studies of bacteria in heated rivers have shown that where temperatures were raised 7–15°C above ambient, psychrophilic forms were less common or less active than the thermophilic forms, as might be expected (Stangenberg and Pawlaczyk, 1961; Rankin, et al., 1974). Miller, et al. (1976), showed that maximum bacterial activity occurred at 20–25°C in a thermal discharge canal, and that ΔTs of more than 15°C over summer ambients of 20–25°C were inhibitory. The density of bacterial cells was also higher in the discharge canal than at the cold-water control station. At temperatures of 28°C ($\Delta T = 7°C$) in a heated discharge to the River Seine (France), Appourchaux (1952) found no detectable effects on enteric, faecal, aerobic sulphur-reducing, or non-sulphur-reducing bacteria.

In both large- and small-scale experimental systems, Guthrie, Cherry, and Ferebee (1974), and Cherry, et al., 1974 found that at ΔTs of 3–5°C above ambient, diversity and abundance were increased. At ΔTs of 5–10°C both were decreased. No recovery data were given.

So far, therefore, studies of aquatic bacteria in heated waters have produced reasonably predictable results, with changes in species composition and in activity. Studies of fungi in heated water bodies are scarce, though some species flourish in cooling-tower environments (see chapter 6).

'Sewage fungus', basically a colonial aggregation of filamentous bacteria, fungi, and protozoa, is an inhabitant of flowing, polluted water (Hynes, 1960). It is more obvious in rivers during winter, because in summer it often is obscured by macrophytes. Growth optima are reported to be between 7 and 17°C. In heated rivers, however, it is not eliminated until temperatures reach 34–36°C (Stangenberg and Pawlaczyk, 1961) though Turoboyski (1973) showed that Sphaerotilus natans declined in the heated part of the Skawinka river at 33°C.

Bacteria pathogenic to humans and fish may be encouraged to develop faster in heated waters, but as yet few cases have been demonstrated (De Sylva, 1969). Fish-kills supposedly caused by bacterial infections enhanced by high water temperatures have been reported (Brett, 1956; Stroud and Douglas, 1968; Mihursky, et al., 1970; De Sylva, 1969; Rankin, et al., 1974) though a direct relationship was not shown in any case. The stresses from living in a heated environment may make fish more susceptible to infection.

The fish pathogen Aeromonas hydrophila has optimum growth temperatures ranging between 25 and 30°C in the cooling ponds of the Savannah River nuclear reactors. Lethal temperature was 45°C (Hazen and Fliermans, 1979).

The human faecal bacterium Escherischia coli was found to die faster in parts of the Wabash River heated to 35°C than in colder areas (Silver and Brock, cited in Brock, 1975). Hoadley, et al. (1975) found, on the other hand, that some strains of Pseudomonas (from infected auditory passages) could be isolated from heated, chlorinated swimming pool water and would grow at 41°C. The bacterium *Clostridium botulinum* has been found to be slightly more common in a heated lake than in cold waters.

Recent research on the Savannah River showed that there were no significant differences in thermophilic and thermotolerant fungi in heated and unheated waters, except for the poultry pathogen *Dactyloria gallopava*. This organism was abundant in the most heated areas in foam and in the microbial mats on the substrates (Tansey and Fliermans, 1978).

So far, data are minimal, but in many areas of the world people swim in waters heated by thermal discharges including, in the author's experience, those in British rivers and coastal waters. Where sewage discharges are also adjacent there is a need for much more data on human pathogens than are available at present.

5.4.2. Protozoa

Cairns (1969) could not demonstrate differences between the protozoan communities upstream and downstream of a heated discharge in the Potomac. In experiments, however, he showed that thermal shocks of 12–30°C for 7–8 min. reduced the numbers of species considerably, as did a steady 30°C for 24 h. Recovery was rapid once ambient conditions were restored, taking 24–144 h. depending upon the magnitude of the shock.

In a study of the human pathogenic protozoan which causes primary amoebic meningo-encephalitis (P.A.M.E.), De-Jonckheere, *et al.* (1975) isolated the organism from a thermal-polluted canal near Antwerp. The organism *Naegleria fowleri* was found only in the parts of the canal heated to 25–32°C by the effluent from a lead–zinc factory. There were none found in the thermal plume from a nearby power station, even though temperatures were similar. Chlorination may have been the main reason for their absence. One case of P.A.M.E. was reported in a boy who had swum near the factory outfall. The authors quote a number of cases of *N. fowleri* occurring in naturally warm waters, and there is no doubt some scope for further research here.

5.4.3. Algae

A great deal of our knowledge of the *in situ* thermal tolerance of algae has come from studies of hot springs in the U.S.A., New Zealand, Iceland, and Japan. The blue-green algae of these habitats were probably the first truly *thermophilic* organisms to be noted (Farrel and Rose, 1967). Patrick (1969, 1974), Gessner (1970), and Brock (1975) have defined temperature limits for various algal groups. The tolerance range for the whole group is $-2°C$ up to 73°C, though each major sub-group shows different tolerances. Hawkes (1969) summarized maximum *growth* temperatures as:

Diatoms: 15–25°C.
Green algae: 25–35°C.
Blue-green algae: 30–40°C.

Notable exceptions are the blue-green *Oscillatoria rubescens* which is a cold-water form and the diatom *Achnanthes marginulata* which can tolerate up to 41·5°C (Patrick, 1969, 1974).

The upper temperature tolerance of the larger freshwater species, for example *Cladophora, Nitella, Spirogyra,* and *Oedogonium* are in the range 29–37°C. Similarly, the larger marine littoral species, for example *Fucus* and *Ascophyllum,* have wide tolerance ranges in order to withstand exposure at low tides (Newell, 1970; Gessner, 1970). In contrast, however, the kelps of the Pacific coast of the U.S.A. have much lower tolerances (North, 1969).

There have been a number of studies of algae affected by thermal discharges. Much of the recent work has involved effects of phytoplankton entrained in cooling systems and these are discussed in chapter 6.

(a) Phytoplankton (communities in receiving waters)

Several studies have shown that species diversity of the phytoplankton decreases in areas consistently heated to over 30°C (Trembley, 1960, 1965; Stangenberg and Pawlaczyk, 1961; Patrick, 1969, 1974; Turoboyski, 1973; Simmons *et al.*, 1974) with a tendency for blue-green algae (Cyanophyta) to become dominant if temperatures over 32°C persist for long periods. In lakes and rivers heated to temperatures of 27–29°C, or where temperatures of over 30°C are only intermittent and short-lived, no over-all changes in the communities have been observed (Appourchaux, 1952; Dryer and Benson, 1957; Swale, 1964; Beer and Pipes, 1969; Kresoski, 1969; Patrick, 1969, 1974; Hawkes, 1969; Harmsworth, 1974; Koops, 1974, 1975). However, some diatoms (Bacillariophyta) may be favoured at 27–29°C, and both diversity and primary production may actually increase (Patrick, 1974; Welch and Ward, 1978). Tilly (1974) suggested that increased primary production in the heated water of a nuclear reactor cooling pond was a result of increased nutrient concentrations caused by evaporation, and Foerster, *et al.* (1974) suggested that increased productivity was a result of more efficient use of nutrients at higher temperatures in the heated zones of the Connecticut River. In contrast, Stuart and Stanford (1978) showed that the thermal discharge into North Lake, Texas, ultimately depressed primary phytoplankton production by causing nutrient limitation. This, in turn, was caused by 'long-term subtle, thermally linked precipitation activities'—not however, very clearly described. Welch and Ward (1978) concluded that the optimum temperature for phytoplankton production in a heated reservoir was ∼ 25°C, though carbon fixation rates at 34°C upwards were similar to those at 10–20°C. Marine phytoplankton species have differing optimal temperatures for growth and production, but both light and salinity may also affect these processes (Eppley, 1972). This is little evidence of decreased diversity or productivity of marine phytoplankton around thermal discharges (Patrick, 1974). One reason is that the natural discontinuous distribution of organisms, together with the instability and dispersal of the thermal plumes, make the results of field studies difficult to interpret. The other reason may be that the changes are, in fact, minimal and therefore undetectable in the field. Studies of thermal discharge canals have, however, demonstrated changes in both cell numbers and productivity (see chapter 6).

Large-scale or *in-situ* experiments with freshwater and brackish-water phytoplankton have shown that diversity and productivity may increase with small temperature increments (e.g. 5–10°C) provided that the maximum temperature does not exceed 32–35°C (Carpenter, E., 1973; McMahon and Docherty, 1975).

It can only be concluded, however, that irrespective of experimental studies and the effects of passage through power stations, there are no significant effects of thermal discharges on the phytoplankton of tidal waters, though in heated lakes and rivers some changes have been identified in the vicinity of outfalls.

(b) Benthic algae (periphyton and epiphytes)

There are many studies of attached benthic algae around thermal discharges (Patrick, 1969, 1974; see Coutant and Talmage, 1976; see Hannon, 1978). However, since the most recent work has demonstrated the effects of chlorination on algae (see chapter 7), earlier studies which interpreted changes solely in terms of temperature must be reviewed with caution. Several studies have demonstrated increases in total standing crop and productivity of benthic algae in heated waters (Fig. 5.3), though species-diversity generally decreased. Increased production was in most cases a result of enhanced growth of blue-green algae (Trembley, 1960; Patrick, 1969; Fleming, 1970; Hickman, 1974; Hickman and Klarer, 1975; Kolehmainen, *et al.*, 1975; Squires, *et al.*, 1979). Similar data have been obtained from natural thermal streams (Boylen and Brock, 1973). Blue-green algae are generally most tolerant to thermal discharges, though they only become dominant when temperatures exceed 32°C for long periods. At slightly lower temperatures, i.e. 27–30°C, some diatoms may be favoured in heated waters, for example *Navicula cuspidata* (Hickman, 1974); *N. mutica*, *N. viridula*, and *Surirella ovata* (Turoboyski, 1973), while others may be reduced in numbers or eliminated, for example *Diatoma elongatum* (Turoboyski, 1973) and *Amphora ovalis* (Hickman, 1974; Milner, personal communication).

Where ice-cover was removed by a thermal discharge in parts of Lake Wabamun, Alberta, increased light-penetration and water temperatures of 7–18°C may have accounted for increased winter-standing crops of epipelic algae (Hickman, 1974). Studies of the larger, filamentous algae in heated waters are scarce, though *Cladophora glomerata* has been observed to increase in abundance below heated effluents (Langford, 1974; Squires, *et al.*, 1979). Tolerance limits for *C. glomerata* are supposedly about 23–25°C (Wong, *et al.*, 1978), though *C. prolifera*, a marine species, may tolerate up to 46°C for 90–120 min (Gessner, 1970). There are a number of records of marine

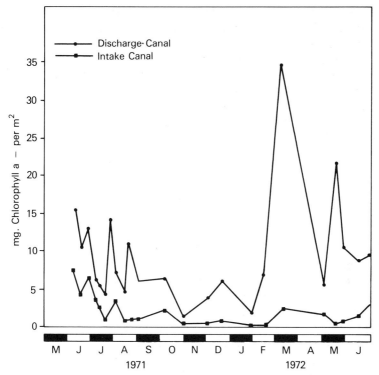

Fig. 5.3. Standing crops of the epipelic algae in the intake and discharge canals of the Wabamun Power Station, Wabamun, Alberta, Canada (ΔTs 7–18°C in the discharge). Redrawn after Hickman (1974).

macro-algae affected by thermal discharges. At Turkey Point Power Station in Florida, the green macro-algae *Halimeda* sp. and *Penicillus* sp. initially declined in abundance in areas heated 3–5°C above winter ambients. Increased scour also removed nutrients and the finer sediments. The substratum then became colonized by blue-green algae (Thorhaug, 1974). In recent years, however, fifteen species of algae are reported growing abundantly in the cooling-water channels at Turkey Point (Wilcox, 1977). Kolehmainen, *et al.* (1975) found that where sea-water temperatures varied from 35°C in winter to 40°C in summer, blue-green algae were among the most tolerant organisms, while the large brown algae, notably kelps and fucoids, were least tolerant. The brown seaweeds *Ascophyllum* spp. and *Fucus* spp. were eliminated from a rocky shore heated to 27–30°C by the discharge from a power station in Maine, U.S.A., though *Enteromorpha intestinalis* increased significantly in abundance near the outfall (Arndt, 1968). In contrast, Vadas, Keser, and Rusanowski (1976) found few over-all changes in the littoral macro-algae on another thermally affected Maine shore, though the distribution and abundance patterns of *Ascophyllum nodosum* and *Fucus vesiculosus* altered near the discharge. Early growth occurred in heated areas, but few viable propagation organs developed in the most heated communities. After relocation and design changes at the outfall had enhanced heat dispersion *Ascophyllum* recovered fully and the original community was re-established. The *F. vesiculosus* population did not recover however (Vadas, *et al.*, 1978). The commercially and ecologically important kelps of the Pacific coast are apparently sensitive to higher temperatures (North, 1969). It was predicted, therefore, that thermal discharges from power stations would eliminate the kelps from large areas of Californian coastal waters. Studies have shown however that at the Diablo Canyon Nuclear Power Plant, only an area of about 0·7 ha (1·5 acres) were affected (Adams, 1969). In the studies quoted no mention was made of chlorine in the discharge and its contribution to the effects on these algae is not fully known.

5.4.4. Angiosperms

The thermal responses of aquatic angiosperms have not been studied by many workers, though some data are given in reviews (Langridge and McWilliam, 1967; Brock, 1970, 1975; Coutant and Talmage, 1976). Several studies in relation to thermal discharges have been published.

Potamogeton spp. have been noted growing below power station outfalls in fresh-water at temperatures of 30–40°C (Trembley, 1965; Langford, 1971b; Anderson, 1969). Anderson (1969) concluded that *P. perfoliatus* was replacing *Ruppia maritima* in heated areas of the Patuxent River, Maryland. *Vallisneria spiralis*, a sub-tropical species, has also been noted inhabiting several thermal discharge canals and heated rivers in temperate regions (Hynes, 1960; Khalanski, 1975; Massengill, 1976a).

At the Savannah River Plant the macro-flora in the cooling ponds was found to be much less abundant than in cooler waters (Parker, *et al.*, 1973). The reeds *Typha latifolia* and *T. domingensis* did not flower at the highest temperatures, which ranged up to 50°C (Gibbons, *et al.*, 1975). The thermal discharges were also found to have killed large areas of forest in the swamp areas, as a result of heating of the tree roots (Sharitz, *et al.*, 1974). Similarly, Banus and Kolehmainen (1976) noted that mangrove trees in a heated bay in Puerto Rico were stressed by water temperatures above 37–38°C and that seedlings from stressed trees were smaller than those from unheated areas.

In Lake Wabamun the increased winter production and standing crop of submerged macrophytes (notably *Elodea canadensis*) in a heated area was a result of decreased ice-cover and higher water temperatures. *E. canadensis* became the dominant species in heated areas of the lake and replaced other plants, notably *Chara globularis* (Haag and Gorham, 1977). Winter temperatures reached 20–25°C in the discharge canal, and summer maxima often exceeded 30°C. Young (1974) found that production and metabolism of *Spartina alterniflora* in a thermally affected Florida saltmarsh were increased in comparison with control areas, though individual stems were smaller in the heated zone. Vadas, *et al.* (1978) noted a decrease in biomass in a similarly heated marsh area.

Turtle-grass (*Thalassia* spp.) is a dominant plant in many tropical coastal waters, and contains a diverse fauna (Thorhaug, *et al.*, 1978). Heated discharges from several power stations have been found to destroy large areas of *Thalassia* and in some cases its associated fauna (Kolehmainen, *et al.*, 1975; Thorhaug, *et al.*, 1978). Temperature rises of more than 5°C were found to be damaging in summer, when ambient temperatures were 30–35°C. At lower ΔTs growth was reduced and the critical ΔT is believed to be 1–1·5°C. At the Turkey Point Power Plant in Florida, which discharges into a shallow bay, 9·3 ha of *Thalassia* bed were destroyed and 30 ha reduced in growth, though scour and high concentrations of metal ions may have been additional stresses in this case (Thorhaug, 1974; Thorhaug, *et al.*, 1978). It seems, therefore, that although some larger plants may be able to tolerate temperatures up to 45°C, changes in resident communities are readily observable in thermal discharge areas at temperatures over 30°C and at high winter temperatures. There may be increases or decreases in production depending upon temperature and nutrients or effects on reproductive processes. The contribution of chlorine to the effects of discharges on macrophytes is not yet understood. Turbidity and scour are probably significant factors where changes have occurred.

5.4.5. Zooplankton and micro-crustacea

There have been many observations of zooplankton passing through power station cooling-water systems (see chapter 6), but studies of resident communities in receiving waters are less common and largely inconclusive. Several studies in temperate waters where thermal discharges had ΔTs of 6–10°C have showed little significant effect on the zooplankton. At most, slightly increased production in spring and summer was indicated (Raymont and Carrie, 1964; Churchill and Wojtalik, 1969; Heinle, 1969; Whitehouse, 1971b; Kititsina, 1973; Harmsworth, 1974; Goryaynova, 1975; Khalanski, 1975; Koops, 1976).

Where temperatures were higher, for example reaching 35–38°C in receiving waters, zooplankton abundance declined and mortalities occurred (Coutant, 1970a,b; Polivannaya and Sergeyeva, 1971a,b; Kolehmainen, *et al.*, 1975; Miller M., *et al.*, 1976; Bowen, 1976).

Effects on distribution were demonstrated by Brauer *et al.* (1974) in Lake Monona, Wisconsin (U.S.A.). Here, however, density in the outfall area was 2–7 times higher than at control stations, mainly because of animals from the dense populations at the intake circulating through the power station's cooling-water system (see also Goryaynova, 1975; Coutant and Talmage, 1977). In the heated areas of Par Pond, in the Savannah River region, the low diversity of ostracod populations was correlated with high minimum temperatures. The elimination of some species from the heated area was believed to be related to the elimination of rooted vegetation, which is their preferred habitat (Bowen, 1976). Early seasonal development of some species was related to increased water temperatures. In similar studies of marine ostracods in Tampa Bay, Florida, increased population densities were related to higher spring temperatures, though at the highest summer temperatures abundance decreased. In the heated area, calcium carbonate deficiencies occurred in the carapace of individuals (Stiles and Blake, 1976).

Anderson and Lenat (1978) noted an increase in warm-water stenothermic rotifers in a heated North Carolina Lake, but several planktonic species declined in abundance. These declines also occurred, however, in cooler parts of the lake and were believed to be the results of natural factors.

5.4.6. Benthic macro-invertebrates

Along with benthic algae, the macro-invertebrate communities of both fresh and tidal waters have been studied extensively in attempts to assess the biological consequences of thermal discharges. The main reasons for such emphasis are the apparent relative ease of sampling, ready identification, and some prior knowledge of the response of invertebrates to environmental disturbance. However, critical reviews of sampling methods and sampler performance together with indications of effort required to provide statistically significant data have suggested that benthic invertebrates are probably no more conducive to quantitative studies than other communities (Needham and Usinger, 1956; Hynes, 1960, 1970; Newell, 1970; Holme and McIntyre, 1971; Elliott, J., 1971; Chutter, 1972; Elliot and Tullett, 1978; Hellawell, 1978).

(a) Community responses and diversity

Invertebrate community and diversity studies around thermal discharges have usually involved either direct sampling of organisms from the substratum (Trembley, 1960; Coutant, 1962; Whitehouse and Aston, 1964; Mann, 1965; Naylor, 1965a; Davies, D., 1967; Langford, 1967, 1971b; Churchill and Wojtalik, 1969; Barnes and Coughlan, 1970; Merriman and Thorpe, 1976; Lenat, 1978; Locan and Maurer, 1975; Thorhaug, et al., 1978; Oden, 1979) or the use of artificial substrata such as fouling panels or trays and containers filled with stones, brushwood, or some other material (Markowski, 1960; Naylor, 1965b; Profitt, 1969; Dahlberg and Convers, 1974; Young and Frame, 1976).

Studies in British rivers have shown both increases and decreases in diversity below power station discharges (Whitehouse and Aston, 1964; Mann, 1965; Langford, 1967, 1971b, 1972; Langford and Aston, 1972). Changes in substratum and current velocity or the presence of biocides were believed to be mainly responsible. In the polluted River Trent, heavily used for power station cooling, the diversity of the bottom fauna decreased sharply below the entry of a grossly polluted river, but changes downstream of power stations did not relate to temperature increases (Fig. 5.4) (Langford, 1972).

One of the earliest and most significant river studies in the U.S.A., at

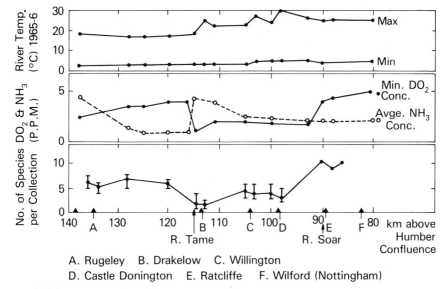

Fig. 5.4. Average number of invertebrate species collected from reaches of the River Trent in relation to temperature and chemical factors in 1965, redrawn after Langford (1974).

Martin's Creek Power Stations on the Delaware, showed distinct reductions in species—diversity when temperatures in the discharge reached 37°C. At the extreme temperatures of 40–42°C in summer very few organisms survived. Recolonization of the heated zone occurred in about 1–2 months (Coutant, 1962). Very few other studies have shown such significant changes. For example, Wurtz and Dolan (1960) found hardly any reduction in diversity below a heated discharge where maxima reached 38–40°C for short periods. Where changes in diversity (usually reductions) were recorded below thermal discharges in the U.S.A., substrate changes, current velocities, or biocides were also believed to be contributory factors (Churchill and Wojtalik, 1969; Profitt, 1969; Merriman, 1973; Merriman and Thorpe, 1976; Cole and Kelly, 1978).

One possible reason for the notable changes at Martin's Creek (Coutant, 1962) was that the effluent was discharged across shallow riffle areas and there was little vertical stratification. In deeper rivers an effluent may not be in such close contact with the substrate because of stratification. In the Severn at Ironbridge (U.K.) no changes in diversity occurred under circumstances comparable to those of the Delaware, though maximum temperatures rarely exceeded 30°C (Langford, 1972; Langford and Howells, 1976).

'Control' stations upstream of discharges are often difficult to establish. Changes in flow or substrates, the presence of weirs, or introductions of other pollutants often intervene between upstream controls and downstream sampling points and produce their own particular effects on the fauna. In lakes and cooling ponds localized benthic faunal changes are more likely caused by scour than heat (Becker, et al., 1979).

In temperate tidal waters little evidence of major changes in diversity caused by thermal discharges has been found (Markowski, 1959; Naylor, 1965a,b; Hawes, *et al.*, 1975). In the studies at Bradwell Power Station in the U.K. surveys taken at intervals over 20 years have indicated that some changes have occurred since the thermal discharge began (Davies, D., 1967; Barnes and Coughlan, 1970; Hawes, *et al.*, 1975) though scour may be responsible for those nearest the discharge. Young and Frame (1976) found no differences in diversity between heated and unheated areas of Barnegat Bay, New Jersey, near the discharge from the Oyster Creek Nuclear Power Station.

Decreased faunal diversity was, however, demonstrated within 300 m of the Yorktown Power Station discharge canal on the York River, Virginia. Lowest diversity was in summer, at a point 100 m from the canal mouth. There were also marked seasonal changes in diversity at other stations during the surveys which were not discussed by the authors (Warinner and Brehmer, 1966). Temperatures in the thermal plume exceeded 35°C for long periods. The eel-grass *Zostera* spp. beds were seriously depleted, probably by scour or biocides and this was also a significant factor in the decline of faunal diversity. In contrast, Locan and Maurer (1975) noted a marked increase in diversity on the fauna of the discharge canal at an estuarine power plant on the Indian River even though temperatures exceeded 36°C in summer.

Distinct depletion of the benthos has been noted in the vicinity of heated discharges in tropical and sub-tropical waters, notably in Biscayne Bay, Florida, and Guayanilla Bay, Puerto Rico (Thorhaug, *et al.*, 1974; Kolehmainen, *et al.*, 1975; Blake, *et al.*, 1976; Thorhaug, *et al.*, 1978, 1979). Here, discharge temperatures reached 40–45°C in summer and few species can survive long periods at such temperatures. The consensus of opinion was that tropical organisms are living near their critical thermal limits and that even small temperature rises, i.e. ΔT of 2–3°C for short periods may be lethal. The hypothesis is not always verified by the field data, however, as many species have been found living at much higher temperatures than expected (Kolehmainen, *et al.*, 1975). Time of exposure may ultimately be more significant than absolute maximum temperature (Thorhaug, *et al.*, 1978).

Even in these higher temperature systems it has not always been possible to relate changes in the fauna to temperature alone, as other pollutants, biocides, scour, and indirect effects on plants such as *Thalassia* spp. may all be contributory factors (Thorhaug, *et al.*, 1978, 1979).

(b) Species composition and relative abundance

Even though changes in macro-invertebrate diversity, as expressed by any of several methods (Southwood, 1966; Hellawell, 1978), have been generally of little significance, several studies have shown changes in species composition and in the abundance of certain indigenous species. In some cases exotic or introduced species have become established and have bred successfully.

In British rivers species which have shown increases in abundance below thermal discharges include the snails *Potomapyrgus jenkinsi*, the exotic *Physa acuta* (Langford, 1967, 1971b) and the worm *Limnodrilus hoffmeisteri* (Aston, 1973; Lenat, 1978). In heated Eastern European lakes two freshwater shrimps (*Pontogammarus crassus* and *P. robustoides*) showed fivefold increases in biomass over populations in unheated lakes (Kititsina, 1973). Effer and Bryce (1975) also found large numbers of the freshwater shrimp *Gammarus fasciatus* near the thermal discharge from the Pickering Nuclear Power Station on Lake Ontario. After 2 years the numbers of *Gammarus* declined to levels comparable with control stations. The reasons for the sudden increase in abundance of this species are not fully understood but were not believed to be related to temperature.

In a study of dragonflies and damsel-flies in the Savannah River, Gentry, *et al.* (1975) found that twenty-three species existed in the colder areas and only seven in the areas heated by the reactor effluent. Ambient temperatures reached 30°C, the heated areas 45–48°C and here temperature appeared to be the major limiting factor. One common factor in most freshwater studies is the apparent tolerance of some Chironomidae to heated discharges. Several workers have shown the group to be tolerant of water temperatures up to 40°C (Coutant, 1962; Profitt, 1969; Kititsina, 1973; Harmsworth, 1974; Brock, 1975; Merriman and Thorpe, 1976; Lenat, 1978). Both Coutant (1962) and Profitt (1969) considered that some species of Chironomidae were actually favoured by temperatures of 34–35°C. The group also contains species which are tolerant to organic wastes and other pollutants (Hynes, 1960), and as we have seen in chapter 4 are characteristic of other stressed habitats.

Evidence of elimination of species caused solely by temperature changes is rare and inconclusive for more temperate fresh waters (Mann, 1965; Langford, 1967, 1972, 1974; Langford and Howells, 1976; Kititsina, 1973; Effer and Bryce, 1975), though as water temperatures exceed 33–35°C, certain species (taxa) begin to decline or disappear (Coutant, 1962; Lenat, 1978). In British rivers, Mann (1965), Cragg-Hine (1969b), and Langford (1971b) suggested that the planarian *Dugesia tigrina* together with some leech and mollusc species were reduced in abundance below some power station outfalls where temperature reached 25–32°C. Scour and biocides were, however, believed to be major limiting factors as the species occurred at similar temperatures in other places. Two molluscs (*Pisidium* sp. and *Bithynia tentaculata*) disappeared from the St. Lawrence River near the Gentilly Nuclear Power Plant discharge, where the temperatures reached 29–30°C. However, the author (Langford, 1971 and unpublished) has found both genera in waters at 30–32°C in Britain and scour again may have been the major factor at Gentilly. Möller (1978b) also considers scour to cause most disturbance to benthic invertebrates near power station outfalls. At Martin's Creek on the Delaware caddis-flies (Trichoptera) and mayflies (Ephemeroptera) were first to disappear as temperatures reached 35°C. At over 37°C only Chironomidae were present. The natural riffle fauna was almost completely eliminated at temperatures over 37°C (Coutant, 1962). Near the Savannah River Nuclear Plant the compositions of the insect faunas were different in hot- and cold-water areas. In the cold zones Diptera and Hemiptera were less abundant than in the hot or recently cooled streams (Howell and Gentry, 1974). Only three species were common to all three habitats.

In marine environments similar changes in species-composition and relative abundance have been recorded, even where diversity has altered only slightly. In Britain, Markowski (1960) found massive growths of algae and hydrozoa on artificial substrates in the Roosecote Power Station outfall area. Smaller crustacea, for example *Sphaeroma hookeri* and *Gammarus zaddachi*, used these growths for shelter and food and consequently increased in abundance as did the snail *Potomapyrgus jenkinsi*. At Bradwell several unexpected species notably the crab *Pilumnus hirtellus*, the anemone *Cereus pedunculatus*, and the worm *Cirratulus cirratulus* were recorded only at the power station outfall (Barnes and Coughlan, 1970).

Animals which appear to be common to several thermal discharge areas are the amphipods *Corophium* spp. and the polychaete worm *Nereis succinea* (Naylor, 1959; Hechtel, *et al.*, 1970; Young and Frame, 1976). At Oyster Creek (U.S.A.) *Corophium acherusicum* was the dominant amphipod in the discharge area but *C. tuberculatum* was dominant in the intake areas. Thus, although diversity remained unchanged, species composition differed between cold and warm areas. Similar differences in species dominance between intake and outfall faunas occurred among the molluscs, polychaetes, and barnacles (Young and Frame, 1976). The over-all conclusion from this study was, that of ninety species recorded at Oyster Creek, the distribution of nineteen species was related to temperature changes. However, much of the work was done in the discharge canal which was a somewhat unnatural environment in that it was dredged. At temperatures over 27°C, in another thermal discharge canal blue-mussels (*Mytilus edulis*) were killed and feeding stopped at temperatures over 25°C (Gonzalez and Yevich, 1976).

In heated tropical waters most species of crabs, sponges, molluscs, and crustaceans were eliminated where temperatures exceeded 37°C (Koleh-mainen, *et al.*, 1975; Thorhaug, *et al.*, 1978), though some benthic Foraminifera were found in a cove heated to 40°C. At 35–37°C several species of crabs, molluscs, and worms survived among mangrove roots, though these roots contained intertidal species which may be more tolerant than fully aquatic species (Newell, 1970).

(c) Growth and production in heated waters

Roessler (1971) showed that the least productive area in Biscayne Bay was inside the +4°C isotherm, i.e. in the warmest zone near the Turkey Point Power Station outfall. However, the next *least* productive zones were not the next *most* heated, though all were in areas where *Thalassia* (eel-grass) was absent. Similar patterns were found at the Yorktown site (Warinner and

Brehmer, 1966). At Guayanilla Bay, Puerto Rico, benthic biomass was inversely related to the ΔT in the plume area, though this was not true for other tropical sites studied (Thorhaug, *et al.*, 1978). In the mangrove root communities biomass was not affected by temperatures of 34°C and was at maximum at ΔTs of 6–7°C above ambient, i.e. 31–35°C. In the discharge canal of the Connecticut Yankee Nuclear Power Plant the standing crop of benthos increased by 10–40 per cent over a full year, though it was lower in summer and higher in winter than at the control stations (Massengill, 1976a). Similar increases were recorded for other discharge canals (North and Adams, 1968; Young and Frame, 1976). In the heated lakes studied by Kititsina (1973) the biomass and production of freshwater shrimps increased fivefold, though only where temperatures stayed below 25–26°C. Over 28°C biomass decreased. Similar patterns were shown by Chironomidae, the snail *Valvata piscinalis*, and the worm *Limnodrilus hoffmeisteri*. Faster growth and shortened life-cycles accounted for the rapid turnover.

Given sufficient food, increased growth may be related to increased temperature up to an optimum, after which the relationship is inverse (Rose, 1967; Newell, 1970; Kinne, 1970). Several studies have demonstrated such effects in the field. For example, Kititsina (1973) noted faster growth of shrimps and mysids in heated lakes, though the breeding adults were ultimately smaller than in colder lakes. Similarly, breeding leeches (*Erpobdella octoculata*) and hog-lice (*Asellus aquaticus*) were found to be smaller in heated parts of the River Trent than in the colder parts (Aston and Brown, 1973; Aston and Milner, 1978). In this river, however, adults of both species were larger in the most polluted reach than in cleaner or heated reaches (Fig. 5.5). In contrast, Aston (1968) found that mature adults of the exotic worm *Branchiura sowerbyi* were larger below a power station discharge than in a colder river. However, food may also have been less abundant in the latter habitat. In the Thames the freshwater mussel *Anodonta anatina* was found to grow faster in the first 2 years in a heated area, but after that no differences in size were noted between warm and cold areas (Mann, 1965; Negus, 1966). Gentry, *et al.* (1975) found that dragonfly nymphs from thermally affected ponds and streams were larger than those from control habitats and they suggested that increased size may assist thermo-regulation.

In marine environments similar phenomena have been noted. The clam *Mercenaria mercenaria* grew faster in the heated discharge from Marchwood Power Station in the U.K. (Ansell, 1963). Coughlan (personal communication), however, noted that the flesh condition was poorer than of those in cold water. This also applied to oysters growing on the Bradwell Power Station outfall (Barnes and Coughlan, 1970). Naylor (1959, 1965a) in his extensive work around Tir John Power Station found, however, that adult sea-squirts (*Ascidiella aspersa*) were smaller in the heated areas than he expected.

The fact that many aquatic invertebrates attain larger final body sizes in cold water than in warmer water is well known (Hynes, 1970; Kinne, 1970).

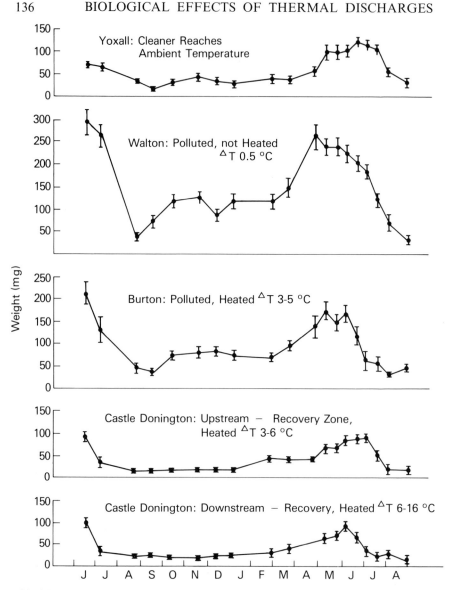

Fig. 5.5. River Trent, 1971–2, seasonal variation in the mean weights of leeches at each sampling with 95 per cent confidence limits are shown, redrawn after Aston, *et al.* (1976).

They tend to have a longer life-span, longer growth-phase, and mature later. At higher temperatures the onset of maturity tends to be earlier and breeding adults are generally smaller. The phenomenon is well known in insects which have overwintering and summer generations of nymphs. Later summer adults are generally smaller than those in the spring hatch (Macan, 1957; Hynes,

1970). The author, in studies of mayflies upstream and downstream of Ironbridge Power Station, found this natural pattern, though the 'size of adults in heated and unheated reaches were similar, during both early and later emergence periods' (Langford, 1975). The increased temperatures did not cause early maturity or smaller adults. Thus, although growth of some species may initially be faster in heated areas, ultimate size and size at maturity may be limited by food availability and gonad development.

(d) Development, breeding-cycles, and life-histories

The rates of development of eggs, duration of breeding seasons, and life-histories of many aquatic invertebrates are altered by temperature changes under experimental conditions, or by temperature differences in different regions (Macan, 1963; Hynes, 1970; Kinne, 1970). It might be expected, therefore, that increases in water temperature near thermal discharges would either speed up or slow down these processes (Naylor, 1965b; McWhinnie, 1967; Precht, et al., 1973).

In the heated reaches of the River Trent young leeches (*Erpodella octoculata*) and young hog-lice (*Asellus aquaticus*) hatched up to 1 month earlier than in the colder reaches (Aston and Brown, 1973; Aston and Milner, 1978). On the other hand, although Elliott (1972, 1978) showed clearly that the rate of egg hatching in the mayflies *Baetis rhodani* and *Ephemerella ignita* was related to water temperature, it was not possible to demonstrate significant changes in the timing of the appearance of nymphs of these species in the heated area of the River Severn (Langford, 1971a). The downstream temperature regime may not, however, have been stable enough to stimulate different hatching rates (Langford, 1971a).

The sex-ratios of some animals are also temperature dependent (McWhinnie, 1967) and an increase in the proportion of male *Asellus aquaticus* was noted by Mann (1965) in a heated area of the River Thames, though Aston and Milner (1978) did not note similar occurrences in the Trent. In a heated South African pond, females of the prawn *Upogebia africana* became more abundant that in a normal population, though breeding cycles did not alter (Hill, 1977).

Advanced emergence of adult insects in warmer waters has been noted under experimental and field conditions (Gaufin and Hern, 1971; Nebeker, 1971a,b; Coutant, 1968b; Rothwell, 1971; Kititsina, 1973; Fey, 1977) and the potential danger of eliminating species by causing early emergence of adults from heated water to cold air has been noted (Langford, 1971a; Langford and Daffern, 1975). There is, however, as yet little evidence of advances of more than 2 or 3 weeks, though Rothwell (1971) and Kititsina (1973) suggested that some species of caddis-fly and Chironomids respectively may emerge from heated lakes up to several months early.

In the Severn, emergence of mayflies and caddis-flies was found to continue during periods when the river was heated to 28–29°C in summer (Langford and Daffern, 1975; Langford, 1975). Early emergence was not generally obvious in any species except in spring 1973 when a small number of mayflies emerged 1 week earlier downstream than upstream. The likelihood of a species being eliminated may be small because of the residual of eggs, larvae, or adults developing outside the normal periods (Langford, 1975). Extended breeding seasons of several marine species have been reported in heated waters. For example, Naylor (1959, 1965a) showed that sea-squirts (*Ciona intestinalis* and *Ascidiella aspersa*) bred throughout the year in the Tir John effluent, instead of only in summer. Similarly, female *Corophium acheriscum* were ovigerous all the year round. Young and Frame (1976) showed similar patterns for *C. acherusicum* and *Jassa falcata* in the Oyster Creek Power Plant discharge canal. These authors point out, however, that Naylor (1965a) showed some confusion about the normal breeding season of *Corophium acherusicum* as he quotes peaks as February–March on one page and three pages later June–July. Extended breeding seasons in heated waters in Britain have also been reported for the immigrant barnacles *Balanus amphitrite* (Crisp and Molesworth, 1951) and *Elminius modestus* (Pannell, *et al.*, 1962), and for several other invertebrates including the clam *Mercenaria mercenaria* (Ansell, *et al.*, 1964) and the bryozoan *Bugula* spp. (Ryland, 1960), the sea-squirts *Botryllus schlosseri* and *Diplosoma listeranum* (Naylor, 1965a). In the thermal discharge area at Hunterston Power Station in Scotland the reproductive periods of the gasteropod snail *Nassarius reticulatus*, the copepod *Asellopsid intermedia*, and the shrimp *Chothroe brevicornis* were found to be advanced by up to 2 months (Barnett and Hardy, 1969; Barnett, 1971, 1972). Two species of mussels (*Mytilus edulis* and *M. californianus*) in a thermal discharge canal in California did not show such advances (Hines, 1979). A reverse reaction was shown by the shore crab *Carncinus maenas*. The species did not breed at all in the hot water at Tir John Power Station in 1960, though in later years as the power station operated less and temperatures were lower, breeding resumed (Naylor, 1965a). Species which have shown clear advances in breeding cycles in heated fresh waters include *Asellus aquaticus* (Mann, 1965; Kititsina, 1973; Aston and Milner, 1978), *Anodonta anatina* (Mann, 1965), the worm *Limnodrilus hoffmeisteri* (Aston, 1973), and the leech *Erpobdella octoculata* (Aston and Brown, 1973). Kititsina (1973) and Aston and Milner (1978) showed that small crustaceans tended to have shorter life histories in heated waters and some mysids had 'all year round' instead of 'summer-only' breeding seasons. Also, in a study of the snail *Physa virgata* in a heated Texas reservoir, McMahon (1975) found that although growth was faster in the second annual generation there were no changes in numbers of eggs per egg-mass laid.

Ginn and O'Connor (1976) exposed shrimps (*Gammarus* spp.) to simulated thermal discharge conditions. They found that 60-min exposures to 34·3°C had no delaying effect on reproduction. Ovigerous females released young normally on exposure to 37°C for 30 min. However, females did not produce young for several weeks. It is evident from these examples that although species may not be eliminated from heated areas, the increased temperatures can have significant effects on their life and breeding cycles. There is little evidence, however, that these changes have led to elimination of any of these species.

(e) Feeding and food

There is ample evidence of relationships between temperature and feeding rates of both freshwater and marine invertebrates (Macan, 1963; Hynes, 1970; Kinne, 1970; Newell, 1970). Feeding activity may cease well before lethal temperatures are reached in some species (Gonzalez and Yevitch, 1976). Other factors such as food availability, size, and acclimatization may also affect feeding (Sheader and Evans, 1975; Fedorenko, 1975; Calow, 1975).

There are no published records of changes in food selection among invertebrates in heated discharges or the indirect elimination of species through elimination of a food item.

Many species are probably opportunistic in their feeding habits and where excessive growths of algae, bacteria, or prey animals occur in thermal discharges other organisms would be expected to feed upon them, possibly in preference to the normal organisms in their diet.

(f) Migrations, mass-movements, and temperature preference

A number of animals are known to migrate to or from deeper or shallower waters during particular seasons. Preference for warmer or cooler water has been suggested as one reason for such migrations (Yonge, 1949; Naylor, 1965b; Hynes, 1970; Newell, 1970). An apparent mass migration of *Gammarus fasciatus* into a thermal discharge was noted previously (see 5.4.5.b), though the reason could have been increased water movement in the discharge zone rather than temperature, as *Gammarus* spp. are known to migrate upstream in running water, probably to compensate for normal downstream drift (Hynes, 1970).

Temperature preferences for invertebrates are not well known. Gentry, *et al.* (1975) demonstrated some temperature selection by acclimatized dragon-fly nymphs, though there was a considerable range of temperature preference shown by each acclimatized group. Newell (1970) also cites a number of references to temperature selection by intertidal invertebrates which may explain their choice of habitat.

(g) Invertebrate drift in streams and rivers

Invertebrates drifting downstream in running water constitute an important food resource for fishes and in most rivers there is a diel pattern of drift with peak abundances at dawn and dusk (Waters, 1962; Elliott, 1967; Hynes, 1970). Water temperature has been considered to be a factor influencing the behaviour and activity of organisms and hence their occurrence in the drift fauna (Müller, 1963; Pearson and Franklin, 1968; Wojtalik and Waters, 1970; Durrett, 1972). Further, the possibility of drifting organisms being carried into heated effluents and hence suffering acute or chronic effects of sudden temperature rises has been suggested (Sherberger, *et al.*, 1977), though any such effect is unlikely to remove the organism from the drift fauna.

The disappearance of the benthic fauna below Martin's Creek Power Plant (Coutant, 1962) must have been because of the death of some resident organisms, which presumably would then pass into the drift. Studies using artificial substrates in heated rivers also record colonization by drifting organisms (Profitt, 1969). Two studies have shown relationships between the numbers of drifting *Gammaridae* and temperature, though the basic diel drift pattern was unchanged (Müller, 1963; Wojtalik and Waters, 1970). In the study by the latter authors, incremental temperatures of up to 8·3°C produced no changes in either the diel pattern or abundance of drifting mayfly nymphs, though higher increments did. Durrett and Pearson (1975) collected drifting fly-larvae (*Diptera*) and phantom larvae (*Chaoborus*) from the discharge canal of a Texas power station. Drift was sparse in winter but abundant in summer. The organisms were almost all derived from the cooler parts of the lake, however, and no mention was made of their survival in the cooling-water channel.

In an experimental study, designed to simulate effects of heat on organisms drifting into a thermal discharge, Sherberger, *et al.* (1977) found that larvae of the mayfly *Isonychia* and of the caddis-fly *Hydropsyche* were not killed by long exposures to 26·5°C and 28°C. Lethal temperatures were 33–35°C and 36–38°C respectively. Neither species showed any behavioural change after exposure to the sub-lethal ranges and thermally shocked specimens were not more susceptible to predation that normal larvae. King and Mancini (1976) showed that mayflies entrained by a power station from the drift fauna suffered high mortalities but other insects were less affected.

In spite of forecasts of 'catastrophic drift' (Wojtalik and Waters, 1970) in heated rivers, it would appear that unless an effluent causes direct mortalities of benthic organisms its influence on drift patterns and drifting organisms is unlikely to be significant.

5.4.7. Biodeterioration, fouling, and wood-borers

Early in the development of cooling-towers, increased bacterial and fungal activity in the warm water caused rapid deterioration of timber packing (Rippon, 1971) (see chapter 6). The possible extended breeding seasons and rapid development of sessile organisms, barnacles, sea-squirts, bryozoans, mussels, etc., in thermal discharges also caused predictions of increased fouling problems (Crisp and Molesworth, 1951; Patel, 1959; Ryland, 1960; Pannell, *et al.*, 1962; Naylor, 1965b). Early settling of fouling organisms in discharge canals has been reported by several workers (Markowski, 1960; Naylor, 1965b; Cory and Nauman, 1969; Nauman and Cory, 1969; McCain, 1975) (see chapter 6).

The possible extension of the breeding season and feeding activity of the destructive wood-boring molluscs and crustaceans, notably the 'shipworms' *Teredo* spp. and *Lyrodus* spp. and the gribbles *Limnoria* spp. and *Chelura* spp. in heated waters, were at one time causes for concern (Bell, 1949; Naylor, 1959, 1965a; Pannell, *et al.*, 1962; Hockley, 1965; Coughlan, 1977). Bell (1949), for example, reported that increased abundance of *Teredo* spp. in Queen's Dock, Swansea, was the result of the heated effluent from a power station, and the species became so prolific that wooden structures had to be replaced by concrete. The gribble *Limnoria tripunctata* was also found to have an extended breeding season in the effluent (Naylor, 1965b).

However, apart from this situation, there is little evidence of massive increases in wood-borers in heated waters. In spite of comprehensive surveillance in some British coastal waters, heated by power stations, there has been no increase in the incidence of these organisms. Other factors, such as removal of old wood and the increased use of metal piling have, in fact, probably led to a marked decline in some species, notably *Teredo* spp. (Coughlan, 1977).

It is likely that although increased temperatures enhance activity and reproduction in enclosed or limited areas, in the further field the temperature patterns are not sufficiently stable to maintain such effects. Further, timber structures are used much less than previously, which has doubtless led to declines in most borer populations.

5.4.8. **Exotic invertebrates**

As we have seen in previous sections very few species have been eliminated by heat alone from areas influenced by thermal discharges. There are, however, a number of records of exotic or introduced species found in heated waters. In the marine environment, Naylor (1965a,b) recorded three species of exotic crabs: *Brachynotus sexdentatus*, *Neopanope texani sayi*, and *Rithropanopeus harrisi*. Eighteen other exotic species are listed by Naylor (1965b), including representatives of Polyzoa, polychaetes, crustaceans, molluscs, and fish.

In fresh waters the exotic oligochaete worm *Branchiura sowerbyi* has been reported from heated rivers in Europe and in the U.S.A. (Aston, 1968, 1973). The Mediterranean and American species of snails of the genus *Physa* (namely *acuta* and *heterostropha* respectively) have been reported from engine cooling ponds and heated rivers in Britain and in Eastern Europe (Macan, 1960; Langford, 1971b, 1972; Turoboyski, 1973). *Physa acuta* was found to be very common and widespread in cooling-tower ponds and rivers in Britain (Langford, 1971b and unpublished). The species was also found to colonize new cooling ponds. For example, Ironbridge 'B' Power Station, commissioned in 1970, included cooling towers. Collections showed no molluscs in these ponds until 1973, when *Limnea pereger* was recorded. Early in 1975 large numbers of adult *Physa acuta* were found, though records of the species in the locality were previously non-existent (Langford, unpublished). The nearest known population was in the River Trent over 30 km away.

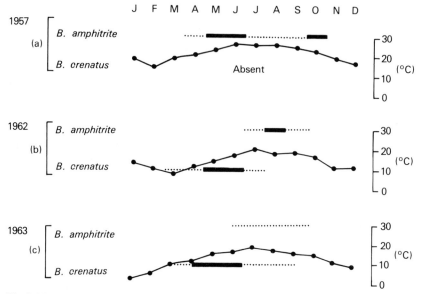

Fig. 5.6. Results of experiments carried out in a heated dock showing settlement on test panels of the warm water immigrant barnacle *Balanus amphitrite* and the native barnacle *B. crenatus*, redrawn after Naylor (1965b).

Many exotic species have also been recorded in unheated waters, to which they have been introduced either by collectors or by shipping or with imported goods (Macan, 1960, 1974; Naylor, 1965b). The clam *Mercenaria mercenaria* (Ansell, *et al.*, 1964) has bred so successfully in Southampton Water that it is the basis of a thriving commercial fishery.

Records of complete replacement of an indigenous species by exotic species in heated areas are scarce (Naylor, 1965b; Macan, 1974), though in a reverse example Naylor (1965b) demonstrated the gradual replacement of an exotic barnacle by an indigenous species in a heated dock, during a period when a power station was going out of use (Fig. 5.6).

5.4.9. Synergistic effects of heat and chemicals

It has already been noted (chapter 3) that the intensity of temperature stress may be magnified by the exposure of aquatic organisms to other environmental stresses such as low oxygen levels, salinity changes, desiccation, etc. (Hawkes, 1969; Newell, 1970). Cairns, *et al.* (1975, 1978) have reviewed the data on synergistic effects on algae, invertebrates, and fish.

Data are almost entirely from experiments, and while it is evident that many organisms exposed to other pollutants may be more sensitive to thermal shocks, the effects are difficult to illustrate in the field. It is possible, however, that some of the organisms, which are apparently eliminated below some power station discharges but appear in similar temperature conditions at others (Langford, 1967, 1971b), may be responding to synergistic effects of heat and pollutants or biocides (see chapter 7).

5.4.10. Fish

Fish occur naturally over a temperature range from $-2 \cdot 5°C$ in polar seas up to 38–40°C in thermal springs and streams (Brett, 1960, 1970; Hynes, 1970; Brock, 1975). Nikolsky (1963) quotes temperatures at which some species occurred at 52°C in thermal springs, though this seems unlikely.

Few species exist successfully in habitats with an annual temperature range of more than 30°C. Some stenothermal species can tolerate annual fluctuations of only $+6°C$, and a few tropical species have minimum tolerance temperatures as high as 16 or 17°C (Brett, 1970). In temperate waters, however, most species are eurythermal, though their optimum temperature range may vary with season or stage in their life-history (Brett, 1970; Coutant and Talmage, 1976, 1977) (Fig. 5.7). Thermal death points of selected species are shown in Table 5.2. Data for many other species may be found in the literature (Coutant and Talmage, 1976, 1977).

Table 5.2. *Upper temperature tolerances of some marine and freshwater fishes*
  A. Marine and anadromous species (from de Sylva, 1969)

Species	Acclimatization temp., °C	Tolerance limit temp., °C	Duration, h.
Alosa pseudoharengus	15	23	90
Aspidophoroides monopterygius		24·4–25	
Atherinops affinis	20	31	24
Caranx mate			
prolarva and postlarva		30	
Clinocottus globiceps		26	
Clupea harengus, larva	15·5	23·0	24
Cyclopterus lumpus		25·5–26·9	
Cynoglossus lingua			
prolarva and postlarva		30	
juvenile		23	
Dussumieria acuta		31	
Enchelyopus cimbrius		27·2	
Fundulus heteroclitus		40	2
F. heteroclitus	28	37	
F. parvipinnis	30	37	24
Gadus morhua		19·8–24·4	
embryo		10	
Gasterosteus aculeatus, adult		31·7–33	
larva		37·1	
Girella nigricans	20–28	31	72
Hemitripterus americanus		28	
Hippoglossoides platessoides		22·1–24·5	
Hypomesus olidus		10	
Limanda ferruginea		24	

Table 5.2 (*Cont.*)

A. *Marine and anadromous species* (from de Sylva, 1969) *(Cont.)*

Species	Acclimatization temp., °C	limit temp., °C	Tolerance Duration, h.
Liopsetta putanami		31·6–32·8	
Macrozoarces americanus		26·6–29	
Megalaspis cordyla			
prolarva and postlarva		31	
Melanogrammus aeglefinus		18·5–22·9	
Micorgadus tomcod		29	
2 cm		19–20·9	
14–15 cm		23·5–26·1	
22–29 cm		25·8–26·1	
Mugil cephalus			
prolarva and postlarva		32	
Myoxocephalus aeneus		26·3–27	
M. groenlandicus		25	
M. octodecemspinosus		28	
Oncorhynchus gorbuscha			
juvenile	20	23·9	168
O. keta, juvenile	20	23·7	168
O. kisutch, juvenile	20	25	168
O. masou, embryo		13	
O. nerka, juvenile	20	24·8	168
O. tschawytscha	20	25·1	168
Osmerus mordax		21·5–28·5	
Pagrosomus major		21	
Petromyzon marinus			
prolarva and postlarva	20	34·0	1·5
Plecoglossus altivelis		22	
Pleuronectes platessa, embryo		14	
Pollachius virens		28	
Polynemus indicus			
prolarva and postlarva		31	
yearling	9	23	40
Pseudopleuronectes americanus		27·9–30·6	
Salmo salar			
prolarva and postlarva		28	1
S. trutta trutta			
prolarva and postlarva		28	1
alevin	20	26	7
Saurida tumbil			
prolarva and postlarva		31	
Scomber scombrus, embryo		21	
Solea elongata			
prolarva and postlarva		32	
juvenile		23	
Tautogolabrus adspersus		29	
Triacanthus brevirostris			
prolarva and postlarva		30	
Ulvaria subbifurcata		27–29	
Urophycis chuss		27·3–28	

Table 5.2 (*Cont.*)

A. *Marine and anadromous species* (from de Sylva, 1969) *(Cont.)*

Species	Acclimatization temp., °C	Tolerance limit temp., °C	Duration, h.
U. tenius		24·5–25·2	
Raja erinacea		29·1–29·5	
juvenile		30·2	
R. ocellata		28	24
R. radiata		26·5–26·9	
Squalus acanthias		28·5–29·1	
Hippocampus sp.	30	Davenport and Castle, 1895	
Clupea harengus			
juvenile–adult	20·8–24·7	Huntsman and Sparks, 1924	
Fundulus heteroclitus			
juvenile–adult	40·5–42	Huntsman and Sparks, 1924	

B. *Freshwater and anadromous species* (from Jones, 1964)

Species	Acclimatization temp., °C	Upper tolerance limit
Lepomis macrochirus	15	30·7
Gasterosteus aculeatus	25–26	30·6
Salmo trutta	26	26
fry	5–6	22·5
	20	23
yearling	?	25·9
parr	?	29
Cyprinus carpio	20	31–34
Ictalurus nebulosies	15	31·8
Oncorhynchus tsawytscha, fry	15	25
	20	25·1
O. keta fry	15	23·1
	20	23·7
O. kisutch fry	15	24·3
	20	25
O. gorbuscha fry	5	21·3
	10	22·5
	20	23·9
Notropsis cornutus	15	30·3
Carpiodes carpio	15	29·3
Semotilus atromaculatus	15	29·3
Notemigonus crysoleucas	15	30·5
Carassius auratus	10	30·8
	20	34·8
	30	38·6
Micropterus salmoides	20	32·5
	30	36·4

Table 5.2 (*Cont.*)

B. *Freshwater and anadromous species* (from Jones, 1964)

Species	Acclimatization temp., °C	Upper tolerance limit
Perca fluviatilis	?	23–25
Lepomis gibbosus	25–26	34·5
Salmo gairdneri	?	28
Rutilus rutilus	20	29·5
	25	30·5
	30	31·5
Salmo salar, grilse	?	29·5–30·5
parr	?	32·5–33·8
	?	29·8
Tinca tinca	?	29–30
Perca flavescens	15	27·7

The effects of temperature on the biology and ecological requirements of fish have been extensively studied and reviewed (Fry, *et al.*, 1946; Nikolsky, 1963; Fry, 1967; De Sylva, 1969; Raney and Menzel, 1969; Brett, 1970; Garside, 1970; Hynes, 1970; Coutant and Talmage, 1976, 1977; Richards, *et al.*, 1977). Temperature can affect survival, growth and metabolism, activity, swimming performance and behaviour, reproductive timing and rates of gonad development, egg development, hatching success, and morphology. Temperature also influences the survival of fishes stressed by other factors such as toxins, disease, or parasites (Coutant and Talmage, 1976, 1977; Cairns, *et al.*, 1978; Groberg, *et al.*, 1978; Sanders, *et al.*, 1978).

(a) Recorded fish mortalities

In 1968 the Federal Water Pollution Control Administration compiled reports from forty-two states which showed mortalities of 15,236,000 fish, of which 1·8 per cent were reported as due to electricity generating stations. However, mortality of 250,000 fish at one site made up the bulk of the 1·8 per cent. Without this, the percentage kill was 0·12 per cent. Truly 'thermal' mortalities in 1968 were estimated as less than 0·1 per cent and these from 3 out of over 400 operating power plants.

Between 1962 and 1969 only eighteen 'voluntary reported' fish-kills were associated with thermal discharges in the U.S.A. involving 700,000 fish, though the relationship to temperature changes was not clear in all cases (I.A.E.A., 1972). Barber (1972) suggested however that most fish-kills at power plants were not recorded; though in such a 'pollution conscious' country this seems improbable.

Mass mortalities of Atlantic menhaden (*Brevoortia tyrannus*) have been recorded near at least three power plants, namely Millstone (Conn.) (I.A.E.A., 1972), in the Cape Cod Canal (Fairbanks, *et al.*, 1968), and at Northport on Long Island Sound (Young and Gibson, 1973). Only in the

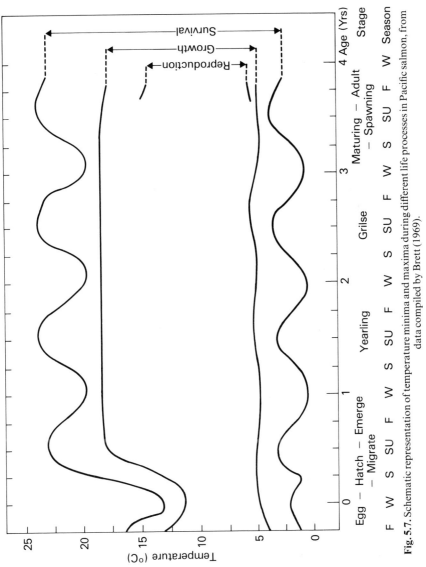

Fig. 5.7. Schematic representation of temperature minima and maxima during different life processes in Pacific salmon, from data compiled by Brett (1969).

third case did free-swimming fish encounter temperatures of 37–38°C which may have been lethal. At Millstone maximum temperatures in the discharge canal were only 19–23°C, though ΔT rose by 13–17°C in a short period and fish had no time to avoid the discharge. The incident in the Cape Cod Canal was similar and trapped fish succumbed.

An important fact to note, however, is that massive *natural* annual mortalities of menhaden occur in near-shore areas after spawning and it is likely that these may occur near a power station and be wrongly associated with it. Impingement mortalities on intake screens have exceeded 100×10^6 at one site alone (Sharma and Freeman, 1977) (see chapter 6).

In the U.K. and Europe recorded fish mortalities related to increased temperature are rare. Alabaster (1962) noted small mortalities of free-living roach (*Rutilus rutilus*) and gudgeon (*Gobio gobio*) where river temperatures reached 36°C during experiments. In the author's experience the only relevant mortalities recorded since that time were of about 100 small cyprinids in the Great Ouse during 1969 at temperatures of over 34°C, and again small numbers of cyprinids in the River Trent in 1976 at temperatures of over 35°C.

One of the more significant reported fish-kills in the U.S.A. was a result of a *reduction* in temperature. In January 1972 there was a 'shutdown', i.e. a rapid, partial load-reduction at the Oyster Creek Power Station, followed by a full closure for over $6\frac{1}{2}$ hours. The discharge canal temperature fell from 22°C to 15°C over a very short period, followed by a fall to about 2°C over the next $6\frac{1}{4}$ hours. Up to 200,000 Atlantic menhaden (*Brevoortia tyrannus*) were found dead in the channel on 29 January. The theory was that menhaden had been attracted and retained by the thermal plume instead of migrating out of the bay as was more usual. The loss of up to 200,000 fish was regarded as insignificant to the massive population of the species in the region (U.S.A.E.C., 1971; I.A.E.A., 1972). At the Turkey Point Power Station in Florida fish mortalities at the outfall were attributed to temperatures of 33–35°C, together with sharp decreases in salinity, high levels of copper and low oxygen levels (U.S.A.E.C., 1971). Recorded mortalities are much rarer than might be expected, especially as we have seen that effluent temperatures can often exceed 35°C and sometimes over 40°C. The possible reasons for the low mortalities are discussed later in this chapter.

(b) Community responses (distribution, abundance, and diversity)

Anglers have believed for many years that fish actually congregate in thermal discharge canals or near outfalls, particularly in winter. If so, such congregations could alter the natural distribution, movements, and ecology of the fish population in a locality. Many studies have been made near operating power stations in fresh and tidal waters.

The factor common to most of these studies is the establishment of marked seasonal fluctuations in distribution, abundance, and diversity of the fish fauna in both natural situations and in thermally affected areas. The basic seasonal patterns of diversity are mostly similar in heated and unheated waters, though differences in species composition and relative abundance may have been noted (Trembley, 1960; Alabaster, 1962, 1969; Grimes and Mountain, 1971; Brown, D., 1973; McErlean, et al., 1973; Neill and Magnusson, 1974; Stauffer, et al., 1974; Storr and Schlenker, 1974; Targett and McLeave, 1974; Carr and Giesel, 1975; Gallaway and Strawn, 1975; Kelso and Minns, 1975; Homer, 1976; Marcy, 1976a; McFarlane, 1976; Teppen and Gammon, 1976; Yoder and Gammon, 1976; Hillman, et al., 1977; White, et al., 1977; Wilkonska and Zuromska, 1977; Sadler, 1979).

The diversity of the fish fauna in any locality may vary as a result of a reduction in the number of species present or a change in the relative abundance in one or two species, for example massive immigration (see references above). Several studies have suggested that some species migrate into and out of thermal discharges and discharge canals seasonally, while others may remain all the year round. Emigrations occurred in the various studies over a temperature range of 28–40°C depending upon the species (Alabaster, 1962, 1969; Moore and Frisbie, 1972; Brown, D., 1973; Moore, et al., 1973; Landry and Strawn, 1973; McNeely and Pearson, 1974; Leynaud and Allardi, 1975; Stauffer, et al., 1975a; Marcy, 1976a; Sadler, 1979).

Brown, D. (1976) showed that juvenile cyprinid fishes in British rivers congregated in heated areas during autumn and winter, but were more evenly distributed in spring and summer over both heated and unheated reaches (Fig. 5.8). On the Florida coast, however, Carr and Giesel (1975) found that there were fewer species and much lower densities of juvenile fish in heated tidal creeks all the year round than in those at ambient temperature, though in both ambient and heated creeks there were similar patterns of seasonal variation.

Marcy (1976a), in an extensive study on the Connecticut River, showed that after the Connecticut Yankee Power Station began operating, few species of fish occurred in heated mixing-zone areas of the river, but he concluded that this was a result of immigration of these species into the discharge canal. At the hottest times of year, however, fewer species were found in the discharge canal than at unheated stations and most fish left the canal when temperatures exceeded 35–36°C. Stauffer, et al. (1975) found that a thermal plume in the New River, Virginia, caused temporal and spatial changes in the distribution of some species but that others were 'indifferent' within the temperature range up to 35°C. White, et al. (1977) in another study of Virginia river, found that although diversity and abundance changed seasonally at the control sampling stations, both were much lower below in the most heated zone, all the year round. McFarlane (1976) found that most fish were absent from thermally stressed streams near the Savannah River Plant where temperatures exceeded 37–40°C, but many species recolonized as the water cooled in winter or when operations ceased.

To summarize the many studies of fish community responses in the field, it appears that there is evidence of immigration and emigration of fish between

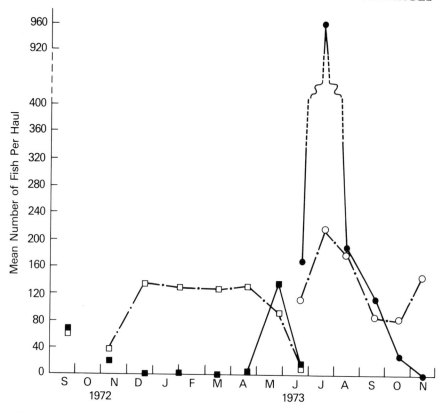

Fig. 5.8. Mean numbers of juvenile fish caught per unit effort, upstream (solid symbols) and in the effluent channel (open symbols) of Peterborough Power Station, redrawn after Brown (1975).

heated and unheated waters which cause changes in diversity and abundance though the reactions of the various species differ. Emigration of all but a very few species occurs when temperatures exceed about 35°C in most U.S.A. habitats. In some discharge canals certain species have been caught at temperatures up to 40°C (Marcy, 1976a). In European rivers evidence suggests peak abundances of cyprinid fish in autumn or early spring in heated areas at temperatures of 25–28°C, and emigration may occur in summer (Brown, D., 1973; Sadler, 1979) or winter (Leynaud and Allardi, 1975).

Although there seems to be consistent seasonal patterns in thermally affected areas, which could be related to temperature selection and acclimatization (Alabaster, 1962; Brett, 1970; Richards, et al., 1977), there has been some doubt expressed as to the true reasons for such seasonal fluctuations and their relationship to water temperature (Neill and Magnusson, 1974; Kelso and Minns, 1975; Spigarelli, 1975; Brown, D., 1976; Langford, et al., 1979).

Brown, D. (1976), for example, showed that juvenile fish stayed in a discharge canal, even when there was no heated water being discharged.

Leynaud and Allardi (1975) showed that there were fewer fish in a discharge canal in both winter *and* summer than in control areas, even though temperatures were very different. In other studies where good control stations have been used, decreased abundance in the discharge areas has been paralleled by similar changes at the control stations (Neill and Magnusson, 1974; Kelso and Minns, 1975; Spigarelli, 1975; Hillman, *et al.*, 1977).

The tendency for non-diadromous fishes to migrate to specific areas for feeding or spawning is well documented (Nikolsky, 1963; Harden-Jones, 1968; Hynes, 1970; Johnson and Hasler, 1977), and the importance of shore-zones to these feeding or spawning movements is also well known (McErlean, *et al.*, 1973; Carr and Giesel, 1975; Spigarelli, 1975; Hillman, *et al.*, 1977; Moore, R., 1978). The natural migrations into such areas would, therefore, bring fish automatically into the discharge area of a power station.

The fact that freshwater fish congregate in certain areas in winter or prior to the spawning season is also documented (Stott, *et al.*, 1963; Johnson and Hasler, 1977; Langford, *et al.*, 1979). Some of these fish may also make extensive migrations within a lake or river system, probably using water currents to navigate (Harden-Jones, 1968; Hynes, 1970; Arnold, 1974).

Such migrations would bring fish into contact with the strong currents created by thermal discharges. These currents may 'attract' fish more than the temperature increases. This hypothesis is supported by the fact that at some sites fishes are strongly attracted to discharges at temperatures which are potentially lethal (Reutter and Herdendorff, 1974; Effer and Bryce, 1975). Also at several sites various species have been found to migrate into and out of thermal plumes when the temperatures were neither within their preference ranges nor exceeding their tolerance limits. Usually these migrations were natural in control areas at that time of year (Storr and Schlenker, 1974; Spigarelli, 1975; Otto, *et al.*, 1976). Minns, *et al.* (1978) concluded that shore fish communities were migrating into discharge areas at two Canadian power stations entirely in response to water currents and topography. On the other hand, Coutant (1975) showed that gizzard shad (*Dorosoma cepedianum*) only entered a discharge canal when the water was heated and left when the canal discharged only cold water, even though there was still a current (Fig. 5.9). Temperatures selected in the field matched temperature preferences found in the laboratory. Alabaster (1962, 1969) found similar relationships for common bream (*Abramis brama*).

The phenomena of congregation near, and emigration from, thermal discharges is as yet not clearly explicable either on temperature or water-current relationships. It is probably that different species respond to different stimuli at different times of the year or life-history stage and, therefore, movements may occur for various reasons. Increased food supply from organisms passing through the station is probably a major factor. There is little doubt, however, that fish in thermal discharge areas remain active in winter and this renders them more liable to capture than torpid fishes in colder

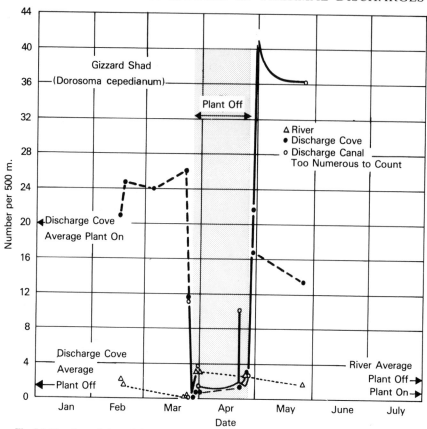

Fig. 5.9. Numbers of gizzard shad, *Dorosoma cepedianum*, per 500 m of shoreline near the Bull Run Steam Plant, Clinch River, Tennessee, U.S.A., February–May 1971, redrawn after Coutant (1975)

water. There are obviously a number of areas for further research concerning fish movements and migrations in thermal discharge channels. Temperature selection is discussed later in this chapter (see 5.4.10.).

(c) Exotic species and elimination of indigenous species

There are no records of the permanent elimination of any species from discharge areas of power stations, though at many of the sites discussed in the previous section species may be excluded from the warmest areas for periods varying from several hours up to a few weeks. Marcy (1976a) found, for example, that the species caught in the effluent channel at Connecticut Yankee could vary almost daily with small increases and decreases in temperature.

Although some exotic species have been introduced to most countries of the world by now, there are few which are entirely associated with thermal discharges. The carp *Cyprinus carpio* introduced to the U.S.A. from Europe seems to be favoured by warm water and has been recorded from heated lakes and discharge canals (McNeely and Pearson, 1974; Marcy, 1976a), though it is also common in many small unheated eutrophic lakes as it is in Europe.

In Britain there are reports of small populations of exotic warm-water species in the St. Helen's Canal near a heated effluent from a glass works, and Bowers and Naylor (1964) recorded the exotic sand-smelt *Atherina boyeri* breeding in the heated area of Queen's Dock, Swansea. This species has, however, also been reported from unheated areas (Palmer, *et al.*, 1979).

(d) Growth

Given adequate food and no other stresses the growth of fish under experimental conditions increases with temperature up to an optimum. This optimum varies with species (Fry, 1967; Brett, 1970). Other stresses such as low oxygen levels, presence of toxins, disease, parasitism, or excessive temperature may depress the growth-rate.

Field studies of growth in heated discharge areas can be complicated by the mixing of populations and migrations discussed in the previous section. There are, therefore, few significant data.

Brown, D. (1973, 1976) found that under-yearling roach (*Rutilus rutilus*), bleak (*Alburnus alburnus*), bream (*Abramis brama*), and chub (*Leuciscus cephalus*) grew faster during their first 6 months of life in heated British rivers than in colder reaches (Fig. 5.10). Growth continued through the winter in heated rivers instead of stopping in November as was normal. The mixing of 'heated' and 'unheated' populations in spring obscured further growth differences. Cragg-Hine (1971) showed that adult roach and bream began to grow up to 2 months earlier in heated reaches than in the colder ones. Langford (unpublished) has also noted that sea bass (*Dicentrarchus labrax*) appear to grow all year round in a thermal discharge canal on the Medway Estuary, whereas growth normally ceases in November. Temperatures in these heated British waters rarely exceeded 30°C and in winter ranged from 7 to 15°C.

In contrast, Marcy (1976a) found that tagged brown bullheads (*Ictalurus nebulosus*) lost on average 0·2 per cent of their body weight per day when over-wintering in the discharge canal at the Connecticut Yankee Nuclear Power Plant. The actual range of weight change among individuals was from + 50 per cent to − 37·5 per cent over the whole winter. At the end of winter the fish migrated out of the canal and made up their weight loss during summer. It was concluded that brown bullheads lost the equivalent of 1·5 to 2 years' growth over their life. Food supply was believed to be the limiting factor because of overcrowding in the canal. The increased metabolic rate caused by higher temperatures may also have resulted in greater food demands for maintenance. Winter temperatures varied from about 5 to 15°C. This study of growth by Marcy (1976a) is probably the most comprehensive published for a thermal discharge canal and serves to illustrate possible deleterious effects of heated discharge, even in winter if the food supply is inadequate. Graham

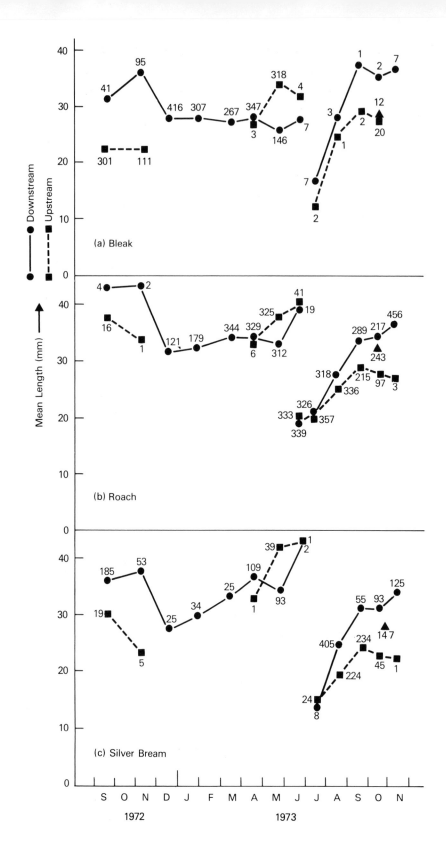

(a) Bleak

(b) Roach

(c) Silver Bream

Mean Length (mm)

Downstream

Upstream

S   O   N   D   J   F   M   A   M   J   J   A   S   O   N
1972   1973

(1974) found similar effects on sunfish (*Lepomis* spp.) in a cooling impoundment in North Carolina as a result of high temperatures and overcrowding.

Sappo (1975) found that growth of bream (*Abramis brama*) in Russian lakes
heated to 31°C was slower than in colder lakes after the first few years, mainly
because of increased metabolism and food scarcity. Fish matured at an earlier
age and a smaller size then normal. Where temperatures rarely exceeded
23–25°C growth was better throughout life. Marciak (1977) found a similar
pattern for the same species in the heated Konin Lakes where in the first 2
years growth was better than in colder lakes. The condition factors (i.e.
weight–length ratio) of the older bream were lower in heated than unheated
lakes and this also applied to the bullheads at Connecticut Yankee (Marcy,
1976a).

Thus, although as we will see later (chapter 5), thermal effluents may be used
to stimulate growth under culture conditions if food is inadequate and
temperatures exceed growth optima, fish condition may deteriorate even in
winter and more so in summer.

(e) Reproduction and larval development

Brett (1960, 1970) showed that Pacific salmon reproduce successfully over a
temperature range which is much narrower than that which the adults tolerate
(Fig. 5.7). Coutant and Talmage (1976, 1977) have listed temperature ranges
over which a number of species reproduce in nature. Artificial increases in
water temperature may lead to changes in the timing of spawning in some
species and also to advanced hatching of eggs. The possible early emergence of
larvae and early development of fry may be harmful as the necessary food for
the fry may not be available. Field studies of reproductive cycles in heated
areas are rare, inconclusive, and even contradictory.

Cragg-Hine (1969a) and Brown, D. (1973) suggested that roach, bream,
and bleak did not necessarily spawn early in thermally affected British rivers,
the former based the suggestion on studies of gonad condition, the latter from
studies of fry distribution. However, Brown, D. (1976) in a further study
suggested that roach (*Rutilus rutilus*) may have spawned early in one river in
Britain. Bray (1971) also suggested that roach in the River Witham spawned
6–8 weeks earlier in the heated than in the unheated reaches.

Marcy (1976a) found earlier development of ovaries in the heated
Connecticut Yankee discharge canal than in the colder areas, but there was no
evidence of early spawning. The conclusion was the fish migrated into cooler
water prior to spawning and that day-length actually stimulated the onset of
spawning. Advanced and prolonged spawning has been reported for several
species affected by heated discharges, including: silver bream (*Blicca
bjoernka*), tench (*Tinca tinca*), rudd (*Scardinius erythrophthalmus*), pike-perch

Fig. 5.10. Mean lengths of juvenile cyprinids caught in trawls upstream, and in the effluent channel
of Peterborough Power Station, redrawn after Brown (1975).

(*Stizeostedium lucioperca*) in Lake Lichen, Poland (E.I.F.A.C., 1968b; Wilkonska and Zuromska, 1977); white suckers (*Catastomus commersoni*) in the Delaware (Trembley, 1960); large-mouth bass (*Micropterus salmoides*) in a Missouri reservoir (Adair and De Mont, 1971; Witt, 1971); and sauger (*Stizeostedium canadense*) in a Tennessee river (Dryer and Benson, 1957). The periods of advancement varied from 1 to 5 weeks in the different habitats.

The spawning of perch (*Perca fluviatilis*) and pike (*Esox lucius*) in Lake Lichen were unaffected, though spawning conditions generally deteriorated when grass-carp (*Ctenopharyngodon idella*) were introduced to crop aquatic weeds (Wilkonska and Zuromska, 1977).

In the Savannah River studies, large-mouth bass (*Micropterus salmoides*) showed early gonad development but not advanced spawning in areas heated up to 10–15°C above ambient (Bennett and Gibbons, 1972).

Kaya (1977) found that brown trout (*Salmo trutta*) in the naturally thermal Firehole River did not reproduce successfully in the heated reaches. Gonad maturation was poor and degeneration of ova occurred. Few juveniles developed. In contrast, rainbow trout (*S. gairdneri*) reproduced successfully but changed their spawning period from spring to autumn, avoiding the hottest period of the year for hatching and fry development.

Development and hatching rates of fish eggs are often time-temperature dependent (E.I.F.A.C., 1968b; De Sylva, 1969; Brooke, 1975; Guma'a, 1978; Alderdice and Velsen, 1978), but there are no recorded observations of eggs actually in heated discharges. Nakatani (1969) showed that Chinook salmon spawned regularly in the heated waters below the Hanford reactors on the Columbia River. Between 1950 and 1967 rudds increased in number from 300 to 3,300 in spite of the thermal plume and there was no evidence of destruction of eggs and larvae.

Increased temperature like other stresses may cause physical deformities or alterations in meristic characteristics of young fish on hatching (E.I.F.A.C., 1968b; Coutant and Talmage, 1976, 1977). Twenty per cent of deformed individuals have been found among carp hatched at over 20°C and roach hatched at over 16°C (E.I.F.A.C., 1968b). Brooke (1975) also showed that abnormal fry of Coregonids were most abundant after eggs hatched at 10°C and least at 4°C. Boitsov (1974) found that body shape and size of fins differed in bream and roach taken from heated and unheated reaches below the Konakovo Power Station, though he suggested that higher current velocities in the heated zone were more responsible for the changes than higher temperatures. There are no published field studies of larval deformities, but Sappo (1976) suggested that morphologically separate local populations of the bream *Abramis brama orientalis* had formed in the area of a reservoir heated by the Konakova Power Station and that this separate population numbered 1.3×10^6 fish aged 1–13 years. It has been demonstrated that genetic changes may also occur in populations of fish in heated waters (Mitton and Koehn, 1975), but these may be reversible in some situations (Ames, *et al.*,

1979). This is obviously an area where much more research is needed, though the separation of morphologically and genetically distinct populations caused by heated effluents in most habitats will doubtless be obscured by the natural mixing of cold and warm water groups.

(f) Food and feeding

Feeding, digestion, and food evacuation rates in a number of fish species are clearly related to temperature (Hathaway, 1927; Fry, 1967; De Sylva, 1969; Brett, 1970; Peters, *et al.*, 1974; Elliott, J., 1976). Species which have fasting periods during the year usually begin to feed when water temperatures reach a threshold (Holmes, *et al.*, 1974; Randolph, 1975). Food selectivity may also decrease as temperature and hence metabolic rate increases (Edwards, D. J., 1974). We have already seen that increased growth or losses in body weight may occur in some heated areas (see 5.4.10.d). Food availability and feeding rates are critical factors determining the extent of these increases or decreases in weight.

Evidence from stomach analysis suggests that many fish are opportunistic feeders, taking food items which are most available provided they can be ingested (Nikolsky, 1963; Langford, 1963, 1966; Hynes, 1970). In several field studies of fish, differences in diet have been observed between heated and unheated waters.

Cragg-Hine (1969b), for example, found that roach and bleak in an effluent channel consumed more vegetable material than in the main river. The effluent channel contained a much more abundant flora than the river, though macro-invertebrates were also more abundant.

In the cooling ponds of a Lithuanian power station roach and silver bream fed on zebra mussels (*Dreissenia polymorpha*) and algae in heated zones, and on insects in the unheated zones, though the diet of rudd (*Scardinius erythrophthalmus*) was similar in both zones (Rachyunas, 1973). Polivannaya (1974) also showed differential selection of zooplankton by perch (*Perca fluviatilis*) in the warmer and colder areas of heated lakes. Spot-tail shiners (*Notropis hudsonis*) in the Connecticut Yankee Discharge Canal consumed 30 per cent more insects that in the river (Marcy, 1976a). In turn, resident brown bullheads (*Ictalurus punctatus*) ate spot-tail shiners in winter instead of their more normal invertebrate diet (Merriman and Thorpe, 1976). A similar change occurred in the heated channel at a Texas power station (McNeely and Pearson, 1974). In most cases the composition of the diet reflects the availability of food organisms in the heated area.

As a contrast, Moore, *et al.* (1975) found no major differences in the diet of white perch (*Morone americana*) between the effluent canal at Chalk Point Power Plant and the nearby Patuxent River. Also diets of large-mouth bass (*Micropterus salmoides*) near the Savannah River Plant were similar in composition in both warm and cold areas though, over a year, more food was consumed in the warm area (Bennett and Gibbons, 1972).

High water temperatures may also deter fish from feeding, though this has not been evident from most field studies. Carp (*Cyprinus caripio*) showed a marked reduction in feeding rate at 29–30°C, under laboratory conditions. Lethal and disturbing temperatures were 34–36°C and 31·3°C respectively (E.I.F.A.C., 1968b). On the other hand, Coregonids (whitefishes) were found to be still feeding in July at about 3°C below their determined lethal temperature in Ukranian rivers and lakes (E.I.F.A.C., 1968b).

There is, therefore, evidence that fish will continue to feed in warm water and alter their feeding habits to meet any change in the flora and fauna of a heated area though they may not always have sufficient food to meet the increased metabolic requirements.

(g) Activity and swimming speeds
The activity and potential swimming speeds of fish are directly related to temperature (Fry, 1967; Brett, 1970). This is an important factor determining the potential for growth in thermal discharge areas because the higher the activity, the more food is required to maintain metabolic requirements, and the scope for growth is reduced. The swimming speed of a fish may also affect its susceptibility to impingement on intake screens (see chapter 6) and temperature here may be a critical factor (Hocutt, 1973) (Fig. 5.11).

Swimming speeds are difficult to determine in the field, though some attempts have been made to estimate swimming speeds and behaviour in thermal plumes using ultrasonic or radio-tracking techniques (Kelso, 1974, 1976a). Brown bullheads, yellow perch, and white suckers were all found to turn more frequently and move more slowly over the bed in a thermal plume than in the colder water outside the plume. However, the reason for the apparent slower speed was that the fish were moving against the currents caused by the discharge. Two fish tracked when only cold water was being discharged showed the same behaviour as those in the thermal plume.

The author and co-workers in studies of sonic-tagged common bream (*Abramis brama*) could show no difference in the normal swimming speeds of fish between summer and winter (Langford, *et al.*, 1979). The fact, therefore, that fish are able to swim faster at higher temperatures does not mean that higher speeds are in fact used for normal movements.

(h) Migratory and diadromous fish
The changes in diversity and abundance of fish in thermally affected areas of water (see 5.4.10.b) have in many cases been the result of migrations of diadromous or non-diadromous species, into or out of the relevant localities.

Naylor (1965b), Nakatani (1969), Hawkes (1969), and Bush, *et al.* (1974), all suggest that the migration of diadromous fishes, particularly salmonids and shads, may be adversely affected by thermal discharges. The hypothesis is that a thermal plume will cause avoidance reactions in migrating fish and if occupying the whole width and depth of a river will deter or at least delay both

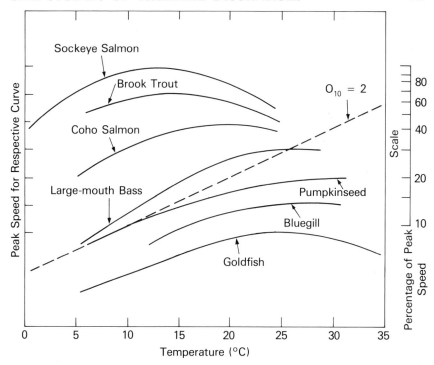

Fig. 5.11. Relative cruising speeds of various fish in relation to temperature when they are acclimated to the test temperature, redrawn after Fry, F. in Rose (1967).

upstream and downstream migrants. The reluctance of upstream migrants to enter tributaries with higher temperatures than others has been noted (see chapter 4), but the relationship between migration and temperature is by no means clear.

There is little evidence to support the 'thermal block' hypothesis. For example, in Britain the River Severn was heated in its middle reaches by the effluent from Ironbridge Power Station for over 45 years. Fully mixed temperatures across the whole width and depth of the river reached between 27 and 31 °C in most years, with ΔTs of 5–9 °C (Langford, 1970b). The salmon fishery upstream did not decline during the life of the power station, therefore fish must have continued to migrate successfully, even though considerable migrations occur in summer.

Coutant (1967) and Becker (1973) describe tracking studies of upstream migrant sockeye salmon (*Oncorhyncus nerka*), steelhead trout (*Salmo gairdneri gairdneri*), and Coho salmon (*Oncorhyncus kisutch*) in the Columbia below the Hanford thermal discharges. Most fish avoided the thermal plumes during the hottest period in summer, migrating along the opposite shore. The migration rates of fish which did encounter the plume were similar to those which did not.

In the Connecticut River migrating adult shad (*Alosa sapidissima*) showed some reaction to a power plant discharge. There was a tendency for the fish which encountered the plume to meander slightly. However, Leggett (1976) concluded that the shad selected certain channel routes and utilized these despite the thermal plume, probably migrating below the warm water where stratification occurred. A few fish did appear to hesitate for several hours near the discharge canal, but continued to migrate after the short delay.

There are few data on smolt migration through thermal discharges, though some experimental studies of survival of caged fish in the Columbia were undertaken (Nakatani, 1969; Becker, 1973).

There is a strong possibility that the current velocities and turbulence produced by large thermal discharges in estuaries, lakes, and coastal waters may attract migrating salmonids or shads, irrespective of the elevated temperature, and there is evidence of this occurring in some species (see 5.4.10.b).

(i) Disease, parasitism, and predation
A number of viral, bacterial, fungal, and protozoal diseases are endemic in fish throughout the world (Nikolsky, 1963). Parasites include flatworms, round worms, tape-worms, and other invertebrates, while predators include invertebrates, fish, and other vertebrates (Nikolsky, 1963).

Generally, the occurrence of disease, parasites, and predators, is at a relatively low level, that is, high enough to maintain the organism without eliminating the host or prey. However, the spread and intensity of a disease or a parasite may be dependent on the size and density of the host population. The extent of predation may also depend on the availability and density of the prey as well as on those of the predator.

Thus, the influence of thermal discharges on the localized abundance of certain fish species may increase densities sufficiently to enhance the transmission of diseases and parasites or to attract predators. Further, some diseases and parasites are known to develop faster at higher temperatures (E.I.F.A.C., 1968b). Also, the increased stresses on fish in warmer water may make them more prone to infection. Finally, disorientation and loss of equilibrium as a result of heat or cold shock by fish make them more susceptible to predation, even though the effects are not in themselves ultimately lethal.

There is some experimental evidence to support these theories, though the field studies are generally inconclusive. Musselius (1963), cited in E.I.F.A.C. (1968b), showed that the parasite flatworm *Bothriocephalus gowkongenis* in the gut of grass-carp reached sexual maturity twice as quickly at 22–25°C as at 16–19°C. Similarly, the eggs of the ecto-parasite *Dactylogyrus vastator* developed faster at higher temperatures.

In the heated Lake Lichen there are reports of high *Ligula intestinalis* infestations, and this was thought to be because of vigorous development of the tapeworm at higher temperatures (E.I.F.A.C., 1968b), though the parasite is also common in unheated lakes and reservoirs (see chapter 4). However, *Ligula* is notoriously sporadic in occurrence and has been found to be abundant in unheated British reservoirs but not present in nearby ponds or rivers (Langford, unpublished).

The numbers of parasitic worms (flatworms, round worms, and tapeworms) taken from large-mouth bass (*Micropterus salmoides*) were significantly higher in the heated areas of the Savannah River cooling-pond system than in the cooler areas, except in summer, although parasite burdens in all parts showed distinct seasonal cycles (Eure and Esch, 1974).

The effects of bacterial infections vary with temperature. For example, laboratory mortalities of sockeye salmon infected with the bacterium *Chondrococus columnaris* were found to increase as temperatures rose (Nakatani, 1969), but recent work has suggested that the relationship of this disease to temperature is complex (Becker, 1973).

Coutant and Talmage (1977) reviewed studies of viral, bacterial fungal, and parasitic invertebrate infections of fish and invertebrates. Temperature relationships were not clear, but in several studies optimal temperatures for successful infection were noted, which varied with the infecting organism from about 7 to 27°C.

Two contrasting effects of temperature were shown in Coho salmon and steelhead trout infected with the causal organism of bacterial kidney disease. Both suffered greater percentage mortalities at temperatures between 6·7 and 12·2°C, than at 20·5°C, though at higher temperatures time between infection and death was shorter (Sanders, *et al.*, 1978). The reverse trend was shown by the same species infected with furunculosis bacteria (*Aeromonas* spp) and higher percentage mortalities occurred at the higher temperatures. However, time from infection to death also decreased with increased temperature (Groberg, *et al.*, 1978). The effect was similar to chemical synergisms where fish exposed to double or multiple stresses succumb earlier than under a single stress.

Predation of spot-tail shiners which congregated in the thermal discharge canal at Connecticut Yankee increased when brown bullheads utilized them as food instead of the usual invertebrates (Marcy, 1976a). Coutant (1973) also showed that under laboratory conditions thermally shocked young sockeye salmon (*Oncorhyncus nerka*) were more heavily preyed upon after exposure to 28°C, though the increase was only by 10 or 11 per cent and may not have been significant in wild populations.

Exposure to cold-shock may or may not cause increased vulnerability to predation (Coutant, *et al.*, 1976; Cox and Coutant, 1976; Wolters and Coutant, 1976; Deacutis, 1978). Predators were found to select fish exhibiting abnormal behaviour after sudden temperature reductions of 3–9°C, even though the fish were not killed by the shocks. Cooling rates of over 2°C per min caused significant increases in predation. In contrast, Deacutis (1978) found that thermally shocked larvae of *Paralicthys dentatus* were less susceptible than controls, though the reverse was true for *Menidia menidia*. No explanation was offered for the differences though the behaviour of shocked fish may be important. Increased gull activity has been noted at several power station outfalls (Young and Gibson, 1973; Prentice, 1969), and in such places the birds may be feeding both on dying or dead fish, or on those which may be sub-lethally affected by the discharge. In the author's experience, however, they are usually feeding on small fish killed after passage through the cooling system.

(j) Indirect biotic effects

Changes in the ecology of one species as a result of changes in another are not always easy to demonstrate in the field. Thermal discharges, however, do create such changes in small areas.

For example, we have already seen that the disappearance of sea-grasses (*Thalassia* spp. and *Zostera* spp.), from some thermal plume areas was a major factor in the disappearance of the invertebrate and fish fauna (Thorhaug, *et al.*, 1978) (see 5.4.6.b).

The change from one food organism to another, shown by fishes in heated discharge areas (see 5.4.10.f), adds predation pressure to the prey species in addition to other stresses. The increased density of local populations of predatory fish is related to the density of the prey species which, in turn, is related to the thermal discharge. This 'chain' effect may lead to the disappearance of the predator from other areas in the relevant water body and indirectly allow the prey species in the vacated area to increase, or possibly allow the entry of other predators.

The effect of the rapid spread of the weed-eating grass-carp *Ctenophryngo-don idella* in some heated Russian lakes has been to remove the vegetation on which other species spawned. This ultimately must lead either to a decline of these species or to a change in their spawning habits.

(k) Gas bubble disease

The incidence of gas bubble 'disease' has been noted in fish from several heated discharge areas (Demont and Miller, 1971; Adair and Hains, 1974; Miller, 1974; Otto, 1976). Fish in discharge canals are particularly prone to the condition and in studies on Lake Norman, in Carolina, thirteen species of fish were found to show physical symptoms of the disease (Adair and Hains, 1974). Otto (1976), however, found that only carp (*Cyprinus carpio*) which were resident in a discharge canal showed signs of the disease, and no other species caught in the discharge area outside the canal were affected.

Supersaturation with oxygen or nitrogen may cause the condition in fish, though Rucker (1972) suggests that while nitrogen concentrations of around 120 per cent saturation value are dangerous, oxygen concentrations need to be 350 per cent saturation before they are harmful.

5.4.11. **Other vertebrates**

Several observations on the ecology of reptiles and amphibians have been made in heated waters. Fry (1967) has briefly reviewed the responses of some amphibia and reptiles to temperature, and noted the presence of acclimatization processes, and more advanced methods of behavioural and physiological thermoregulation than in fishes.

Eggs of the toad *Bufo terrestris* laid in the warmest areas of the Savannah River Nuclear Plant reactor cooling ponds, suffered higher mortalities than in cooler water. However, larvae which survived grew and metamorphosed faster in the warm areas (Nelson, D. H., 1974).

Female turtles (*Pseudemys s. scripta*) in the heated areas were also larger, and more fecund than those in cooler areas. Over 20 years these turtles in heated areas would produce 10^6 offspring instead of only about 10^3 in a nearby normal population (Christy, *et al.*, 1974). Some species of parasitic nematode worms and *Acanthocephala* were also more plentiful in turtles from heated waters, but trematode parasites were less abundant than in colder waters (Bourque and Esch, 1974).

American alligators also tended to congregate in the heated zones near the Savannah River Plant and did not go into a winter dormancy period. Radio tracking showed that at 20–25°C in the cooler lake, during the autumn, alligators were stimulated to move at random. If they came into contact with a thermal area they remained there, but in spring when the main lake reached 20–25°C they dispersed (Murphy and Brisbin, 1974).

The waterfowl distribution differed in heated and unheated areas around the Savannah River Plant. Generally, the diversity and abundance of species was lower in the heated ponds, possibly owing to a decline in certain types of vegetation and food. Coots (*Fulica americana*) were able to utilize the algal mats in the hot water much more than 'diving' or 'dabbling' species. The latter were more common in colder areas (Brisbin, 1974). During a freeze in the area (December 1972), waterfowl decreased in abundance in the unheated areas but not in the heated areas.

There are implications for the transmission of parasites in this situation. For example, winter aggregations of waterfowl in a heated discharge area increased the availability of bird-borne parasite stages of such species as *Ligula intestinalis* or *Diplostomum spathaceum* which use birds as primary or intermediate hosts (E.I.F.A.C., 1968b). If fish aggregate in the same places infection rates may be increased markedly.

5.4.12. **Biological factors modifying thermal effects in the field**
(a) **Temperature preference, avoidance, and behavioural responses**
The fact that many fishes and higher vertebrates when presented with temperature gradients will select a preferred range is well documented. Preference may depend on species, acclimatization temperatures, size, and season (Alabaster, 1962, 1969; Fry, 1967; Brett, 1970; Garside, 1970; Meldrim

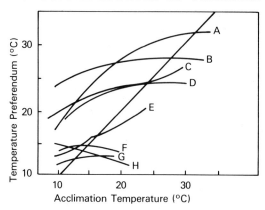

Fig. 5.12. Temperature preferenda of various fish, redrawn after Fry, F. in Rose (1967).

and Gift, 1970; see Esch and McFarlane, 1976; Richards, *et al.*, 1977; Ehrlich, *et al.*, 1979; Spigarelli and Thommes, 1979; Peterson, *et al.*, 1979). Coutant and Talmage (1976, 1977) have listed and reviewed preferred temperatures for many species.

Figure 5.12 shows the relationship between preferred temperature and acclimatization temperature for eight species. With the possible exception of the rainbow trout, prefered temperature increases with acclimatization temperature, though maxima differ with species.

The migrations and movements of fish into and out of heated areas have been related to temperature preferences found in experiments (Alabaster, 1962, 1969; Stauffer, *et al.*, 1975a; Cherry, *et al.*, 1975; Coutant, 1975; Otto, *et al.*, 1976) and both Bull (1937) and Alabaster (1962) showed that some fish would react to changes in temperatures as small as 0·05°C.

However, temperature may not be the only stimulus which attracts fish to and deflects them from certain areas (see 5.4.10.g). Cherry, *et al.* (1975) showed that some fish would select their acclimatization temperature in experimental troughs. They suggested that the fish selected not only on the basis of temperature but that ecological, social, and physiological factors may also be important. Beitinger and Magnusson (1975), for example, found that juvenile bluegill sunfish (*Lepomis macrochirus*) usually selected 31°C in preference to 27 or 34°C when left to select undisturbed. If, however, a socially dominant fish was introduced to this preferred temperature area, juveniles avoided it and selected 'non-preferred' temperatures.

Neill and Magnusson (1974) suggested that fish exhibited a thermoregulatory behaviour pattern, maximizing their exposure to preferred temperatures in the vicinity of a heated discharge. However, they concluded that the presence of most species in an outfall area was directly related to their presence in unheated areas and that a general migration was more responsible for accumulation near the outfall than temperature selection.

Nyman (1975), Kelso (1974, 1976a), Spigarelli (1975), and Langford, *et al.* (1979) have shown that, while certain species may be found congregating near outfalls, residence times of individual fish are not very long, usually a few days at most. Movement away from the outfalls appears to be a result of natural mobility rather than avoidance or temperature selection. Wren (1976) found, however, that a sonic-tagged walleye (*Stizostedion vitreum*) stayed in the effluent channel at the Colbert Power Station in Alabama for 30 days, during which time the temperature rose from 19·2–27·4°C.

The question of temperature selection, avoidance, and thermoregulatory behaviour is vital to the prediction of effects of any discharge on the relevant fish populations. As Neill and Magnusson (1974) state 'it matters little whether in the laboratory yellow perch (*Perca flavescens*), can grow at 30°C or survive at 33°C for 1,000 min, if yellow perch never occur in an outfall area where temperatures exceed 29°C. What matters is, whether given a choice, the perch will invade water of a particular temperature, if so for how long and for what reasons, ecologically prudent or otherwise. This applies to all species potentially affected by any discharge. The attraction of fish to a 'home range' may also influence the readiness with which they avoided adverse tempera-tures and re-colonize. Stott and Cross (1973) showed that it was more difficult to dislodge 'homed' than 'non-homed' fish from an area by presenting them with low oxygen levels to stimulate avoidance reactions. Also, once the adverse conditions passed, homing occurred rapidly. This type of effect could also apply to temperature avoidance. Similarly, Langford, *et al.* (1979) showed that fish displaced by floods tended to home as flows subsided.

Other behavioural changes may occur if fish are exposed to near-lethal temperatures. For example, Power and Todd (1976) showed that the social groups of pumpkinseed (*Lepomis gibbosus*) remained stable as temperatures increased, though ritualized behaviour increased. Between 31 and 38°C stress signs appeared and the social groups broke down. Golden shiner (*Notomi-gorus crysoleucas*) showed no such effect, but only physiological symptoms of stress at near-lethal temperatures. Aggression increased in yellow bullheads (*Ictalurus natalis*) at 30°C, some 9°C below their lethal temperature (McLarney, *et al.*, 1974).

(b) Internal temperature control

Although it is generally assumed that 'poikilotherms' have little or no internal temperature control, the term really means literally 'many temperatured' and does not necessarily imply that internal temperature is always the same as environmental temperature (Fry, 1967).

Usually, a fish has a body temperature similar to that of the environment. Some species, however, notably the skipjack tuna *Katsowonus pelamis* and the yellowfin tuna *Thunnus albacores* can have muscle temperatures of 4–8°C above ambient (Fry, 1967; Dizon, *et al.*, 1977).

Neill and Magnusson (1974) measured body temperatures of fish caught in thermal discharges but found little evidence of thermoregulation. McCauley and Huggins (1976) found that body temperature of rainbow trout fluctuated much less than that of the water in experimental channels, though this was regarded as a passive buffering result of slow heat conduction to the internal sensor. Stauffer, *et al.* (1975b) found that there was very little time lag between heating the water and an increase in body temperature of bluegills. They concluded that the body did not buffer effects of temperature on internal organs.

Bennett (1971) and Spigarelli (1975) both concluded that large-mouth bass (*Micropterus salmoides*) and salmonids respectively showed some internal and behavioural thermoregulation in heated effluent plumes, and Spigarelli, *et al.* (1977) found that the rate of internal temperature change in fishes was closely related to body weight and followed first-order kinetics.

(c) Other factors modifying thermal effects and the relevance of experimental and field data in prediction

It should be evident by now that although there have been many changes in animal and plant ecosystems in the vicinities of thermal discharges large-scale permanent catastrophic ecological effects of heat have not occurred. Where there are measurable and deleterious effects such as at Turkey Point, Florida (Thorhaug, *et al.*, 1978), the Delaware River at Martin's Creek (Coutant, 1962), or in parts of the Savannah River system (Gibbons, *et al.*, 1975), they are mainly fairly localized and seasonal. Even the worst affected areas where organisms disappear tend to be re-colonized at some times of year, usually when water temperatures are lower. Also, the causal agent of some ecological changes is, in some places, in doubt (U.S.A.E.C., 1971).

Why then are the thermal effects not as harmful as were forecast, even though temperature rises at most power stations are 8–15°C in temperate waters and maxima may reach 40–45°C in tropical waters?

The reasons are, in general terms, the instability of the thermal plumes from power stations both temporally and spatially (chapter 2), together with the physical, physiological, and behavioural mechanisms which most organisms use to withstand adverse conditions (Fig. 5.13). In many of the cases quoted in these chapters the major ecological changes recorded in heated waters have actually been observed in discharge canals or cooling ponds and not in the aquatic environment outside those places. Needless to say, the discharge canal and cooling pond are usually specialized, artificial habitats. Once discharged into the larger environment the warm water as a plume often becomes extremely unstable because of the factors described in chapter 2.

Although many of these physical factors were known, and there was a great deal of information on the adaptations of organisms to temperature, there seemed to be little attempt to 'temper' predictions of thermal catastrophe, particularly in the U.S.A., in the 1960s. Extrapolations were, therefore, made often on the basis of relatively simple tolerance data for a small number of organisms.

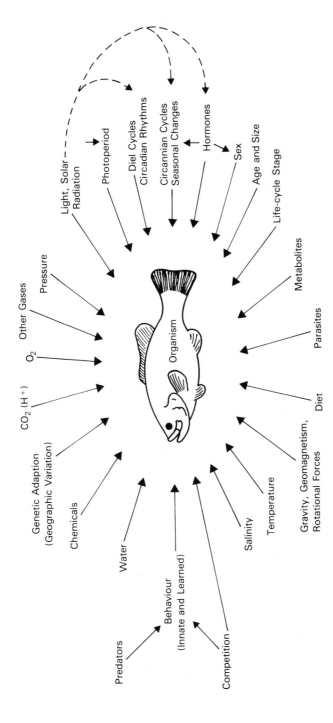

Fig. 5.13. Some factors that may influence the thermal tolerance of organisms, redrawn after Hutchison (1976).

Since that time, however, much more has become known about the physical and chemical aspects of thermal discharges and about the adaptation of organisms to temperature, and other stresses.

To predict effects of any discharge all these factors must be taken into account. The effects of actual passage of organisms through power stations must also be predicted and these are reviewed in chapter 6. The emphasis of this book is mostly on observed effects of power stations in field situations. However, predictive data on acclimatization, exposure times, effects of size and life-history phase, seasonal tolerance behaviour, internal temperature control, temperature preferenda and avoidance have come from comprehensive experimental studies. These are reviewed extensively by many of the authors already quoted (Fry, 1967; Kinne, 1970; Brett, 1970; Hynes, 1970; De Sylva, 1969; Krenkel and Parker, 1969b; Precht, *et al.*, 1973; Coutant and Talmage, 1976, 1977), to whom the reader should refer for data on individual species. Prediction of the ecological consequences of a cooling-water discharge on any habitat is therefore a complex exercise involving:

(i) A thorough knowledge of the distribution, species composition, and ecology of the plant and animal communities of the habitat prior to any kind of construction work, because dredging, silt discharges, and scour from cold-water pumping trials may all cause major changes.

(ii) An accurate prediction of the spatial and temporal behaviour of the thermal plume.

(iii) An assessment of temperature tolerances of sessile organisms or those with no or low mobility such as plants, worms, molluscs, and other small invertebrates, either from experiments or, preferably, from studies of operating power plants in similar situations.

(iv) Information on the natural mobility, migrations, temperature preferenda and avoidance reactions of organisms such as fish, preferably from a combination of site studies and experiments.

Operating-site studies are, in the author's opinion, probably of higher priority than experimental or prolonged before-and-after studies. It may be argued that sites are unique and that data from one site may not be applicable to another. While this may be generally true for physical and hydrographic conditions it may not be true biologically. In any one region of the world the number of species is relatively limited. Thus, provided 'generic' studies of different major types of habitat, for example rocky shore, muddy estuary, river riffle, lake shore, etc., are made with respect to operating power stations, they will include many species common to other potential sites with similar basic physical characteristics, and field data can be used for predictive purposes at these potential sites once plume behaviour at the new site is modelled. Prolonged studies at a specific site are perhaps justified where multi-stage developments, i.e. several power stations are envisaged. Obviously, some follow-up study is necessary to check predictions and these should be planned to try to illustrate seasonal and over-all changes in the major ecosystems and to put these changes into perspective against the natural background.

Given the basic data on enough species, prolonged studies at any one site may only be justified where an important commercial species or very rare species is endangered.

5.5. SUMMARY

The perception of a thermal discharge merely as an agent of temperature rise is a simplistic view. Most discharges produce physical and chemical changes in the immediate environment of the outfall. Scour removes fine sediments and their associated fauna and the resulting substrata are colonized by 'hard-bottom' or 'coarse sediment' species. At some sites these changes have almost certainly in the past been attributed to heat. Similarly, changes probably caused by chlorine have been related to temperature rises. For these reasons many of the conclusions about 'thermal effects' in the field are contradictory or confusing. There are very few places in the world where a detectable permanent change in a plant or animal community can be attributed to heat alone.

Detailed studies of single species have shown that life-histories may be changed in the area of heated discharges. To date, this does not appear to have eliminated any species.

At a number of well-studied sites the discharge canal or cooling pond has been used as a habitat for comparison with natural control habitat. The validity of data from such man-made habitats is doubtful because of their specialized characteristics.

In the environment at large the ability of organisms to acclimatize to temperature and to avoid adverse conditions, combined with the often transient nature of thermal plumes both spatially and temporally, means that the dramatic consequences extrapolated from results of experimental or short-term exposures do not often occur.

Chapter 6 Problems with organisms at intakes and in cooling-water systems

6.1. INTRODUCTION

So far we have been mainly concerned with ecological consequences of power generation on the aquatic environment beyond the immediate perimeter of the power station. However, water entering hydro-electricity turbine intakes or cooling-water circuits may contain many millions of floating or free-swimming organisms including bacteria, algae, other plants, invertebrates, and small fishes. Some of these organisms actually cause considerable operational and economic problems owing to fouling or blockages within the system. Others, subjected to thermal, mechanical, and chemical stresses may be killed in transit from intake to outfall (Langford, 1977). Figure 6.1 shows the major problem areas and stresses in a thermal power station cooling-water circuit. Mussels growing in cooling-water circuits have resulted in losses in thermal efficiency and even total shutdowns at power stations over many years. Financial losses have been extremely large at some British sites (Coughlan and Whitehouse, 1977).

Similarly, ingresses of weed or fish have clogged both cooling-water and turbine intakes in several parts of the world (Langford, 1977; Balon, 1978; Moazzam and Rizvi, 1980).

To counteract settling and actively growing fouling organisms, cooling-water circuits and some turbine intake tunnels are usually dosed with biocides (usually chlorine, in some form) in very large amounts. This causes mortalities of both the fouling and the non-fouling organisms in the circuit. This discharge of the resulting chlorinated effluents, together with dead organisms, may in turn have effects on the habitat beyond outfalls (see chapter 7). There is, therefore, a close relationship between operationally necessary antifouling procedures within systems and the wider ecological consequences of the discharge.

6.2. OPERATIONAL PROBLEMS

6.2.1. Intake screening and location

At both hydro-electric and thermal power stations the primary intake protection is provided by either a coarse mesh metal screen or a simple vertical

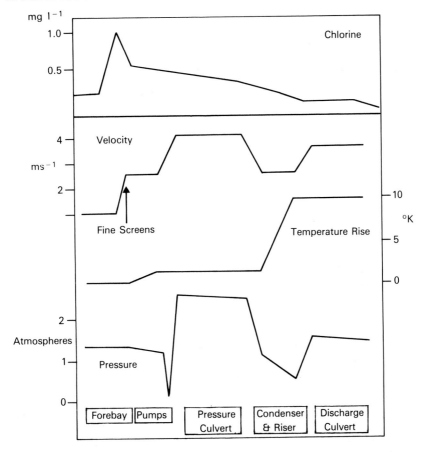

Fig. 6.1. Typical changes in velocity, temperature, pressure, and chlorine levels in the cooling-water system of a thermal power station.

grid of bars spaced about 15 cm apart. This keeps out large debris such as oil drums, tree-stumps, and large weed masses. At turbine intakes a fine-mesh static screen may also be installed to keep out fishes or small debris.

At thermal power stations the cooling water is drawn through a secondary screen, usually a moving fine screen, either of the 'drum' or 'band' type (N.T.I.S., 1973). Mesh aperture is usually about 50 per cent of condenser tube diameter, i.e. between 5 and 9 mm, depending upon design. Many smaller organisms can, of course, pass through these meshes into the cooling-water circuit.

The fine screens usually rotate in a large chamber or screen well, from where the water passes to the condensers. There are many variations on basic designs, but a modern British 2,000 MW power station (4 × 500 MW units) usually has four drum screens, though band screens are more common in other countries. Smaller debris and fish collect in the screen-wells and are removed via ledges on the inner side of the rotating screen, from where they are washed by water jets into channels and large metal baskets.

Intakes may be located in one of two basic positions, namely:

Onshore: either at the end of a channel or parallel with the bank or shore, or;

Offshore: incorporated into a structure some distance from the pumps and screens and connected by tunnels or culverts.

6.2.2. Causes of screen blockages

There are two major types of material which commonly block intake screens: organisms (living or dead) and man-made debris.

Collins (1976), concerned with the design of deflection screens for cooling-water intakes, classified material caught on screens into six categories, as follows:

'Stringlike': flexible—strings, ropes, hydroid fronds, grass.

'Sticklike': long, stiff-straw, reed stems, sticks.

'Sheet': stiff-polythene, kelps, paper, leaves, etc.

'Flake': stiff, platelike—plastic, glass, crab shells, mussel shells, etc.

'Lumps': (stiff) coal, concrete, rock fragments, etc.

'Jellylike': flexible-jellyfishes, siphonophores, etc.

Although these categories are somewhat over simplified they all require different types of screeens or mechanisms to collect, deflect, or align the materials so that they are either prevented from entering the cooling-water system or allowed to pass through without stopping.

The one category which Collins does not include in this list is 'fish', which have caused serious screen blockages at some British power stations (Langford, *et al.*, 1978). There have doubtless been many screen blockage problems in power stations throughout the world, though there are few published reports. The removal of screen debris is a normal part of daily power station operating schedules and the volume of debris varies at many sites. Daily fluctuations in weight of weed from 20 to 800 kg have been recorded from drum screens at one coastal power station, and weekly totals ranged from 300 to 5,000 kg (Holmes, 1975). The amounts of material, including fish and weed, varied with season and climatic conditions. Storms, high tides, and wind direction were all found to influence catch-patterns (Holmes, 1975; Langford, *et al.*, 1978; Turnpenny and Utting, 1980). At Wylfa Nuclear Power Station, in North Wales, the detachment of kelps (*Laminaria* spp.) during autumn gales has caused serious intake blockages in several years. Similarly, at other power stations, the autumn sloughing of seaweeds such as *Griffithsia* sp. and *Ulva* spp. causes problems (Savage, 1975; Holmes, 1975). At Poole Power Station (U.K.), increases in town sewage discharged to the intake area resulted in massive growth of *Ulva lactuca* which blocked intake screens when it became detached.

In British rivers autumnal spates dislodge the long-fronded water weeds such as *Potomageton pectinatus* and *Ranunculus* spp., and these can cause blockage problems in some river intakes (Langford—unpublished observation).

Animals may also cause problems. For example the colonial hydroid known as white-weed (*Sertularia* sp.) has resulted in screen and condenser blockages in the Thames Estuary (Board, 1975). In the tidal reaches of the Trent, massive populations of the amphipod *Gammarus zaddachi* have been known to invade the cooling system at a power station, blocking intake screens and spray nozzles. Problems are, however, usually short-lived.

Fish have always been prominent in intake screen catches and in some places the screens have been used as samples for both fish and invertebrates (Hardisty, *et al.*, 1974; Hardisty and Huggins, 1975; Claridge, 1976; Van den Broek, 1978, 1979). During the past few years, however, the potential effects of screen mortalities on local and migratory populations have become of more concern.

Shoals of migrating sprats (*Sprattus sprattus*) became impinged on the band screens at Dungeness Nuclear Power Station on the Kent coast (U.K.) to such an extent that the station was forced to shut down, owing to jamming of screens and the resulting breakdown in cooling-water supply. This kind of occurrence has been recorded sporadically at other power stations in the U.K., though not usually with such drastic results. At Dungeness estimated costs of closure and cleaning and repairs was over £1m sterling ($2m). The blockages were reportedly associated with the presence of the usual migrating shoals together with strong tidal currents, spring tides, and gale-force onshore winds.

A similar occurrence was recorded at the Millstone Power Plant on the Connecticut coast where more than 2×10^6 menhaden clogged intake screens during the late summer of 1971, again causing complete closure of the power plant (U.S.A.E.C., 1971; I.A.E.A., 1972). However, it is well known that menhaden have huge natural mortalities and dead fish have clogged water intakes at other locations (Wisconsin Electric Power Company, 1976).

6.2.3. The fouling of intake culverts

There are many different species of fouling organism throughout the world, and the effects of fouling on structures and ships has been known for many hundreds of years (Woods Hole Oceanographic Institute, 1952; Ray, D., 1959; Crisp, 1965; De Palma, 1971; Mangum, *et al.*, 1973). It is surprising, that faced with the experience of fouling in the world, that engineers building coastal power stations did not foresee the problems. In some cases fouling was so unexpected that no provisions were made, and during some cleaning operations in culverts, hundreds of tons of fouling material had to be removed via small inspection hatches (Beauchamp, 1967, 1969).

In most regions of the world mussels (*Mytilus* spp.) are a common fouling organism in power station intakes with massive growth occurring on culvert walls and in screenwells. In the 1960s closure of power stations for culvert cleaning was, in Britain and other countries, a regular part of the annual or biennial maintenance schedules. At some British power stations 300–400 tonnes of living mussels were removed from some culverts. In some cases mussels growing in condenser end-boxes or blocking tubes caused pressure to build-up so much that leakages occurred, and sea-water (used for cooling) entered the boiler water causing serious problems. Mussel larvae (spat or plantigrades) enter cooling-water systems as part of the normal plankton and settle in crevices or areas of low current velocity. As they grow they create small areas of back-eddy where other spat can settle or grow. Settlement usually occurs in June–September, though spat may be found all year round in some waters (Dare, 1976; Jensen, 1977).

Mussels are not, however, the only fouling organisms and marine fouling communities often contain representatives of several groups including barnacles, tube-worms, hydroids, sea-squirts, sponges, and other molluscs (Woods Hole Oceanographic Institute, 1952; Ray, D., 1959; Beachamp, 1967; Holmes, 1970a; Coughlan and Holmes, 1972; McCain, 1975; Menon, *et al.*, 1977; Jensen, 1977). Various methods of monitoring fouling communities are outlined by Coughlan and Holmes, 1972; Richards, 1977; Hillman, 1977.

Holmes, (1970a) described briefly the sequence of events in the development of a typical fouling community in a cooling-water culvert at a new British coastal 2,000 MW station. The leaching of chemicals from the concrete prevented colonizing animals from settling for the first few months. After that a bacterial slime developed, followed by the other larger organisms. In fouled culverts the settled communities included hydroids (*Sertularia* and *Tubularia*), barnacles (*Balanus* spp.), tube-worms (*Pomatoceros* and *Mercierella*), and the inevitable and ubiquitous mussels (*Mytilus*). Free-living organisms also existed, for example crabs (*Carcinus*), dog-whelks (*Nucella*), amphipods (*Jassa*), and polychaete worms (*Nereis*). The tube-building amphipod *Jassa falcata* can itself cause considerable fouling problems, particularly on screens and in condensers (Board, personal communication). In the cooling-system at Sizewell Nuclear Power Station on the Suffolk coast large quantities of starfish (*Asterias*) were present in collecting bins after the first chlorination of the year. There was little doubt that the starfish were feeding on the resident mussel population in the intake culverts.

In fresh waters, intake culvert fouling is not usually as important a problem as at marine sites, although the author has noted blockages of cooling-tower sprays and condenser tubes by shells or living molluscs, notably snails (*Bithynia* sp. and *Physa* sp.) and the zebra mussel *Dreissenia polymorpha*.

The zebra mussel causes problems at water supply intakes and in European reservoirs, and the species is spreading actively in Britain (Macan, 1974). At some power stations in the U.K. colonial hydroids are found in profusion in cooling-tower ponds and may eventually cause problems at some sites (Milner and Langford, unpublished). In the U.S.A. the freshwater Asiatic clam

Corbicula manilensis has become a serious fouling problem at many hydro and thermal power stations, particularly on the Tennessee River, and drastic control measures and cleaning are necessary (Goss and Cain, 1977).

6.2.4. Condenser fouling

The most common form of organic fouling within condenser tubes is a bacterial slime, and this occurs in both freshwater and marine systems, though in the latter other organisms such as hydroids, tube-worms, and molluscs may also invade (Board, 1967; Cole, 1977; Draley, 1977).

Rippon (1971) and Rippon and Wood (1975) reported the presence of bacteria and other micro-organisms in many parts of a power station cooling and boiler feed water systems. Out of seven inland power stations sampled, three had heavy deposits of condenser slimes, but all seven showed evidence of some kind of deposit. A wide variety of bacterial groups were recorded from these condensers, most of which originated in the polluted river water used for cooling. There were changes in the cooling-water populations throughout the year and certain species became more prominent in the cooling-water system than they were in the original river water. Most of the work was, however, at power stations using closed-circuit (cooling-tower) systems in which the bulk of the water is recirculated continuously. Many of the bacterial strains isolated produced ammonia, and it was believed that eventually the action of these organisms could cause corrosion of condenser tubes (Board, personal communication).

Cole (1977) described the composition of condenser slimes as 15–30 per cent by weight of organic matter (bacteria, fungi, etc.), and the remainder a mixture of clay, fibrous, and siliceous material. Occasionally, organic content may be as high as 80 per cent. Where sewage effluent is used for cooling-augmentation the growth of bacterial slimes may be greatly enhanced (Humphris and Rippon, 1978).

6.2.5. Biodeterioration and wood-borers

One of the most troublesome problems in early cooling-tower systems was the deterioration of unprotected wood-packing in the towers, resulting in structural collapses. Increased temperatures enhanced the growth and activity of bacteria and fungi (Rippon, 1971; Ross and Whitehouse, 1973). Deteriorated packing has to be replaced, though fungicide treatment of new wood inhibits the development of organisms. The growth of wood-boring organisms has been discussed in chapter 5.

6.2.6. Fouling prevention and control

Antifouling procedures fall into three main categories: physical, mechanical, and chemical.

Physical methods include culvert construction, heat, and to some extent antifouling surfaces such as paints or gels. Mechanical methods include the cleaning with rubber balls, iron balls, brushes or straightforward manual scraping and removal. Chemical methods include dosage of cooling-water with acids, biocides, or the use of copper and paints as coatings which are partly chemical and partly physical in their action (Burton and Liden, 1978). The possible use of radiation to control clams has also been investigated (Tilly, et al., 1978), though the system was rejected because of problems with effluent disposal.

To prevent or reduce fouling in culverts the walls and surfaces should be so smooth that organisms cannot find a 'foothold'. At the same time the velocity of water should always exceed 2·5 m s^{-1} (Board and Collins, 1965). Lewis, B. (1964) quoted data to show that on true smooth surfaces velocities of over 2·4 m s^{-1} would prevent settling, but Burton and Liden (1978) noted that velocities of 1·5 m s^{-1} were successful in preventing settling at Calvert Cliffs Power Plant in Maryland. Established communities may, however, withstand over 3·9 m s^{-1} (Lewis, B., 1964). Construction and operational constraints do not, at present, allow either the smoothness of surface required or the velocity criteria in all situations. To some extent paints, gels, or self-cleaning flaking surfaces may keep surfaces free of fouling organisms, at least for a time.

Ritchie (1927) found that raising water temperatures to 42°C killed established mussels very effectively, and at Portobello Power Station in Scotland reversal of the cooling-water flow once every 4 weeks achieved some degree of control. This process required extra heating, and was costly. Fox and Corcoran (1957) reported that mussel fouling in some Californian power stations has been reduced by thermal methods, at temperatures over 40°C.

Fouling studies in other regions, particularly in the tropics where ambient temperatures are high, have shown that the epifauna grows less successfully at temperatures over 35°C (McCain, 1975; Thorhaug, et al., 1978). As a result of Fox and Corcoran's work a number of thermal stations, both nuclear and fossil-fuelled, in California now use heated water recirculation to prevent mussel and other fouling in culverts (Stock and Strachan, 1977). At the San Onofré nuclear power stations the cooling-water system is arranged so that recirculation of effluent water can occur usually at 4–6 week intervals (Fig. 6.2), reaching temperatures of 40–42°C at a rate of 30°C h^{-1}. The temperature is maintained for about 2 hours and 60–70 per cent of the cooling-water flow is recirculated. The final discharge temperature is about 52°C, though the period of discharge is reported to be down to 30 min. There are problems in the procedure arising from thermal stresses on tunnels, condenser tube expansion, and environmental effects of discharge temperatures. Also, the method does not control condenser slimes. At least two European stations are known to have similar facilities and in the light of growing pressures against the use of biocides such processes may come into much more use.

Mechanical methods using iron balls to crush molluscs in culverts were proposed by Ritchie (1927), but the size of culverts renders this method impractical. Condenser cleaning may be carried out, however, using brushes

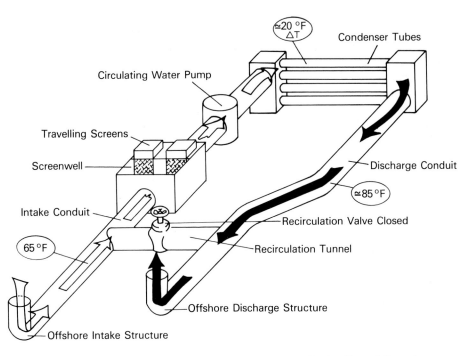

(a) Schematic Diagram of Circulating Water System During Normal Operation

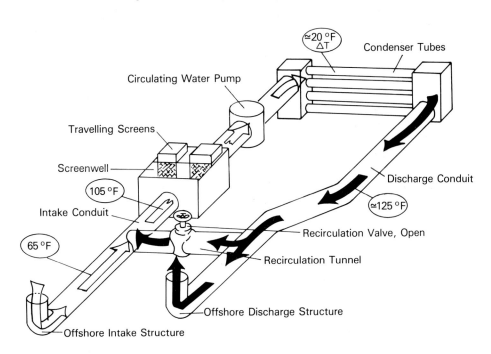

(b) Schematic Diagram of Circulating Water System During Heat Treatment Process

Fig. 6.2. Water circulation during heat treatment and normal operation at San Onofré Nuclear Generating Station, U.S.A., redrawn after Stock and Strachan (1977).

or sponge rubber balls, forced through the tubes under pressure. These are known as the M.A.N. and Amertap systems respectively (Burton and Liden, 1978). Other methods such as irradiation or ultrasonic vibration have not been successful to date (Burton and Liden, 1978).

For marine culverts a chemical and physical control method has been used which combined osmotic shock with anoxic conditions. The culvert is filled with fresh-water and kept closed until the water goes anaerobic. The dual physiological stress kills most organisms. The anoxic method alone can also be used in fresh-water or salt-water systems (Holmes, 1970a; Burton and Liden, 1978). Like other non-biocidic methods the process requires part of the plant to be shut down, but this can coincide with annual maintenance or overhaul of other equipment.

By far the cheapest, most convenient, *and* most effective method of fouling control in cooling-water systems has been by the use of biocides, mainly chlorine in some form or other, i.e. either liquid, gaseous, or as hypochlorite (White, 1972). In the U.S.A. estimates of the total chlorine used in power station cooling-water systems have varied between 50×10^3 and 200×10^3 tonnes per annum, though Hamilton (1978) suggested that the real figure is more like 26×10^3 tonnes. In the U.K. the annual usage is about 8.5×10^3 tonnes (Coughlan and Whitehouse, 1977). Intermittent dosing with chlorine has been used for many years in cooling-water circuits to remove bacterial slimes, but it is not always effective in controlling mussels. Increasing the frequency of dosing does not enhance control and Beauchamp (1967, 1969) found that mussels at Hayle Power Station in the U.K. had their shells bleached by intermittent heavy doses of chlorine, but were still very much alive. He concluded that under intermittent dosing the mussel responds by closing its shell and waiting until the chlorine passes. Doses of up to 20 p.p.m. were found to be ineffective. Lewis, B. (1961) showed that continuous low-level chlorination of culverts during spring and summer would prevent settling and kill off very young, recently settled, spat. Dosing at the rate of $0.5–1.0$ p.p.m. (mgl^{-1}) in combination with current velocities of $1.5–2.0\ m\ s^{-1}$ has been a very effective control procedure in the U.K. and other countries for a number of years. Usually, the aim is to produce about 0.2 p.p.m. residual at the condenser inlet. The actual dose-rate may be higher in some regions where intake water has a high organic content and the chlorine demand is high (Holmes, 1970b; White, 1972; Jensen, 1977; see Jolley, *et al.*, 1978) (see chapter 3).

Chlorine is extremely cost effective. It removes condenser slimes and controls fouling very efficiently if applied correctly. However, its toxicity, its capacity to form toxic residual organic compounds (e.g. chloramines), and its enhanced residual toxicity in sea-water where bromine is liberated may be hazardous to organisms outside the cooling-water system. Further, while disturbing or killing-off fouling organisms, it may cause mortalities of other organisms in transit such as planktonic algae, crustacea, larval molluscs, and larval fish. This potential environmental hazard has in the past 5 years led to whole new areas of chemical and ecological research and to a series of new and proposed constraints on power station discharges, particularly in the U.S.A. and Europe (Coughlan and Whitehouse, 1977; see Jensen, 1977; see Jolley, *et al.*, 1978). The environmental effects are reviewed later in this chapter and in chapter 7.

The electricity industry in Britain has a code of practice with regard to the use of chlorine which notes 'that chlorine discharged from the system is both a waste and a potential hazard'. In North America rigid constraints are being applied to the use of chlorine in cooling systems (see Jolley, *et al.*, 1978; Schubel and Marcy, 1978) which makes the finding of alternatives of some high priority (see chapter 9).

It is not within the scope of this book to give details of the efficiency of chlorine and other biocides as antifouling agents and White (1972), together with various authors in Jensen (1977) and Jolley, *et al.* (1978), deal with this in some detail. Other biocides have been tried or considered for power station cooling-water systems including various chlorine-based compounds, acrolein, 1.2 benzisothiocylate, ozone, bromine compounds, and hydrogen peroxide (Coughlan and Whitehouse, 1977; Jensen, 1977; Burton and Liden, 1978). To date, none has been found to be as convenient, efficient, or as economic as chlorine. Toxic paints have been used in culverts at some sites and have been reasonably effective, but large-scale applications are economically impractical as yet, and repainting still necessitates closure of the culvert. This may be feasible, however, during normal annual maintenance.

6.3. ECOLOGICAL PROBLEMS

6.3.1. Impingement of fishes on cooling-water intakes

As we have seen from reported screen blockages (see 6.2.1) impingement mortalities may involve millions of individuals.

In many areas of the U.S.A. intake mortalities of larger larvae, juvenile, and adult fishes are believed by many ecologists to be reaching proportions which may cause serious population declines (Van Winkle, 1977). The potential effects of the mortalities and the forecasts have led to rigid and costly constraints being applied to intake design, location, and operation (see chapter 9).

As a result of legislative conditions, data on the impingement of fish have been collected from many operating sites in North America (Adams, 1969; Marcy, 1971, 1976b; Edsall and Yocum, 1972; Jensen, 1974; Landry and Strawn, 1974; Grimes, 1975; Grotbeck and Bechthold, 1975; Edwards, *et al.*, 1976; Wisconsin Electric Power Company, 1976; Freeman and Sharma, 1977; Martin Marietta Corporation, 1977; Mathur, *et al.*, 1977; Sharma and Freeman, 1977; Stupka and Sharma, 1977; Hannon, 1978; Uziel and Hannon,

1979; Uziel, 1980). Other data are available from sites but are not always published (Sharma and Freeman, 1977). In the U.K. and most of Europe fish impingement has not, as yet, been regarded as a general danger to stocks, though several studies are in progress or completed (C.E.G.B., South-West Region, 1975; Holmes, 1975; D'Arcy and Pugh-Thomas, 1978; Langford, *et al.*, 1978; Hadderingh, in press; Turnpenny, 1981).

At several power stations intake screens have been used as convenient fish sampling devices for studies on pollution recovery, heavy metal uptake (Hardisty, *et al.*, 1974; Van den Broek, 1978, 1979), and the biology of selected species (Hardisty and Huggins, 1975; Claridge, 1976; Claridge and Gardner, 1978).

In most freshwater and tidal habitats, larvae, juveniles, and to a lesser extent adults, of most migratory and resident fishes are potentially susceptible to impingement and entrainment. Generally, larvae and small juveniles tend to pass through the cooling-water circuits. However, in the author's experience, spawning adults of smaller species such as gobies have been found to pass through screens at some power stations in the U.K. The numbers of fish killed by impingement varies with many factors. At some power stations on the Great Lakes annual mortalities of over 19×10^6 fish (mostly alewives) have been recorded (Sharma and Freeman, 1977). In Britain, Holmes (1975) recorded weekly catches of up to 60×10^3 fish (mainly smaller species and under-yearlings) at Fawley Power Station (Fig. 6.3). Hadderingh (in press) reports that up to 10^6 small fish per day were entrained or impinged per day at Bergum Power Station in Holland.

The mortalities of striped bass larvae (*Roccus saxatilis*), at the Indian Point power plants on the Hudson River (U.S.A.), led to an important legal action which is discussed in chapter 9. At most places the mortality rate of young fish impinged on screens is 90–100 per cent, though where rescue schemes are in operation this may be reduced almost to nil (Hadderingh, in press; C.E.G.B., South-West Region, 1975).

6.3.2. Factors causing fish impingement and mortalities

At most sites where long-term regular surveys have been carried out there was considerable variation in screen catches related to season, swimming speeds, tides, climatic conditions, and operating schedules at the relevant power station (Grimes, 1975; Holmes, 1975; Claridge, 1976; Edwards, *et al.*, 1976; Freeman and Sharma, 1977; Martin Marietta Corporation, 1977; Mathur, *et al.*, 1977; Sharma and Freeman, 1977; Stupka and Sharma, 1977; Langford, *et al.*, 1978; Van den Broek, 1979).

(a) Seasonal fluctuations

Screen catches at most power stations fluctuate markedly with season (see references above). These seasonal fluctuations are probably a function of two major factors, namely seasonal migrations into and out of intake areas, and

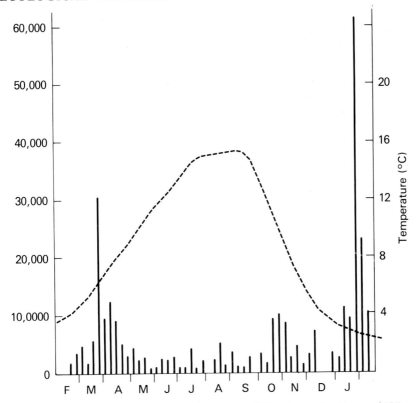

Fig. 6.3. Fish on Fawley Generating Station screens, 1973, redrawn after Holmes (1975).

the effects of temperature changes on the ability of fish to withstand water currents. Seasonal fluctuations in power demand, and hence intake volumes and velocities may also be of some significance. Many species migrate into shallow water to feed or spawn and juveniles are commonly found in such areas (Nikolsky, 1963). Estuaries are noted nursery areas for marine species and populations fluctuate seasonally in size and diversity (Perkins, 1974; McErlean, *et al.*, 1973; Hillman, *et al.*, 1977).

Different species tend to have different migration patterns and their seasonal abundance may or may not be reflected in their prominence in screen catches. The occurrence of salmon smolts on the intakes at Uskmouth Power Station in the U.K. and at Hanford on the Columbia were mainly during the known smolt-running periods (C.E.G.B., South-West Region, 1975; Freeman and Sharma, 1977). At Fawley Power Station on Southampton Water, herring (*Clupea harengus*) were mainly caught in mid-winter, sand smelt (*Atherina presbyter*) peaked in autumn and winter, and pout (*Trisopterus luscus*) were at constant levels all the year round (Holmes, 1975; Langford, *et al.*, 1978). Diversity and total numbers of fish were much lower in screen catches in summer (Fig. 6.3). Similar patterns of catch were found at power stations on the Medway and Severn Estuaries and at a Florida power station, all showing lowest diversity and abundance in summer and highest in autumn and winter (Grimes, 1975; Claridge, 1976; Van den Broek, 1979).

In non-tidal locations catch patterns differ with situation. For example, Mathur, et al. (1977) found low summer and high winter screen catches in 1973 and 1974 at the Peach Bottom Atomic Power Station in the Susquehanna River, but in 1975 the reverse occurred. Here the catch was apparently directly related to the operating load at the power station and to the resulting intake current velocities. At the Zion Nuclear Power Station, on Lake Michigan, most of the common species were caught all the year round on the intake screens, but trawl and gill net data showed the fish to be abundant off the intake only for relatively short periods of the year (Freeman and Sharma, 1977). However, the fishing effort of a power station is many times greater than that of normal fishing methods and might be expected to collect specimens even when the relevant species are not at their peak abundance. Further, the behaviour of species may also be influenced by their reaction to intake currents, as it is known that species may respond differently to currents at different times of the year (Harden-Jones, 1968; Arnold, 1974).

(b) Swimming performance and environmental conditions
Once any fish is in the vicinity of an intake its ultimate impingement may depend on a number of factors including its response to water currents and its swimming performance. Swimming performance may in turn depend on the size and physiological condition of the fish (i.e. fatigue, disease, and parasitism) and on environmental factors such as water temperature, oxygen concentrations, and chemical stresses (Bainbridge, 1960; Brett, 1964, 1970; Beamish, 1966; Fry, 1967; Blaxter, 1969; Oseid and Smith, 1972; Hocutt, 1973; N.T.I.S., 1973; Kutty and Sukumaran, 1975; Wardle, 1976; Rulifson, 1977; Berezay and Gee, 1978). The swimming speeds of fish may be conveniently divided into three categories:

Cruising speed: The speed which the fish is able to maintain for long periods.

Sustained speed: The speed which may be maintained for shorter periods, usually a few minutes.

Darting or *'burst' speeds*: The speed which may be maintained by a single effort, or during a period of only a very few seconds.

As a simple guide, cruising speed is used in normal movements such as migrations or 'browse' feeding. Sustained speeds are used to avoid hazards or stem a fastish current. Darting speeds are used for panic avoidance, or by predators such as pike to catch their prey. Cruising speeds tend to be about half the sustained speeds. Darting speeds are greater than cruising speeds by about a factor of six (N.T.I.S., 1973). By far the majority of fishes caught on intake screens are larger larvae, juveniles (young-of-year), or adults of smaller species, though larger specimens are often impinged. At Fawley Power

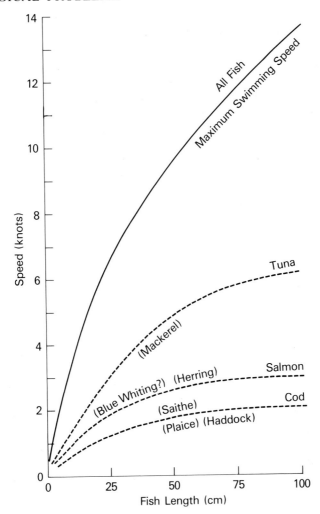

Fig. 6.4. Fish swimming speeds, in knots, related to fish length, redrawn after Wardle (1976).

Station, for example, adult salmon, sea-trout, and rays have been collected together with large bass and mullet (Holmes, 1975). Current velocities at intakes vary widely from about 0·5 m s^{-1} to 6 m s^{-1} (N.T.I.S., 1973; Sharma and Freeman, 1977), but at most sites the velocities exceed the sustained cruising speed or even the ultimate burst speed of many juvenile fishes (Fig. 6.4).

At coastal power stations, with an intake canal, the intake velocity may vary with tide height. At high tide incoming velocities may be only a fraction of those at low tide. Thus, fish swimming within the intake areas may maintain position against high tide intake velocities but not at low tide when these exceed their maximum sustained speeds and the fatigued fish are drawn into screen-wells and on to the screens. Turbulence is often less in screen-wells at high tide than at low tide and 'quiet' areas for fish are also more numerous. Clear tide-related catch patterns have been demonstrated for two British power stations. Over 70 per cent of the catch occurred within the 3 or 4 h spanning low tide (Fig. 6.5). At offshore intakes with no approach channels the same pattern was not as obvious, and factors other than intake current velocity may have been influential (Langford, et al., 1978; Turnpenny and Utting, 1980).

Early offshore intakes were open-topped vertical pipes, protected by a large grid. In many cases large ingresses of fish occurred as shoals passing over the grid were drawn downwards in a direction to which they were not orientated (N.T.I.S., 1973). Subsequently, many intakes were capped so that the orifices were in the horizontal plane (Fig. 6.6), to which fish were better able to orientate.

Tidal influences do not operate in rivers or lakes and here it is probable that small fish actively follow intake currents or drift passively and eventually become trapped in screen-wells.

Storms and high winds have been shown to increase screen catches of weed and fish at Fawley Power Station, particularly in autumn and winter (Holmes, 1975; Holmes, personal communication).

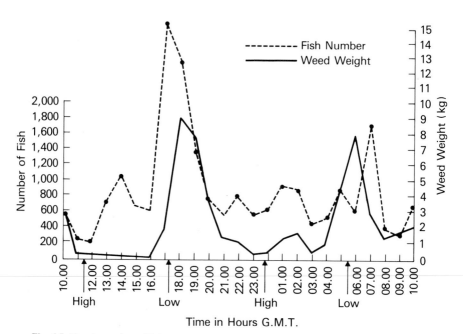

Fig. 6.5. Hourly catches of fish and weed on intake screens at Fawley Power Station, U.K., 20–21 November 1975, redrawn after Langford, et al. (1978).

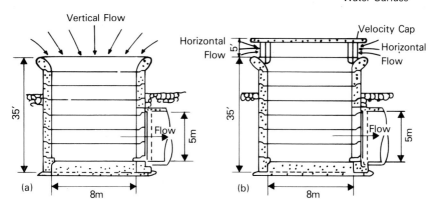

Fig. 6.6. Conventional offshore intake structures, redrawn after Downs and Meddock (1974).

The impingement of larger fish is difficult to explain as fish over 20 cm are often observed swimming in and out of intake areas searching for food. Random movement may explain their occurrence in screen-wells and chlorination may either kill or stress fish which then become impinged. Fish over 10–15 cm long should, in theory, be able to evade intake currents at most British power stations, though some do become impinged.

Swimming performance is directly related to water temperature (Brett, 1970; Hocutt, 1973; Rulifson, 1977) (see Fig. 5.11). Thus, ability to withstand intake velocities should be lowest in winter at natural temperatures and this may explain high winter impingement rates (Grimes, 1975; Edwards, *et al.*, 1976). However, in many of the inland power stations studied in the U.S.A., the seasonal fluctuations are not always as obvious as at estuarine stations (Sharma and Freeman, 1977; Stupka and Sharma, 1977).

The impingement of smolts and other migratory species at hydro-electric stations and thermal stations occurs mainly in the migration seasons when the fish are using downstream currents to migrate to the sea.

Prediction of fish catches on intakes has been attempted using theoretical models and empirical data from site studies (Mathur, *et al.*, 1977; Murarka, *et al.*, 1978; Turnpenny, 1981). Most suffer from a lack of data on fish density in intake areas but generally a major part of the catch can be explained by and related to water temperature, water use, and fish abundance.

As fish orientate and generally maintain position by visual stimuli (Harden-Jones, 1968; Arnold, 1974) the loss of vision at night would be expected to reduce this capacity and cause fish to fall back into screens. There was no evidence of this at Fawley, where tidal influences were dominant. However, both Grimes (1975) and Turnpenny and Utting (1980) have shown some increases in catch during hours of darkness which could be related to orientation problems.

6.3.3. **Entrapment and entrainment of invertebrates**

Apart from the fishes many other animals are drawn on to the screens and impinged. At Fawley Power Station most of the larger locally occurring crustaceans (crabs, lobsters, shrimps, and prawns) are found regularly in screen catches. Seasonally, squids, cuttlefish, and other organisms such as jellyfishes and sea-squirts are found in large numbers. Hardisty, *et al.* (1974a) and Claridge (1976) used power station intake screens on the River Severn to collect specimens of gammarids, mysids, and larger shrimps (*Crangon crangon*) for various studies. There have been relatively few detailed studies of invertebrate impingement, however, or of its significance to the local populations (N.T.I.S., 1973). With animals such as *Carcinus* (shore-crabs), and larger crustacea, some migratory or behavioural mechanism is again probably responsible for their seasonal occurrence in large numbers in intake areas (Naylor, 1965a; Newell, 1970).

6.3.4. **Effects of impingement mortalities on populations**

The large mortalities of fishes and invertebrates impinged or entrained and killed at power stations may have some effects on local populations, but the significance of such effects has not actually been quantified from field data. Successive migrations of animals into the intake areas make it difficult to estimate localized effects, and the size of the whole habitat (lake, river, or tidal water) often makes sampling and population estimates impractical, if not impossible (see 6.5.1.f). Usually, total weights of fish or invertebrates caught are a small fraction of the commercial catch. Also, power stations mainly crop the young stages of most species while commercial fishing crops adults, often around spawning size (Martin Marietta Corporation, 1976; Hadderingh, in press.

The major approach to the estimation of potential damage to species populations has been through mathematical models (Englert, *et al.*, 1976; Van Winkle, 1977), and this is discussed further later in this chapter (see 6.5.1.f).

Most models so far suffer from lack of data for validation and most rely on random distribution assumptions, particularly of larvae or juveniles in the environment. Wallace (1978) reported that fish larvae are not randomly distributed and may show avoidance of intake areas. Information on behavioural changes and re-distribution with increasing age and, on avoidance behaviour, is still lacking for many species in many areas.

There is, as yet, no conclusive evidence of actual population declines in any species as a result of power station intake losses, though long-term studies are still in progress in the U.S.A. (Van Winkle, 1977; Swartzman, *et al.*, 1978) and in Europe (Turnpenny and Utting, 1980).

6.3.5. **The prevention of impingement and intake mortalities**

Over the past five decades a number of devices have been used to try to prevent fishes entering turbine intakes and cooling-water intakes.

The importance attributed to migratory fish is indicated by the fact that the U.S. Corps of Engineers spent over $4 million on specific research during 1960–5. At the present time a great deal of money is being spent on similar research around thermal generating stations, particularly in the U.S.A. (N.T.I.S., 1973; Ray, et al., 1976; Hannon, 1978; Cada, et al., 1979; Uziel and Hannon, 1979).

In the U.K. research to prevent impingement and mortalities of smolts at hydro-stations has also been in progress for many years (Pyefinch, 1966). The major deflection methods at both hydro- and thermal stations have involved visual stimuli such as louvres or air-bubble screens, lights, water velocity and pressure changes, and electrical shock (N.T.I.S., 1973; Downs and Meddock, 1974; Ray, et al., 1976). Other exclusion devices have involved the use of fine-screens, or clinker bunds, surrounding an intake pond from which cooling-water is drawn. The alleviation of turbine mortalities at hydro-stations by actually passing fish through suitably designed and operated turbines has been discussed in a previous section (see 4.7.2.c) as has transportation of downstream migrants around dams and turbines (see 4.7.2.c) (Mills, 1966; Park and Farr, 1972; Arnold, 1978).

Apart from deflection or exclusion structures, rescue or removal schemes may be used to extract fish from screen-wells or congregation areas before they become impinged (N.T.I.S., 1973; Ray, et al., 1976). Ray, et al. (1976) include an extensive bibliography of reports not on general circulation.

(a) Artificial light

The reactions of species to bright light can vary from strong avoidance to strong attraction. Fields (1966) found that the effect of artificial light on salmonids was at times to attract and at others to repel, depending very much on water conditions. He concluded that:

(i) 'Under some conditions, artificial light can repel migrants and divert them from certain areas. In such situations, the problem is one of balancing various environmental stimuli so that light intensity overrides velocity, turbidity, depth and temperature.'

As we have seen fish may swim into 'thermal discharges' under conditions which would seem to be potentially lethal as a result of other stimuli (see 5.4.12). Thus, for light to be an effective deterrent to a fish reacting to such stimuli it would need to produce an extremely strong reaction.

Field's second conclusion was that:

(ii) 'Under other conditions, artificial light may attract migrants and concentrate them in particular areas. Some degree of light adaptation is necessary before attraction will occur.'

Although lights are used to attract or hold fishes in large and small fishing operations throughout the world (Hela and Laevastu, 1961; Woodhead, 1966) the reactions of fish to light depends very much on the species, the life-history stage, season, acclimatization, and water conditions, as well as the intensity of the light. Woodhead and Woodhead (1955) found that activity of herring larvae could be dependent on light intensity. The lower intensity threshold for activity was 20 lux, maximum activity was at 4,000 lux, decreasing at higher intensities up to 65,000 lux. Light also appears to be vital to the schooling behaviour of many marine species (see Loukashkin and Grant, 1959, for references), mainly because the formation of schools or shoals depends on visual contact. At night, schools tend to disperse. In comprehensive experiments on the Pacific sardine *Sardinops caerulea*, Loukashkin and Grant (1959) tested the reactions of a school of fish to coloured lights of varying intensities. Preference tests showed a significant avoidance of red areas but no obvious preference for green or blue zones.

As far as cooling-water intakes are concerned, particularly at coastal sites, lights may, in theory, be useful either as a deterrent or as an attractant. For example at Huntingdon Beach, California, fish are attracted into an area away from the intake by steady lighting, removed alive with a fish pump, and returned to the habitat. Fouling and repair of lights in the marine environment may, however, cause considerable problems, as underwater lights may be totally obscured by settling organisms in a very few weeks.

Power station intake areas and structures onshore are generally illuminated at night. Although no one has investigated the influence of these lights on the local fish populations the low-level illumination may attract some species, particularly the shoaling, migratory species such as herrings, sprats, anchovies, or sardines. Grimes (1975) mentions the possible effect of the intake lighting on screen catches but provides no data on the subject.

In fresh waters Fields (1966) found that a light barrier thrown across a stream at a 90° angle would block migrants for a time but in swift currents fish would eventually be carried into light areas and adapt. They may then be attracted to lights downstream if the other lights are doused. The brighter the initial light, and the longer the adaptation period, the better the migrants can be controlled.

Banks (1969) has reviewed the information on natural illumination and salmon migration, but there is some disagreement in the literature over the relationship between migration and time of day. Neave (1943), cited in Banks (1969), described unsuccessful attempts to induce fish to use a pass at night by artificially lighting the area. Fields, *et al.* (1964) also failed to attract upstream migrants to *enter* a fishway at night either with lights or spillway manipulation. Fish already in the lighted fishway did, however, continue upstream.

To summarize, therefore, artificial light does produce definite responses, but alone is unlikely to be universally effective as a deterrent unless at very high intensities. In combination with other artificially induced factors such as water currents or visual stimuli lights may be used to deflect and guide fishes into areas which may be safe or from which they may be rescued.

(b) Pressure

The sensitivity of marine and freshwater organisms to pressure changes and the significance of pressure are summarized and discussed by Knight-Jones and Morgan (1966).

Pressure in combination with light has been considered as a guiding stimulus for fish, though there is as yet little evidence of the effects (N.T.I.S., 1973). Smolts of four species of salmonids were found to swim towards a faint light when subjected to pressure (N.T.I.S., 1973). The pressure increase encountered by fish descending from about 7 m to 20 m at a dam site evoked a response of swimming toward light sources of 100–500 W, though there are no data on the effects of water conditions, natural light, currents, etc., or operating techniques. The use of any kind of pressurization to guide fishes is, therefore, little more than speculative.

(c) Temperature

As we have already seen in chapter 5 (see 5.4.12) fish can avoid adversely high temperature zones, but may also be attracted to areas of moderately increased temperature. Heated water as an artificial barrier is therefore of doubtful use. Also, high water temperatures at cooling-water intakes would lead to lower efficiency and operational problems in a thermal station. At hydro-dams the production of heated water would entail use of energy. Costs would be prohibitive and its use doubtful especially as the 'thermal barrier' theory is not really substantiated (see 5.4.10.h)

(d) Sound

Fishes are sensitive to mechanical, infra-sonic, sonic, and also some ultrasonic vibrations (Nikolsky, 1963). Low frequency sound from 5 to 25 Hz is detected by the lateral line, higher frequencies, i.e. up to 13,000 Hz by the auditory labyrinth.

Fishes also produce low frequency sounds, deliberately or accidentally, during their activities.

The fact that fishes react to sound is used in some areas of the world as a fishing aid. For example, grey mullet are in some regions scared by sounds into leaping from the water and are caught in a semicircular mat on which they land (Nikolsky, 1963). Also, pelagic fishes can be frightened away from the mouth of a purse seine by a special bell lowered into the mouth. This prevents the fish from escaping while the net is being closed. Van der Walker (1966) found that salmonids responded to selected frequencies and in some rivers sounds of various kinds have been used to scare other species in some parts of the world.

Attempts to use low frequency sound as a complete deterrent at the Indian Point Power Plant on the Hudson River and at Oldbury in the River Severn were generally unsuccessful (N.T.I.S., 1973; Ray, *et al.*, 1976). The most successful noise at Indian Point was that of a pneumatic 'popper', producing sounds at 2–15 cycles min^{-1} (Schuler and Larson, 1974).

(e) Chemicals

The avoidance reactions of fish to toxic chemicals are well documented (Jones, J. R., 1964; Cairns, et al., 1978). The use of such substances in natural waters to deter fish is not only impractical, however, but illegal in most countries with anti-pollution legislation.

Minute concentrations of an extract from mammalian skin and other substances were found to repel Salmonidae in a freshwater system. On the larger scale, however, the quantities needed would be so vast that such methods are impracticable (Brett and Mackinnon, 1954).

(f) Electrical stimuli and fish barriers

One of the most commonly used and as yet most effective methods of deflecting fish from turbine or cooling-water intakes in fresh-water is electrode arrays forming electrical barriers (Aitken, et al., 1966). The effects of alternating and direct current voltages (A.C. and D.C.) on fish are well known (Vibert, 1967; Smith, E., 1974). In general A.C. tends to 'stun' or repel fish while D.C. tends to 'attract' fish to the electrodes before tetany and loss of equilibrium occurs.

While these barriers may be of use in guiding downstream migrants at hydro-stations, or perhaps repelling fish from outfall areas, there are more problems at cooling-water intakes (Ray, et al., 1976; Sharma and Freeman, 1977). Stunned fish will be impinged, therefore the current must stimulate fish enough to guide or repel but not enough to cause loss of equilibrium. It is likely that the stimulus of the water currents is too great in any case for any barrier except a physical screen to be effective. E.I.F.A.C. (1968b), quoted in Banks (1969), described the use of a tailrace screen in the Shannon hydro-electric scheme. Fish swimming upstream were stunned by an alternating current as they entered the tailrace and were carried downstream. Unfortunately, the attraction to the tailrace was so strong that the recovered fishes swam into the field repeatedly. This persistence may be common where flow is the overwhelming stimulus to migrants.

However, such persistent upstream movement will probably eventually result in the fish being guided toward some bypass, fish-ladder, or 'rescue' area. Ray, et al. (1976) suggest that electrical barriers may have been successful at some cooling-water intakes but that there have been no significant tests carried out. They conclude that, as yet, electrical barriers are not successful in preventing fish impingement, except perhaps for individuals of one or two species within limited size ranges. In the sea, electrical screens have not yet proved effective, mainly owing to the high power consumption needed to produce an effective field around the electrodes. Data are scarce and further research is necessary but the basic behavioural problems would apply as in fresh-water (McK. Bary, 1956; Ray, et al., 1976).

(g) Deflection by visual and flow stimuli

Deflection or guidance schemes based on the reaction of fish to visual and flow stimuli have used air-bubble screens, louvres, travelling screens, water jets, and models of predators such as dolphins.

Air-bubble curtains are in operation at several North American power stations, but the results are not generally convincing (Ray, *et al.*, 1976). Fish may be deterred from passing through the bubble curtain during daylight but not at night when the visual stimulus is removed. At one power station, in Arkansas, an air-bubble curtain seemed to cause an *increase* in the number of impinged fish in spring (Ray, *et al.*, 1976). Some workers have concluded that bubble screens are effective at some sites though there are few convincing data (Ray, *et al.*, 1976).

Preliminary studies on water jets discharged at angles across a main flow have suggested that fish may be diverted, though little research has been done (Ray, *et al.*, 1976).

Louvre screens, i.e. vertical arrays of slats set at angles to the main water flow, have been said to be successful in deflecting fish and guiding them into fishways or collection areas for removal or bypass (Bates and Vinsonhaler, 1957; Aitken, *et al.*, 1966; N.T.I.S., 1973; Bainbridge, 1975; Ray, *et al.*, 1976). Louvre screens work by a combination of visual stimulus and flow direction but, as with other deflection methods the efficiency of installations in all situations has not been universally established. The problem with fixed louvres is that debris other than fish may block spaces and impede flow. To deflect fish and remove such debris travelling louvre screens are used set in the horizontal plane rather than in the normal vertical band or drum arrangement (Ray, *et al.*, 1976).

(h) Physical barriers

Fixed or travelling screens may still cause mortalities owing to impingement. To reduce the velocities at the screen face other types of intake protection have been tried, including rubble or clinker bunds through which water infiltrates to the intake orifice. Rapid sand filters have also been investigated (N.T.I.S., 1973; Ray, *et al.*, 1976; Scotton and Anson, 1977). The major problems are the clogging of sand beds and clinker interstices which require backwashing facilities. However, the low current velocities produced by these filtration methods can be effective in reducing impingement of both larval and older fishes. Some larvae may be actually attracted to the shelter of the crevices in coarse clinker or rock bunds.

(i) Fish removal schemes

Accepting that fish are attracted to intake areas, and eventually become impinged on screens, several methods of removing fish have been used to prevent mortalities. At Uskmouth Power Station, in the U.K., salmon smolts were sluiced from the screens into a collecting trough and returned to the river downstream. Survival was over 90 per cent (C.E.G.B., South-West Region, 1975).

At a number of U.S.A. power stations fish-pumps are used to remove fish from screen faces or collecting areas in screen-wells. At the Monroe and Contra Costa Power Plants 20 cm volute fish-pumps allow fish up to 42 cm long to pass through (Ray, *et al.*, 1976). Impingement mortalities have been reduced by 80 per cent.

Other removal methods involve modification of the normal travelling or drum screens to include either water-filled troughs or mesh baskets which collect fish and debris as normally. Low pressure jets wash the live fish into a collecting tank as the baskets invert. The fish are then returned to the environment. Larval fish may also be removed by this method with at least some survival, though the system may be disrupted by huge amounts of weed or other debris (Ray, *et al.*, 1976).

(j) Reduced intake velocities

The greatest reductions in impingement mortalities have been at sites where water velocities at the screen surface have been significantly reduced (N.T.I.S., 1973; Ray, *et al.*, 1976; Freeman and Sharma, 1977). However, the capital costs of increased pumping capacity, increased size of intakes, and screening, make velocity reductions below the critical velocities for all species and all sizes of fish extremely expensive. Recommended velocities in the U.S.A. are about 0.13 m s^{-1} (0.5 ft s^{-1}) at the screen surface.

6.3.6. The design and efficiency of fish conserving intakes at thermal power stations

As yet no really effective method of keeping all fish out of all cooling-water or turbine intakes has been evolved (Cada, *et al.*, 1979). It is said by some of the authors quoted here that shoreline intakes should not be at the inner end of a canal. However, the major ingress problems in the U.K. have occurred at offshore intakes, though velocity caps do appear to be partially effective in reducing fish ingress at such intakes (Schuler and Larson, 1975). Some recommendations for design criteria for cooling-water intakes to protect fish are given below (N.T.I.S., 1973).

A. SHORELINE INTAKES

In rivers, estuaries, bays, and harbours.
  Criteria:
  1. Establish a uniform velocity across the face of the screen.
  2. Avoid the use of fixed 'skimmer' walls or inverted weirs.
  3. Place circulating water-pumps behind screens.
  4. Do not locate screen-well or intake in highly productive or high population density area.

5. Prohibit recirculation of cooling-water which may attract fishes to the intake.

I. Screen-well flush with shoreline

6. Base total screen area on design approach velocity; minimum size of fish and maximum cooling-water flow rate.

7. Include provision for lateral escape of fishes.

II. Screen-well away from shoreline

8. Include provisions within screen-well for safely returning fish to main stream.

9. Avoid excessive negative pressure in the intake conduits.

10. Do not use intake canals.

B. OFFSHORE INTAKES

Located in oceans, lakes, estuaries.

Criteria:

1. Do not locate in nursery areas.

2. Provide for gravity flow from intake to screen-well.

3. Include provision for safe removal of fish from screen-well.

4. Circulating water-pumps behind screens.

5. Design approach velocities on swimming speeds of resident and migratory fish species.

6. Structure should not impede navigation.

7. No recirculation of cooling-water from outfall.

8. Use velocity caps or accept lower intake velocities.

Many of these criteria and provisions are being applied to plant under construction and in the design phase (N.T.I.S., 1973; Ray, *et al.*, 1976). To conclude, the N.T.I.S. report states that 'the backfitting of fish protection devices to existing intake structures has met with limited success. Although occasionally avoidance and guidance of fish has been accomplished using a combination of stimuli, this approach is not normally sufficiently reliable to completely offset design inadequacies.' In the absence of good data on the effects of screen mortalities on even very local populations the main question remains, i.e. 'Is it economically justifiable to spend large sums of money to prevent fish ingress when it has not been possible to establish any deleterious effects?' At present it is only fair to conclude that, in the absence of evidence either way, it is at least politically 'safer' to make some provision to exclude fishes from cooling-water intake screens.

6.4. ENTRAINMENT AND TRANSIT EFFECTS

Entrainment is defined as 'the dragging along or carriage of materials or particles in a flow of fluid' (*C.O.D.*, 1976). Thus any organism which is propelled into an intake by the in-going water current is said to be entrained. This applies equally to hydro-electricity intakes and thermal power station

cooling-water intakes. Large organisms entrained by the in-going current at
cooling-water intakes are generally trapped or impinged by screens. Plankton,
micro-invertebrates, and larval or juvenile fish small enough to pass through
the fine screens, pass into the cooling-water circuit where they are subjected to
'transit' effects.

Organisms may also be entrained in the thermal plume or discharge canal as
a result of turbulence or behavioural responses, even if they have not passed
through the power station. During the passage from intake to outfall,
organisms will experience marked physical and chemical stresses as a result of
the changes in pressure and temperature, shear and acceleration forces (Fig.
6.1), abrasion, and collision with structures. They are also subjected to dosage
with biocides and anticorrosion agents and finally contact with corrosion
by-products such as copper, nickel, and chromium salts. In addition,
organisms discharged from outfalls may be subjected to heavy predation by
congregated fishes or large invertebrates (see 5.4.10.f). The total transit time
depends very much on the design of the power station and its operating
schedules, and can vary from 2 min up to 1 h. The longer exposure periods
usually occur where the discharge is made via a long effluent channel in which
the organisms may be exposed to both chemical and thermal stresses before
discharge to the receiving water.

Effects of entrainment are likely to be a cumulative result of all the stresses
outlined above, though the relative contribution from physical and chemical
components may vary with the design and operation of the power station and
plant. Field studies and experimental studies have been carried out in a
number of regions of the world (Hirayama and Hirano, 1970a,b; Koops, 1972;
E.D.F., 1977; Jensen, 1977; Van Winkle, 1977; Schubel and Marcy, 1978;
Jolley, et al., 1978; see Carrier and Hannon, 1979).

6.4.1. Site studies

(a) **Bacteria and heterotrophs**

Verstraete, et al. (1975) estimated the abundance of bacterial cells entering
and leaving three industrial cooling-water systems in Belgium. They found
that in spite of temperature increases of 5–10°C the number of cells in the
discharge was increased 1 to 7 times, particularly those of *Escherichia coli* and
other faecal streptococci. There was also a shift in dominance from
psychrophilic to mesophilic and thermophilic organisms (see 5.4.1). The
author concluded that increases were a result of 'culturing' in the cooling
system, though no mention is made of chlorination. Fox and Moyer (1973)
also showed increases in bacterial activity and populations in a power plant
cooling-water system. Davis and Coughlan (1978) using ^{14}C labelled glucose
found that heterotrophic activity was almost completely suppressed in
sea-water which had passed through Fawley Power Station (U.K.), though
only after chlorination. They state that 'bacterial activity virtually ceased
when chlorine was present'. Changes in Vmax (i.e. maximum rate of uptake of

the glucose substrate) suggested that traces of chlorine were present even when they were not detectable colorimetrically. The authors concluded that 'the elimination of bacteria is not surprising since chlorination removes slimes'. When chlorine was known to be completely absent physical and thermal stresses had no effect on activity. Humphris and Rippon (1978) demonstrated that nitrifying bacteria (*Nitrosomonas*; *Nitrobacter*) were more resistant to chlorination than other groups in cooling-water systems and that after a dose of 3 p.p.m. for two hours survivors recovered rapidly.

(b) Phytoplankton

Field assessments of the effects of entrainment on phytoplankton have included direct counts, estimations of physical damage, and measurements of photosynthetic activity. Briand (1975) concluded that 1,700 tonnes (biomass) of phytoplankton was killed in passage through two Southern California power plants and that diatoms were killed in greater numbers than dinoflagellates. Koops (1972) attempted to assess mechanical damage and effects of thermal inputs on colonial diatoms passing through the Flevo Power Station. He concluded that there was no evidence of cell damage, though data were scarce. Kreh and Derwort (1976) also concluded that mechanical damage did not occur in passage through the Oconee Power Station in South Carolina.

Table 6.1 summarizes some of the field data from power station studies. Where biocides were not used some reductions in primary productivity or respiration began to occur when discharge temperatures exceeded about 27°C. Some highly significant reductions occurred at temperatures over 30°C. In winter and spring, temperature increases actually stimulated photosynthesis at some sites. The most marked reductions in phytoplankton metabolism were, however, in the presence of chlorine.

A recent account of the effects of power plant chlorination from field studies shows a clear relationship between the concentration of residual chlorine in the discharge and the rate of carbon-fixation, i.e. photosynthesis (Davis and Coughlan, 1978). Figure 6.7 shows clearly that there was little or no reduction where chlorine was not detectable colorimetrically, but at 0.4 mg l^{-1} the rate of photosynthesis in the discharge was only about 30 per cent of that at the intake. At all levels over about 0.8 mg l^{-1} photosynthesis was virtually completely inhibited, i.e. by 95–100 per cent. The evidence for inhibition of phytoplankton activity by passage through power stations is, therefore, quite clear and unequivocal. Whether the smaller cells (i.e. nano plankton) or the larger cells are most affected is not clear.

Recovery rates of phytoplankton are not well known. Where activity is completely eliminated recovery does not occur (Eppley, *et al.*, 1976; Gentile, *et al.*, 1976; Morgan and Carpenter, 1978). Where only part of a population is killed, however, or where cells are only rendered partially or temporarily inactive, primary productivity may recover rapidly. Dilution of discharged

Table 6.1. *Summary of data from entrainment studies of phytoplankton in the U.S.A. and the U.K.*

Authors	Power plant	Chlorination	Temperature		Effects
			Amb. T°C	*ΔT°C*	
Warinner and Brehmer, 1966	York River, Va.	No	1–10	7–10	Productivity increased in discharge area
Warinner and Brehmer, 1966	York River, Va.	No	15–20	above 5·6	Decreased productivity (O_2 consumption)
Morgan and Stross, 1969	Chalk Point, Md.	No	<16	8	Increased productivity
Morgan and Stross, 1969	Chalk Point, Md.	No	>20	8	Decreased productivity (O_2 consumption)
Hamilton, et al., 1970	Chalk Point, Md.	No	14	7	No measurable effect on photosynthesis
		No	23·5	7	30·5% reduction in photosynthesis
Brook and Baker, 1972	Allen S. King	Yes	25–27	9–11	50–91% reduction in photosynthesis
		No	23–25	9–11	5–15% reduction in photosynthesis
		Yes	23–25	9–11	50–90% reduction in photosynthesis
Carpenter, et al., 1972	Millstone Point	No	9–18·3	13	+11% increase to 26% decrease in productivity
Carpenter, et al., 1972	Millstone Point	Yes (intermittent)	7–19·7	12·1–12·9	0·05–0·25 p.p.m. residual chlorine discharge. 24–86% decrease in productivity at discharge
Carpenter, et al., 1972	Millstone Point	Yes (continuous)	9·5–18·2	10·3–15·7	0·05–0·40 p.p.m. residual at discharge (1·2 p.p.m. at injection point) 56–98% decrease in productivity
Fox and Moyer, 1975	Crystal River	No	23	6·1	2–21% decrease in primary productivity
Fox and Moyer, 1975	Crystal River	No	27	5·5	37% decrease in primary productivity
Fox and Moyer, 1975	Crystal River	Yes	27	5·5	65% decrease in primary productivity
Davies and Coughlan, 1978 (see Fig. 6.13)	Fawley Power Station	No	12–19	3–8	12% increase to 20% decrease in carbon-fixation recorded
		Yes	12–19	3–8	60–80% reduction at 0·2 p.p.m. 85–100% reduction over 0·8 p.p.m.

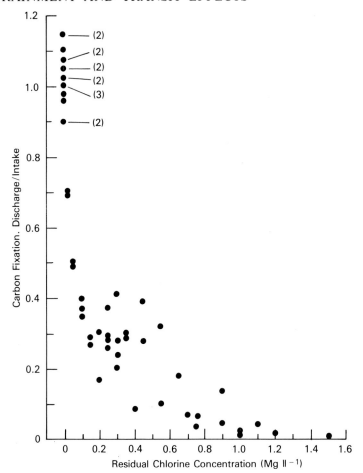

Fig. 6.7. The effect of cooling-water chlorination on carbon fixation by marine phytoplankton at Fawley Power Station, U.K., redrawn after Davis and Coughlan (1978).

chlorinated plankton in the receiving water appears to obscure effects of entrainment mortality quite quickly (Fox and Moyer, 1973; Goldman and Quimby, 1979). At some power stations the numbers of cells passing through the station are lower than in the surface layers of waters from which the cooling-water is taken. This is usually a result of abstraction from depths where phytoplankton is not normally present (see chapter 4). The effects of discharges on the phytoplankton of receiving waters are discussed in chapter 5.

(c) Zooplankton (excluding fish eggs and larvae)

Entrained zooplankton includes small copepods and cladocera, together with the tiny free-swimming larvae of larger organisms such as crabs, oysters, clams, mussels, and polychaete worms. Pelagic fish eggs, larvae, and fry are also found entrained and in transit at some sites, together with larger crustacea such as adult shrimps, mysids, amphipods, and isopods. In fresh water, planktonic crustacea and insects may also be entrained (Markowski, 1959).

Markowski (1959) showed that living freshwater and marine zooplankton could be collected from power station outfalls, but as he used plankton nets trailing in discharges it is likely that some of the animals collected may not have passed through stations but had been 'entrained' by turbulence in the outfall area. He reported that larger animals were damaged, though subsequent work indicates that this may have occurred during sampling (Coughlan and Fleming, 1978).

One of the earlier American studies, on the Green River, Kentucky, produced possibly misleading results (Churchill and Wojtalik, 1969). In May 1964 observations showed fewer zooplankters in the discharge area when compared with the intake, though the abundance increased further down-stream (see 5.4.5). However, as Coutant (1970b) and Coughlan and White-house (1977) point out, losses are difficult to explain as a power station may well cause mortalities but, apart from the possibility of complete mechanical destruction of a few specimens, the plant does not digest or consume carcasses. Therefore, although dead plankters may be discharged, actual numbers should not decrease markedly at the outfall. The viewpoint that the power plant is a 'predator' cropping-off plankton (Bunting, 1974) is, therefore, erroneous.

The discrepancies in numbers between intake and outfall may be explained by several factors. First, plankton is notoriously patchily distributed both temporally and spatially (McLaren, 1963). Marcy (1976b) and Davis and Coughlan (1978) noted marked inhomogeneity of zooplankton with depth and over short time intervals. Thus, samples taken at set depths at the intake where distribution is patchy may not be comparable with samples from the outfall where there is a complete mixing of water after passage through pumps, condensers, and outfalls.

To offer further explanation, Carpenter, et al. (1974) found that copepods sank some 2·5 times faster after passage through a cooling-water system than did control animals. Thus, any sampling which does not take account of both natural and induced vertical distribution may also produce misleading quantitative conclusions.

The more recent work on effects of entrainment and transit has, therefore, tended to concentrate on the comparison of percentage mortalities between intake and outfall. The two main methods used to assess death have been loss of movement (Reeve and Cosper, 1970), or ability to take up a vital stain (Dressel, et al., 1972; Fleming and Coughlan, 1978). Although Davies and Jensen (1975) used movement as their criterion for survival they actually indicated in an earlier paper (Davies and Jensen, 1974) that they found neither movement nor vital staining to be particularly successful criteria. Jensen, *et al.* (1969) suggested that non-motile but 'twitching' plankters may recover while Reeve and Cosper (1970) indicated that once mobility is gone the plankter would not survive. In neutral-red staining methods some specimens may turn out somewhat pink-tinged rather than the positive deep red colour of the living animal, and assessment of their survival potential may tend to be subjective (Fleming and Coughlan, 1978). Thus, data must be viewed critically where small differences in mortality rates are claimed (Table 6.2).

Marcy, *et al.* (1978) and Beck (1978) in comprehensive reviews of entrainment mortalities quoted field studies which showed mortalities of 15–100 per cent of zooplankters at various power plants in the U.S.A. The effects of thermal, mechanical, and chemical components of the system on mortalities are not always clear. However, mortalities of 1·3–10·7 per cent of intake populations at four power plants were recorded by Icanberry and Adams (1974), and they showed a direct relationship between percentage mortalities and temperature. The reliability and significance of such estimates are, however, a little doubtful, especially as maxima occurred at 24·5 and 29·5°C. Subsequently, the authors showed that mortalities were significant only where entrained animals were kept for long periods at discharge temperatures. Those which were returned to ambient temperatures recovered. Thus transit alone was not sufficient to kill the zooplankton. Massengill (1976b) also showed mortalities of 100 per cent when entrained organisms passed down a discharge canal at the Connecticut Yankee Nuclear Power Station. Temperatures here ranged between 31 and 39°C. However, at Turkey Point only 80 per cent of the zooplankters were dead at temperatures of 40°C (Thorhaug, *et al.*, 1974). At 33°C no significant increases in mortalities occurred in the outfalls of either Turkey Point or two other power stations in the U.S.A. (Thorhaug, *et al.*, 1974; Davies and Jensen, 1975). Storr (1974) showed mortalities of 24·8 per cent at Nine Mile Point where ΔT was 17°C and only 14·5 per cent at the Ginna Plant where ΔT was 7°C in summer.

The data suggest that discharge temperatures of below 33°C do not generally cause significant mortalities.

Where chlorine is present the data also conflict. For example, whereas Heinle (1969) reported 100 per cent kill of eggs and larvae of *Acartia tonsa* after passage through the Chalk Point Power Station as a result of chlorine, Carpenter, *et al.* (1974); Gentile, *et al.* (1976); and Davis and Coughlan (1978) found that even at 1·0 p.p.m. residual chlorine instantaneous mortalities following transit were low, ranging from 3–17 per cent of the catch. Mortalities after 48 hours, however, could range from 62·7 to 87 per cent for various planktonic organisms (Davis and Coughlan, 1978). Davies, *et al.* (1976), found no increase in mortality after passage through a thermal discharge canal. Davies and Jensen (1975), using their 'motility ratio' (i.e. per cent immobile at outfall/per cent immobile at intake), indicated 100 per cent kills of zooplankton when total chlorine residual at the Indian River Plant

Table 6.2. *Summary of data from entrainment studies of zooplankton in the U.S.A. and the U.K.*

Author(s)	Power plant	Temperatures Amb. $T°C$	Temperatures Max. $\Delta T°C$	Chlorination	Effects
Thorhaug, et al., 1974	Turkey Point	30	40	? }	80% mortality
Thorhaug, et al., 1974	Turkey Point	20	33	? } ?	12–14% mortality (sampling)
Carpenter, et al., 1972	Millstone Plant			No	No initial mortality Mechanical damage reported
Carpenter, et al., 1972	Millstone Plant			Yes	7·5% mortality at 0·18 mg l^{-1} 7·5% mortality at 0·18 mg l^{-1} (dosed at 1·0 mg l^{-1})
Davies and Jensen, 1975	Marshall Plant, Lake Norman	0–20	33	No	No mortality. Increase productivity in winter
Massengill, 1976	Conn. Yankee	0–20	33	Yes	100% mortality at 1·5 mg l^{-1} 0% at 0·25–0·75 mg l^{-1}
			31·0–39·4*	?	100% mortality after transit in discharge canal
Icanberry and Adams, 1974	Various (California coast)		24·5–29·0		1·3–10·7% mortalities (Linear relationship to temperature)
*Davis and Coughlan, 1978, and personal communication	Fawley, U.K.	10–20	15–28	No	99–100% survival
				Yes	1·0 mg l^{-1} low temperatures. 5% mortality
				Yes	1·0 mg l^{-1} at 20°C+. 85% mortality (Various groups—different mortality rates)

* Approximate figures

reached 1·5 p.p.m. and also when total residuals reached 5·0 p.p.m. at the Chesterfield Plant on the James River. However, at concentrations in the 0·25 to 0·75 p.p.m. range, little change in percentage kill was noted. In two instances when temperatures reached 20°C, and chlorine residual about 1·00 p.p.m., mortality at the discharge was more than at the intake. The authors give no explanation in their paper for the apparent differences in zooplankton reactions to chlorination at the two power plants, but dismiss the point by stating that they 'may have been related to site-specific combinations of factors such as maximum temperature, duration of exposure, temperature rise, wastewater contamination and species composition'. It is possible, however, that the synergistic effects of heat and chlorine are responsible for higher mortalities at higher temperatures (see 5.4.9 and Hiryama and Hirano, 1970b; McLean, 1973; Cairns, et al., 1978; Jolley, et al., 1978; Schubel and Marcy, 1978).

Larger species of zooplankton and other invertebrates are particularly susceptible to damage in transit through a power station. Bunting (1974) noted mechanical damage in up to 90 per cent of larger Cladocera passing through power stations but much lower percentages on smaller species. Icanberry and Adams (1974), on the other hand, recorded 31–43 taxa from the intake and outlet water at the four Californian power stations. They concluded that soft-bodied forms, namely polychaete larvae, trochophore larvae, and phoronid larvae were very resistant to temperature and mechanical stresses in the cooling-water systems. Adams (1969) also showed that many species which settled in the discharge canal of the Humboldt Bay Power Station must have passed through the station in their larval stages, as the net flow in the channel was always outward. Species included oysters, cockles, and several species of clams.

Barnacle oysters and mussel larvae can also survive passage through power station cooling-water systems (Utting, 1975; Coughlan and Whitehouse, 1977; Jensen, 1977; Utting and Millican, 1977) though settlement of mussel larvae may be inhibited by chlorine. Oyster larvae are, however, tolerant to normal chlorine levels in power station systems (Waugh, 1964). Beck (1978) and Marcy, et al. (1978) noted mortalities of several meroplanktonic organisms in transit through power stations including ctenophores, mysid, and polychaete larvae, arrow-worms, and coelenterates, though the percentage mortality varied very much with the site and the organisms involved.

(d) Fish eggs, larvae, and juveniles (ichthyoplankton)
There are a number of published accounts of field and theoretical studies of impingement, entrainment, and transit mortalities of fish eggs, larvae, and juveniles (Coutant and Talmage, 1975, 1976; Van Winkle, 1977; Schubel and Marcy, 1978; Kelso and Leslie, 1979), and many suggested schemes for preventing losses (N.T.I.S., 1973; Ray, et al., 1976). Beck (1978) and Marcy, et al. (1978) have summarized many of the data from field studies in the U.S.A.

Larval mortalities in transit have varied from 27 to 100 per cent, and estimated total numerical losses at some sites have ranged from 20×10^6 to 300×10^6 over the period in which larvae or juveniles were present in the habitat. Various workers have found considerable variation in mortality with species, site, and sampling method (Ney and Schumacher, 1978).

Hadderingh (in press) found that over 0.5×10^6 larval and juvenile cyprinids and northern pike were entrained at Bergum Power Station, but that mortalities were very low, i.e. 0–10 per cent except in the warmest periods. Smaller eggs and larvae generally survived the physical and mechanical stresses better than those which were larger and softer bodied.

Experiments in which fish have been deliberately introduced to intakes have shown variable mortalities, for example Kerr, J. (1953) found 10–20 per cent mortalities of juvenile Chinook salmon and striped bass through the Contra Costa Plant in California, and Knutson, et al. (1976) found over 50 per cent mortality of fathead minnows, 30–60 mm long, after passing through the Monticello Plant. Entrained eggs of various species have been found to suffer large mortalities (Flemer, et al., 1971).

It is clear from the data that very large numbers of young fish and eggs are killed by entrainment in power plants in various countries. In the U.K. no estimates of mortalities have been produced, though the author has observed numbers of dead clupeids being discharged from the outfalls at Kingsnorth Power Station. Also, dead juvenile bass (*Dicentrarchus labrox*) up to 60 mm long have been collected from the condenser system at the same power station (Langford, unpublished data).

(e) Causes of entrainment mortality

A considerable amount of experimental and field research has been carried out recently in attempts to evaluate which of the stresses (physical or chemical) are most responsible for the mortality of various organisms in power station systems (Schubel, et al., 1978).

Thermal stresses can be divided basically into two categories for entrainment assessments:

Thermal shock, i.e. sudden temperature increase as the organisms enter the condenser (Figs. 6.1, 6.8).

Prolonged exposure to gradually decreasing temperatures, i.e. entrainment in effluent channels or thermal plumes following condenser passage (Fig. 6.8).

Most of the data derived from classical constant temperature studies are inappropriate to entrainment work, and specific simulation has been used more recently for predictive studies. Hair (1971) established that the 48 h LT50 for one mysid species was $22.5°C$ when acclimatized to $11°C$ and $25°C$ when acclimatized to $22°C$. However, Burton, et al. (1976) found that there were no mortalities after exposure for 6 min at $25°C$ when animals were acclimatized at $10°C$. Generally, organisms may withstand short-term exposure to temperatures much higher than their LT50, provided that they are

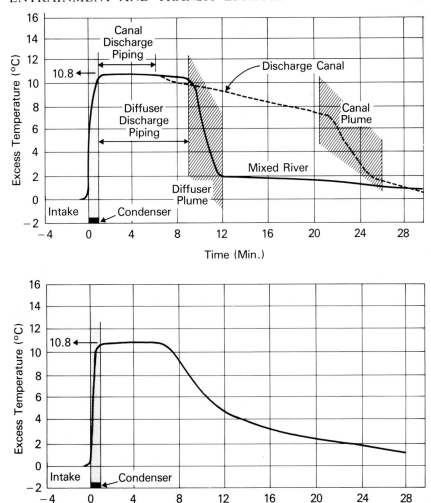

Fig. 6.8. Hypothetical time-course of acute thermal shock to organisms entrained in condenser cooling-water, (a) discharged by canal or diffuser, (b) discharged as a jet; redrawn after Schubel, *et al.* (1978).

returned to cooler water after exposure. Schubel, *et al.* (1978) have summarized most of the available data on effects of temperature shock and time temperature exposures on aquatic invertebrates and fishes. Instantaneous mortalities of common North American species generally occurred when the organisms were exposed to ΔTs of 17–20°C at winter ambient temperatures and ΔTs of 10°C above summer ambient temperatures.

Debilitation of thermally shocked fish occurred to such an extent that predation increased markedly under experimental conditions, suggesting that mortalities may increase in the field because of predation on entrained fishes or invertebrates, though there are few data to confirm this (see 5.4.10). Further, exposure of eggs to temperature shocks caused reductions or eliminated hatching in several species of fish, and deformities or physical abnormalities have also been recorded in fishes hatched from 'shocked' eggs (Schubel and Marcy, 1978).

It has already been noted that chlorine is obviously the major cause of mortalities of entrained bacteria and phytoplankton, and field data show very clear evidence of this (see 6.5.1). Fragmentation of colonial or chain-forming species has, however, been observed as a result of mechanical stresses at at least one power station (Koops, 1972; Gentile, *et al.*, 1976), though recovery of such fragmented colonies has not been studied. Different species of organisms show different chlorine sensitivities within major groups (Morgan and Carpenter, 1978). For example, the diatom *Skeletonema costatum* was killed by 5–10 min exposure to 1·5–2·3 p.p.m. chlorine, but *Chlamydomonas* was much more resistant (Hirayama and Hirano, 1970a). Similarly, significant differences occur in other organisms and different natural assemblages show different mortality curves (Morgan and Carpenter, 1978).

The time of exposure to chlorine is also important. Figure 6.9 shows data for the marine copepod *Acartia tonsa* at different chlorine concentrations. One hundred per cent mortality occurred at time exposures between about 8 and 100 min in concentrations over 1·0 p.p.m. This type of dose-effect also applies in principle to many other organisms, but many are much more resistant, for example oyster and mussel larvae (see 6.3.6). High water

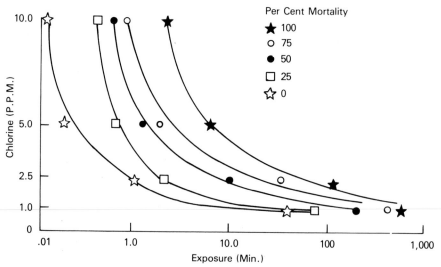

Fig. 6.9. Response isopleths for the marine calanoid copepod *Acartia tonsa*, exposed to chlorine, redrawn after Morgan and Carpenter (1978).

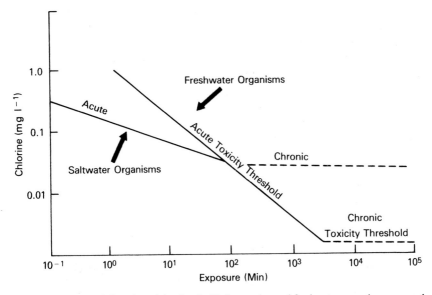

Fig. 6.10. Acute and chronic toxicity thresholds for marine and freshwater organisms exposed to chlorine, redrawn after Mattice and Zittel (1976).

temperatures exacerbate the effects of chlorine (Cairns, *et al.*, 1978). Thus at a temperature of 15°C the 30 min LC_{50} for *Cyclops bicuspidatus thomasi* was found to be 15–16 p.p.m. while at 20°C the LC_{50} was 5·76 p.p.m. (Brooks and Seegert, 1977).

Chlorine concentrations of about 0·2 to 0·4 p.p.m. were usually lethal to 50 per cent of larval fishes in 24–48 h (Morgan and Carpenter, 1978). Again, higher temperatures increased chlorine effects. Freshwater and marine organisms show different toxicity thresholds, probably owing to the effects of the complex chemical changes in sea-water and the toxicity of bromine compounds produced (see 3.5.2a) (Jensen, 1977; Morgan and Carpenter, 1978; Jolley, *et al.*, 1978) (Fig. 6.10).

Chlorine toxicity may also be affected by the presence of other toxins, hardness of the water, and presence of nitrogenous compounds (Katz, 1977; Cairns, *et al.*, 1978). These latter tend to reduce toxicity. Sunlight also stabilizes some organo-chlorine compounds and reduces toxicity (Katz, 1977).

Several authors have shown that organisms can actively avoid potentially lethal physical and chemical conditions. This is not really applicable to transit effects but only to 'external' effects, i.e. in the thermal plumes (see 7.6). Marcy, *et al.* (1978) summarized the data on the mechanical damage found in entrained animals. He states that 'acceleration shear stresses, 3 times that of gravity may be lethal to eggs and larvae of striped bass and white perch'. Pressure changes of 0·1 atm s^{-1} resulted in no mortality among roach (*Rutilus rutilus*) fingerlings, though at 3 atm s^{-1} the mortality rate was 100 per cent.

Changes of about 1 atm s^{-1} occur in power plants. In a comparative assessment of physical and chemical stresses, Beck (1978) concluded that physical damage was mainly responsible for some mortalities among entrained small fish and larger invertebrates, though chlorine and high temperatures might increase mortality rates. Bacteria and phytoplankton are at risk mostly where chlorine is injected. The survival of thermally shocked or chlorinated organisms may also depend on the speed with which they are returned to cool, cleaner water. Thus, discharge via a diffuser or outfall enhancing rapid mixing may be less damaging than via an effluent channel. Intermittent chlorination will probably kill a much lower proportion of the over-all entrained populations than continuous chlorination, particularly if the latter is at levels of over 0·5 p.p.m. Data on chlorine toxicity in relation to the organisms receiving waters are reviewed in chapter 7.

6.4.2. The significance of entrainment mortality

Power plant entrainment removes living organisms from one part of a habitat and returns them, dead or alive, to another. At some power stations intake and discharge points may be within several hundred metres of each other, in the same part of the water body. At other stations intake and outfall may be separated by a land-barrier or may be in waters of different chemical or physical character.

Losses from the 'abstraction' habitat and organic inputs to the receiving water may, therefore, be ecologically important. A number of attempts have been made to assess the effects of impingement and entrainment losses on the resident populations of phytoplankton, zooplankton, and fishes (Van Winkle, 1977; Schubel and Marcy, 1978; Lawler, Matusky, and Skelly, 1979). The increased activity of phytoplankton populations in heated waters has been noted previously in chapter 5 (see 5.4.3) and nutrients released from dead organisms may be partly responsible.

Davies and Jensen (1975) predicted that, given maximum mortalities, the zooplankton population in the receiving water, downstream of the Chester-field Plant on the James River, should have been reduced by 33 per cent, but this was not the case. In fact, as was found by other authors (Churchill and Wojtalik, 1969; Heinle, et al., 1974), there was evidence of increased standing crops downstream of the discharge, though some reduction may have been observed in the immediate discharge zone.

Carpenter, et al. (1974) found that where a cooling pond intervened between the Millstone Power Station outfall and Long Island Sound 70 per cent of copepods entering the intake were not returned to the Sound, as the damaged and dead organisms sank and stayed in the effluent pond. The authors estimated that the loss in secondary production to Long Island from the Millstone Power Station was about 0·1 per cent in the 333 km^2 area adjacent to the plant. Highest daily losses ($1·4 \times 10^6$ g dry weight) occurred in April, the lowest ($45·8 \times 10^3$ g) in November. By rather crude extrapolation they predicted that 0·05 per cent of the secondary production of Long Island Sound would be lost by mortalities in all the power plants discharging to it.

Massengill (1976b) estimated that about 4 per cent of the zooplankton of the Connecticut River was entrained through the Connecticut Yankee Power Station. At high temperatures 100 per cent mortality was recorded, at lower temperatures practically nil. At the Indian Point No. 2 Plant on the Hudson River, 6–31 per cent of the 'passive' zooplankton could be entrained at low water flows, though actual mortalities in transit were small. Heinle (1969) predicted that the population of the copepod *Acartia tonsa* could not survive entrainment mortalities of 20–25 per cent daily, but he still could not measure any detrimental effects downstream of the Chalk Point Power Station even at almost 100 per cent entrainment mortality levels. Shocked or dead zooplankton may be prone to increased predation. Neill and Brauer (1970), for example, found intensive feeding by fish on entrained zooplankton in the outfall area of a power plant on Lake Monona, Wisconsin. Brauer, *et al.* (1974) found that this accumulation at the outfall in the littoral zone was 2–7 times greater than in other littoral zones. The reason was that population density at the offshore intakes was naturally higher than nearer the shore and that animals were transported to the outfall area via the cooling-water system.

Marcy, *et al.* (1978) summarized the few data on percentages of larval fish population entrained by various power stations in the U.S.A. These ranged from about 2 to 12 per cent. In the Connecticut River, Marcy (1976b) estimated that 61 per cent of the larval fish population of the main river in the lower reaches congregated at the Connecticut Yankee Nuclear Power Station intake area. Of these, 4 per cent were entrained on average. Data for winter flounders off the Millstone Power Station showed that up to 20 per cent of the local stock could be entrained (Marcy, *et al.*, 1978). Koops (personal communication) and Hadderingh (in press) found that the Bergum Power Station entrained over 5 per cent of the resident larval coarse fish population, though mortalities were low (see 6.5.1.d).

Mortalities of juvenile fish are naturally high, mainly 70–80 per cent between hatching and the end of the first year. From studies of exploited fish populations it seems clear that many continue to exist with adult cropping rates of more than 25 per cent (McFadden, 1977), though many of these adults will have spawned before being cropped. There are, however, records of over-exploitation (Beverton and Holt, 1957; Cushing, 1974; McFadden, 1977), leading to dramatic population declines.

The total effect of losses from impingement and from entrainment in power stations has been the subject of much speculation and, recently, mathematical population models have been used to try to assess the extent of impacts on various species (Van Winkle, 1977; Muraka, *et al.*, 1978; Swartzmann, *et al.*, 1978; Ogawa, 1979). Different population models of species produce different estimates of their potential vulnerability to power station impacts (Van Winkle, 1977; Muraka, *et al.*, 1978). In some cases larval mortalities as low as

2 per cent above natural mortalities have been predicted as hazardous in the long-term (Saila and Lorda, 1977). In others it is suggested that compensatory factors such as increases in fecundity and survival, together with predator mortality, will more than compensate for high larval mortalities, unless the populations decline to critically low levels (Goodyear, 1977; McFadden, 1977). The major limitation to the use of mathematical models to specific situations is the general inability of field studies to produce accurate estimates of population sizes and the potential areas of influence of an intake or outfall. Also, the marked fluctuations in reproductive success and survival of organisms from year to year makes predictive models of limited use, unless stocks are assessed annually (Van Winkle, 1977).

It is, however, possible to improve both the quality of data and mathematical models and these are the areas of research which need much more work to provide the accurate predictions necessary for future developments.

Plant design and operational criteria which will minimize entrainment effects whilst maintaining the efficiency and economy of electricity generation are difficult to produce. It is likely that these aims are not compatible. Where criteria for minimizing entrainment mortalities are suggested (see Schubel and Marcy, 1978), the suggestions, while making ecological sense, are often economically naive and operationally difficult to achieve (see 6.4.6). Further, in spite of the fact that there have, as yet, been no demonstrably significant effects of entrainment mortalities on planktonic or nektonic invertebrates or fish populations, *and* also that modelling studies generally suggest that there may *not* be such effects from present power stations, the future expansion of power developments and the increasing size of installations may require power stations to incorporate costly features to reduce intake flows and velocities, reduce ΔTs, and use alternative antifouling methods, particularly in more enclosed water bodies.

6.5. SUMMARY

There are considerable operational and ecological problems caused by organisms within, and passing through power station water systems. Operationally, these problems may be costly. The key to the assessment of potential ecological damage still lies in the assessment of the significance of entrainment and impingement mortalities to the populations in the habitat, both local and widespread. It would be, perhaps, more immediate economic and ecological sense to allocate more resources to population studies and predictive modelling rather than make much more costly design, engineering, and operating modifications without such biological data. This would, however, call for substantial advance warning of siting proposals.

Chapter 7 Biological effects of chemical contaminants

7.1. INTRODUCTION

Contaminants which may have significant effects on aquatic biota are noted in chapter 3 (see 3.5.2). Some of these may be discharged in cooling-water and this may be important if synergisms occur with temperature (Cairns, *et al.*, 1978). The volumes of contaminants discharged from power stations are small in comparison with those from other industries. In this chapter possible effects of these contaminants in the wider environment, outside the power station, are described and reviewed. The effects of chlorine on organisms in transit through cooling-systems have been discussed in chapter 6. Radio-activity and its biological consequences are in an anomalous position. The massive increase in nuclear generation in most developed parts of the world has brought the discharge of radio-active materials into public focus. Unlike many other potential aquatic hazards radio-active wastes may have direct consequences on man, through consumption of aquatic organisms or direct exposure of fishermen and water-users. In view of this many of the criteria for the control of such discharges were established early in the development of the nuclear industry (Eisenbud, 1973; I.A.E.A., 1974a,b,c). This book is concerned mainly with the relationships between aquatic organisms and power station operation. It is likely that the effects of radio-activity on man are of far more concern than the effects on the organisms themselves. For this reason the subject is not given the space which the conservationist may expect from the publicity given to the subject at present.

7.2. ASH EFFLUENTS

There are few published biological studies of ash effluent discharges. The components which may be biologically significant are pH, dissolved solids, suspended solids, and high metal ion concentrations (see McGuire, 1977) (see 3.5.2.b).

Cairns, *et al.* (1972) found that the invertebrate fauna of the Clinch River, Virginia, was eliminated immediately downstream of a major ash spill, which occurred when a dyke was breached at settling lagoons of a power station. The pH ranged from 12·0 to 12·7, and the total volume flushed out equalled 40 per cent of the daily flow of the river. An estimated 220×10^3 fish were killed, mainly as a result of high pH combined with low oxygen levels. A survey 10 days after the spill showed that the bottom fauna was completely eliminated for a distance of 5·5–7·5 km below the spill, a reduction in faunal diversity was evident for 140 km downstream, and molluscs (snails and mussels) were eliminated for 21 km below the spill (Fig 7.1). A few months later the river was partially recolonized by the more mobile organisms.

Two years after the spill, recovery was well advanced immediately downstream, though there was still some evidence of decreased diversity.

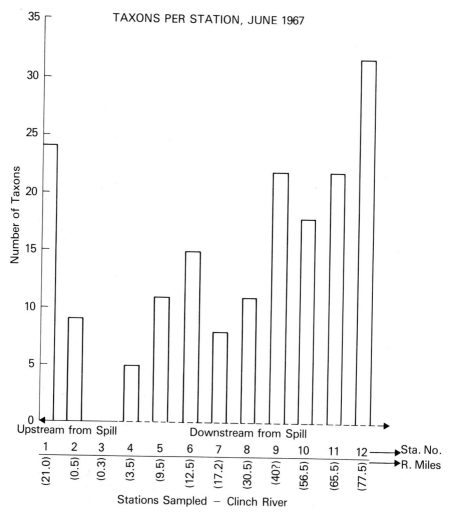

Fig. 7.1. Number of taxons found at each station for June 1967 bottom fauna survey after fly ash spill, Clinch River, U.S.A., redrawn after Cairns, *et al.* (1972).

Molluscs were, as might be expected, slower to recolonize than the more active insects, crustaceans, and fishes.

The author made occasional surveys of a small Shropshire stream in which the flow was largely composed of the effluent from the Ironbridge Power Station ash-lagoons and secondary settling tanks. Before the discharge began the bottom fauna consisted of a few species, mainly because the stream was polluted by farm effluents. In the year following the onset of the ash-lagoon discharge the stream bed was covered by a hard accreted layer of ash and the fauna was completely eliminated.

Surveys 2 and 4 years after the discharge began showed that the stream fauna was more abundant and more diverse than before the ash-lagoon began discharging. The ash-lagoon also contributed large numbers of zooplankters to the stream fauna which encouraged filter feeding invertebrates such as the insects *Simulium* spp. and *Hydropsyche* spp., and the fauna eventually resembled that of a typical lake outflow (Hynes, 1970). Growths of filamentous algae, mainly *Cladophora* sp., were extensive below the discharge. Subsequent surveys (Langford, unpublished) have shown that the fauna has again been partly eliminated as a result of ash solids. Ash-lagoons themselves may develop diverse faunas as they age. Studies of four ash-lagoons at British power stations showed that all had abundant zooplankton faunas, though of restricted diversity.

In the U.S.A., Guthrie, Cherry, and Rogers (1974), and Cherry, *et al.* (1979), found that ash discharges reduced the diversity of macrobenthos and fish faunas in a small river system, though the fauna recovered to almost its original state after the construction of a secondary settling system. Cherry, *et al.* (1976) found that the mosquito fish *Gambusia affinis* 'appeared to successfully function' in a system of swamp drainage channels near the Savannah River Nuclear Plant, which were affected by both ash effluents and thermal discharges. Temperatures reached 44°C and turbidities were high. The authors concluded that temperature was the major limiting factor in certain parts of the channel system and the ash was having no significant effects on the fish population. The fish did, however, accumulate trace metals, particularly selenium (see 7.4).

In a study of bacterial populations in the same area it was concluded again that extremely high temperatures were more responsible for instability and a decrease in diversity than the ash effluent (Guthrie, *et al.*, 1978). Another disposal method for ash is dumping at sea. In the north-eastern part of England ash dumping at sea has been carried out for a number of years (Jones, D., 1973). The only detailed study is that by Bamber (1978), who has surveyed the benthic fauna in the dumping area some 6–8 km off the coast and studied the effects of ash on benthic invertebrates under experimental conditions.

The field study showed clearly that the diversity and biomass of the fauna decreased as ash concentration in the sediment increased. At concentrations of more than 3×10^7 particles of ash g^{-1} the diversity of the fauna was severely suppressed. At 1×10^6 particles g^{-1} there was no apparent effect, though the divisions were not strictly clear cut. The ash formed large aggregates which boring species penetrated much as they would soft rocks. Raw ash contains no organic matter and, although a bacterial epiflora does develop, this takes some time. Extensive faunal colonization of raw ash may take some years once dumping ceases (Bamber, 1978). The groups most affected in the fauna were those of the benthic infauna and particulate deposit-feeders which ingested the sterile ash particles. Tube-building species were found to actively select non-ash particles for tube construction. Epifaunal species colonized ash sediments fairly rapidly as the deposit aged. The effect of the dumping was essentially localized and dependent to a great extent on the thickness of the ash on the sea-bed. Similar changes in marine benthic faunas have been recorded in areas where other finely divided solids have been dumped, for example on china clay wastes in Cornwall (Portmann, 1970). High suspended solids concentrations have been shown to inhibit feeding in some marine filter feeders. For example, 100 p.p.m. of silt caused an average reduction of 57 per cent in pumping rate (and hence feeding rate) in the oyster *Crassostrea virginica*. Long exposure caused death in both this species and the mussel *Mytilus edulis* (Loosanoff, 1962).

In rivers, discharges of other suspended solids, for example sand, limestone dust, and china clay, have also resulted in decreases in faunal abundance or diversity, or both (Herbert, *et al.*, 1961; Gammon, J., 1970; Nuttall, 1972). Where water quality is otherwise suitable for maintenance of freshwater fisheries, suspended solids concentrations of 25 mg l^{-1} would have no harmful effects on fisheries. However, at 25–80 mg l^{-1} yields may be lower, at 80–400 mg l^{-1} *good* fisheries could not exist, and over 400 mg l^{-1} only poor fisheries could exist (E.I.F.A.C., 1965).

Fish may be killed in waters with high suspended solids mainly as a result of clogging of gill filaments and asphyxiation (E.I.F.A.C., 1965). However, it is possible that fish may avoid the worst affected areas. Wallen (1951) found that fishes would tolerate concentrations of 20,000 mg l^{-1} before taking avoiding action, but exposure of up to 100,000 mg l^{-1} could be tolerated for up to 1 week in some species. Fish which died in high suspended solid concentrations had clogged gills and opercular cavities. Bio-assays of ash effluents on fish have shown that exposure to ash effluent leachates did not affect rainbow trout after 16 days' exposure (Zimmerman, *et al.*, 1974; Lynam and Brown, 1978). Some algae actually showed stimulated growth on similar exposure.

Problems with ash discharges are rare in Britain, and where ash is pumped into gravel pits or other water bodies land reclamation is usually the main objective (Gillham and Simpson, 1974). Fish rescue operations have been tried where the gravel pits used for ash dumping and land reclamation have been thought to hold good fish populations (Cragg-Hine, 1968; Brown and Aston, 1972).

7.3. COAL STOCK DRAINAGE

Trout exposed to drainage water from coal stocks did not die after 16-day exposures, though mercury concentrations were high (0.45–9.13 $\mu g\, l^{-1}$) and tissue concentrations were higher than those of normal lake fish (Zimmerman, *et al.*, 1974). The conclusion from these preliminary studies was that the effluents from active ash-lagoons and coal heap drainage were not actually toxic though, as the authors say, much more research is necessary on seasonal fluctuations. Lynam and Brown (1978) showed that ash and coal leachates from a British power station were not significantly more toxic to trout than aerated tap-water over a 28-day exposure. Coal dust in suspension, above 500 mg l^{-1}, has been shown to cause severe abrasive damage to the leaves of mosses (Lewis, K., 1973) after only 1 week. After 3 weeks similar damage was shown by 100 mg l^{-1} of dust. In high concentrations (500 mg l^{-1}) of suspended dust, growth was also inhibited. New shoots were less abundant and chlorophyll *a* production limited, though even at 5,000 mg l^{-1} chlorophyll *a* production was not prevented entirely. Where plant growth is inhibited, either by light extinction or abrasive effects of suspended solids, there is a potential indirect effect on available food and micro-habitats for invertebrates and fishes.

7.4. HEAVY METALS

The most common heavy metal ions to be found in power station discharges and their origins are noted in chapter 3. Most of these are toxic to aquatic organisms if their concentrations are raised artificially, though all are found naturally in small quantities in fresh and saline waters (Hutchinson, 1957; Sverdrup, *et al.*, 1963; Jones, R. E., 1978; Oehme, 1978). Many metal ions are also essential micro-nutrients in surface waters (Burton, 1977). It is well known that heavy metal ions are accumulated in the tissues of aquatic animals and plants during their normal metabolic processes. Such accumulation may result in poisoning in man, where the organisms are consumed in sufficient quantities (Bryan, 1971; Topping, 1973; Perkins, 1974; Romeril, 1972, 1974, 1976; Morris and Bale, 1975; Davis, M., 1977; Jones, R.E., 1978). Specific field data on the bio-accumulation or toxicity of metals in power station discharges are rare. Abalone deaths in the discharge area of the Diablo Canyon Nuclear Power Station in California were a result of high copper concentrations in the discharge (Martin, *et al.*, 1977).

Generally, the concentrations of heavy metals in cooling-water discharges, originating from corrosion of screen surfaces or from tubing and piping in cooling-water systems, are below the levels of detection by the usual methods (Romeril, 1972). The ability of organisms to accumulate metals from such concentrations may, however, be very much enhanced at higher metabolic rates caused by higher temperatures (Pringle, *et al.*, 1968; Cember, *et al.*,

1978). Where cooling is by a once-through system and the effluent is dispersed readily at the outfall there are likely to be few toxic effects on the local fauna, and only low accumulation factors (Romeril, 1974).

However, where dispersion is slower, or where polluted water (from domestic or industrial wastes) is used for cooling, the fauna in the vicinity of an outfall may accumulate metallic elements much more readily. In tower-cooled systems the metal ions may be concentrated in the cooling-water as a result of evaporation, and any animals or plants living in the system will be exposed to levels of 1·5 to 2·5 times that in the cooling-water sources. Further, where this cooling-water is used for aquaculture, metals or other toxins in the water will be transported undiluted to the organism being cultured (Boyden and Romeril, 1974). The ability of shellfish to accumulate metals means that any water source used for their culture requires careful monitoring. Data on trace metal contents exist for many organisms in so-called clean and polluted freshwater and marine environments, but it is evident that in most organisms the rate and degree of accumulation can depend on many factors apart from the concentration of metal in the water or sediment. For example, in the littoral seaweed *Fucus vesiculosus* accumulation may depend upon seasonal variations in growth rate and on tidal exposure (Bryan and Hummerstone, 1973). In cockles and clams there are marked seasonal differences plus size-related differences in trace metal levels in tissues (Romeril, 1974; Davis, M., 1977), which may be related to metabolic variations. In fishes diet may also affect the trace metal levels in tissues (Hardisty, *et al.*, 1974). In addition, different species may show a selective accumulation of different trace metals. Clams (*Mercenaria mercenaria*) show selectivity for zinc, oysters (*Ostrea edulis*) for copper, and cockles (*Cerastoderma edule*) for nickel (Romeril, 1974; Davis, M., 1977). This tendency is apparent in many species of both plants and animals (Perkins, 1974; Leland, *et al.*, 1977). Also, as with other contaminants, different species have different tolerance thresholds to specific metal ions (Jones, J. R., 1964; Roberts, 1977).

In one of the first studies of heavy metals discharged from power stations, Roosenberg (1969) noted 'greening' and copper accumulation in oysters (*Crassostrea gigas*) in the vicinity of a power station outfall which he attributed to corrosion products from the condenser tubes.

Several studies of heavy metals in the shellfish of Southampton Water (U.K.) have included sampling stations within or in the vicinity of cooling-water discharges from the Marchwood and Fawley Power Stations (Romeril, 1972, 1974; Davis, M., 1977).

Southampton Water is a busy shipping route and is bordered by docks, shipyards, and metal structures in many parts. Clams (*Mercenaria mercenaria*), cockles (*Cerastoderma edule*), and mussels (*Mytilus edulis*) were collected at various sites (Fig. 7.2).

Highest levels of some metals (zinc, copper) in clams were associated with the Marchwood cooling-water outfall, though major contributions probably come from a nearby shipyard as there were no detectable levels of these metals in the cooling-water (Romeril, 1974).

Transplanted mussels were used at the Fawley cooling-water outfall in an attempt to measure accumulation of iron resulting from dosage of ferrous sulphate used as an anticorrosive agent in the cooling-system. These mussels showed no evidence of accumulation of iron, copper, and zinc, though mussels actually growing on the outfall did show such accumulations (Romeril, personal communication).

Davis, M. (1977) showed considerable site-to-site and seasonal variation in the concentrations of metals accumulated by cockles in Southampton Water and the Solent. Briefly, the work showed that the metal levels in the tissues could be related to inputs from shipyards, shipping, sewage-works, and possibly small amounts from the two power station discharges. At March-wood Power Station cockles from the outfall contained significantly greater concentrations of copper than those from the intake. However, the major contributions originated from the nearby shipyard. Near the Fawley Power Station outfall cockles contained higher than average levels of iron, copper, and nickel, but these were not as high as in the main shipping area (Fig. 7.2).

The situation in Southampton Water was, therefore, quite complex and although some metals may have possibly entered mollusc tissues via cooling-water discharges, most originated from shipbuilding operations, metal structures, and other liquid effluents. Roberts (1977) has recently reviewed the accumulation and toxicity of metals to marine mussels.

Data from freshwater-cooled power stations are even more scarce than from marine sites. Rothwell (1971) suggested that high metal levels in Lake Trawsfynydd were responsible for the scarcity of the fauna in the sediments of the warm lagoons. She also suggested that high levels of copper originated from corrosion of condenser tubes at the nuclear power station which discharged to these lagoons. However, this was found not to be so and the metals were found to have originated from local deposits. Also, the substrata were particularly poor in nutrients, being mainly peat, and this was probably the main reason for the paucity of the fauna.

The problems caused by trace metals in fish farms using cooling-water from power stations are outlined in Chapter 8.

Data on heavy metal accumulation by organisms exposed to fly-ash effluents are rare. Cherry, et al. (1976) analysed tissues of mosquito fish (*Gambusia affinis*) living in stream and swamp drainage channels on the Savannah River, polluted by both ash effluent from settling ponds, and a thermal discharge. The effluents entered the drainage system from opposite ends so that it was possible to distinguish effects of each before they combined. Selenium concentrations in ash effluents were up to 0.107 mg l^{-1}, exceeding the Environmental Protection Agency's recommended limits for public water supply by one order of magnitude. Sediment concentrations averaged 6.1 mg

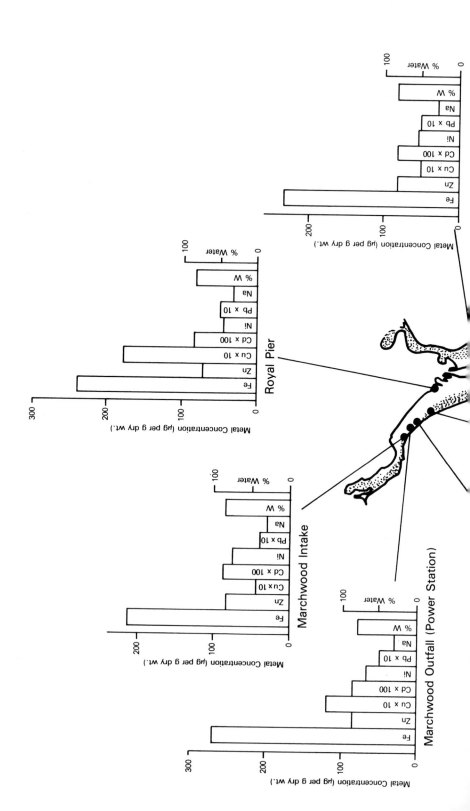

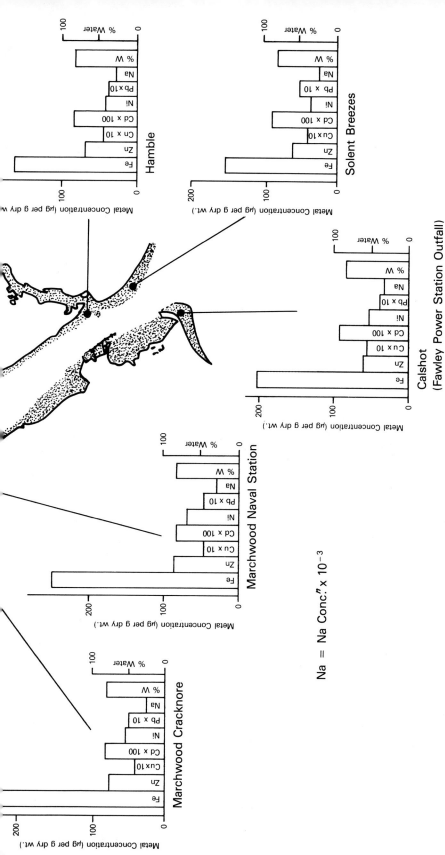

Fig. 7.2. Site-to-site variation of trace metals in the common cockle *Cerastoderma edule* from Southampton Water, U.K., redrawn after Davis (1977).

1^{-1} and those in muscle tissue of the fish 9·4 mg 1^{-1}. The fact that selenium can be concentrated through the food chain of other fishes had been reported by Sandholm, *et al.* (1973). Mosquito fish feed mainly on chironomids, micro-crustaceans, and benthic material.

Selenium, zinc, and calcium were more concentrated in fish nearer the discharge than in those further away. The most abundant elements in the water and sediments were iron, aluminium, and silicon, but these were fifth and sixth most abundant in fish. Copper (8·45 mg 1^{-1}), arsenic (0·55 mg 1^{-1}), mercury (0·22 mg 1^{-1}), and uranium (0·11 mg 1^{-1}) were also concentrated in fish muscle.

The situation in the Savannah River swamps was perhaps unusual in that flows were low in the drainage channels and temperatures were extremely high, but under such circumstances the trace metals abundant in ash effluents may be rapidly and selectively accumulated by resident fishes, especially where benthic organisms are the main food.

Zimmerman, *et al.* (1974) showed that fish kept in drainage-water from coal stock-piles accumulated mercury.

7.5. pH (ACIDS AND ALKALIS) IN LIQUID DISCHARGES

Waste acids and alkaline solutions are discharged from various processes in power stations and are generally diluted and disposed of in the cooling-water (see chapter 3), usually causing no significant changes in pH.

Coal stock leachates are also acidic and bio-assay data from the very few studies were reported earlier (see 7.3).

The fact that most aquatic organisms are tolerant to a limited range of pH is well known (Doudoroff and Katz, 1950; Jones, J. R., 1964; E.I.F.A.C., 1968a). Ash effluents may have pH values of 3·2 to 13·0. The extreme values are lethal to many species of invertebrates and fishes (see 7.2), though much more information exists on the effects of low pH than on the higher pH ranges. At pH less than 3·2 and above 10·5, survival of rainbow trout was less than 100 min (E.I.F.A.C., 1968a). Jordan and Lloyd (1964) showed that 24 h LD50s were 9·86, 9·91 and 10·13 for fish acclimatized at 6·65, 7·50 and 8·50 respectively, which suggested in fact some degree of acclimatization. After acclimatization to pH 8·3, 50 per cent of rainbow trout survived for 15 days at pH 9·5. Roach (*Rutilus rutilus*) survived 10 days at pH 10·15 after similar acclimatization. Evidence suggests that for trout, long-term exposure to pH above 9·0 may be lethal. For more tolerant species, pH 9·5–10·0 may be the threshold values. Values over pH 11·5 appear to be rapidly lethal to all species of fish (E.I.F.A.C., 1968a) (Table 7.1). Again, few data come from studies to ash effluents, though as we have already seen, Cairns, *et al.* (1972) considered high pH (12·0–12·7) to be the main cause of both fish and invertebrate mortalities in the Clinch River spill (see 7.1).

The E.I.F.A.C. Report (1968a) on pH also indicates that little is known about the effects of alkaline discharges, probably reflecting, the report states, 'the lesser importance of the problem' as compared with acid pollution. Further research is necessary, particularly for the power industry.

7.6. OIL

Most aquatic animals and plants are sensitive to oil, usually in fairly high concentrations, but emulsifiers and dispersants are generally much more toxic and have been found to do more damage to the ecosystem in the short-term than oil itself (Carthy and Arthur, 1968; Nelson-Smith, 1970; Cowell, 1971; Reish, *et al.*, 1977). Data from power station spillages are rare in the literature.

Spillages from power stations usually occur via site surface drainage and many discharges contain small amounts.

One recorded oil spill in Britain occurred at Drakelow Power Station, when the amount discharged to the River Trent caused mortalities among a herd of swans some 200 m downstream. The Trent in that area was at that time heavily polluted and fishless, with a fauna consisting only of Oligochaeta and *Asellus aquaticus* (Langford, 1967).

Oil spilled from power stations is not generally regarded as a major ecological hazard, though pollution control authorities often establish maximum allowable concentrations in discharges.

7.7. RESIDUAL CHLORINE AND OTHER BIOCIDES

The biological effects of chlorine and chlorine compounds on entrained organisms have already been discussed (chapter 6). It has also been suggested that some of the biological effects attributed to heat may have been caused by chlorine residuals (see chapter 5). The chemistry of chlorine is complex, both in fresh- and salt-water, though in the latter the complexity is greater. The toxicity of chlorine to aquatic organisms may be enhanced or suppressed by chemical and biological factors, including the presence of organic materials, ammonia and other nitrogenous substances, cyanides, hydrogen, or other halogens, species, physiological condition, time of exposure, and other stresses (White, 1972; Block and Helz, 1976; Wong and Davidson, 1977; Jensen, 1977; Jolley, *et al.*, 1978; Roberts and Gleeson, 1978).

Sunlight is also an important factor in the detoxification of chlorine. (Hostgaard-Jensen, *et al.*, 1977). The biological effects of chlorine in water have been reviewed by Doudoroff and Katz (1950), Jones, J. R. (1964), Evins (1975), E.I.F.A.C. (1973), Brungs (1973), Whitehouse (1975), Coughlan and Whitehouse (1977), Dickson, *et al.* (1977), and various other authors in symposium proceedings (e.g. Block and Helz, 1976; Jensen, 1977; Jolley, *et al.*, 1978; Morgan and Carpenter, 1978). Mattice and Zittel (1976) also listed data on toxicity to many organisms, and compared toxicity to fresh- and salt-water species. In the U.S.A. sewage effluents are often chlorinated heavily and many problems in fisheries have arisen from such effluents (see references above).

Two points concerning chlorinated cooling-water discharges must be borne in mind:

(i) The concentrations of residual chlorine diminish with time and distance from the outfall at a fairly rapid rate (Mattice and Zittel, 1976; Hostgaard-Jensen, 1977; Bender, *et al.*, 1977; Grieve, *et al.*, 1978; Hergott, *et al.*, 1978).

(ii) Chlorine is usually associated with the heated cooling-water discharge and toxicity may be increased synergistically by the increase in temperature (Jones, J. R., 1964; Evins, 1975; Cairns, *et al.*, 1978). At some power stations where chlorination is intermittent, usually at 8 hour intervals with dosage of $0.5–2.0$ mg l^{-1} for periods of 20–30 min, organisms invading outfall channels, or being entrained during non-chlorination periods may suddenly be subjected to heavy dosing, with no possibility of avoidance.

7.7.1. Bacteria, algae, protozoa, and higher plants

The effectiveness of chlorine as a bacteriocide and algicide in water supply systems is well known and the use of chlorine in power station slime-control has been discussed in chapter 6. It is to be expected, therefore, that within power plants and in thermal plumes, mortalities of bacteria, algae, and inhibition of algal activity will take place. Generally, time of exposure affects toxicity, so that the longer the exposure, the more intense the effect. Thus, algae may withstand exposures of 1–5 min to quite high levels, but longer exposure to lower or diminishing levels, say, for up to 70 min, in a discharge canal or plume may be much more harmful (Mattice and Zittel, 1976; Morgan and Carpenter, 1978) (see chapter 6).

Many of the field studies of phytoplankton have been concerned with intake and outfall measurements, i.e. 'transit effects' as discussed already in chapter 6, and few have concentrated on thermal plumes. Eppley, *et al.* (1976), however, observed that photosynthesis of marine phytoplankton was depressed by 70–80 per cent in the thermal plume at the San Onofré nuclear power stations in California, where chlorine levels were 0.04 and 0.02 mg l^{-1}.

There are few data on the combined effect of temperature and chlorine or chlorine compounds to phytoplankton, though Bender, *et al.* (1977) did note a synergistic effect at higher temperatures on the photo-synthesis of natural phytoplankton populations and cultures of *Tetraselmis* sp. Cairns, *et al.* (1978) suggested, however, that the major effect of temperature on chlorine toxicity to algae 'could simply be the effect on the rate of change of toxic residual chlorine into non-toxic compounds', as there were few examples of true synergistic effects. Increased metabolic rate may also increase the rate of chlorine uptake. The presence of ammonia appears to increase the toxicity of chlorine to phytoplankton, and chloramines may be more toxic than free chlorine (Eppley, *et al.*, 1976). All reactions are, however, complicated by the presence of bromine compounds, for example hypobromates and bromamines (Erikson and Freeman, 1978). Cairns and Plafkin (1975) exposed protozoan communities to 1.15 mg l^{-1} free chlorine for 5 min three times in 2 h and 0.66 mg l^{-1} for 20 min. Both types of exposure reduced the species diversity significantly, though a number of tolerant species were noted. Zimmerman and Berg in 1934, showed that after 6 days' exposure to an initial dose of about 3 mg l^{-1} both *Elodea canadensis* and *Cabomba caroliniana* became chlorotic. Chlorine levels were not, however, measured at the end of the experiment.

7.7.2. Zooplankton and macro-benthos

The effects of entrainment chlorination on zooplankton have been discussed in chapter 6. Laboratory bio-assays of chlorine toxicity on zooplankters are summarized by Mattice and Zittel (1976) and by Morgan and Carpenter (1978). As with other organisms, survival is time and dose dependent. Where chlorine is dosed intermittently and rapidly dissipated many species may survive, though metabolism and reproduction may be severely but temporarily depressed even at very low levels (i.e. 0.01 mg l^{-1}) in sea-water (Goldman, *et al.*, 1978; Capuzzo, 1979).

Where chlorine is dosed continuously at levels of more than 0.5 mg l^{-1} and the discharge exits along a canal, the prolonged exposure, even to diminishing levels, may be lethal to many species (Brooks and Seegert, 1977). Also, unlike phytoplankton, some species in the zooplankton may become increasingly more sensitive as temperatures increase in both fresh- and salt-water systems (Morgan and Carpenter, 1978; Goldman, *et al.*, 1978). McLean (1973) showed that mortality among three estuarine invertebrates in chlorinated water did not, however, increase with temperature. Goldman, *et al.* (1978) found that a synergistic relationship between temperature and chlorine did not appear to occur until the test zooplankters neared their thermal tolerance limits. Capuzzo (1979) found that synergistic effects of chlorine and temperature occurred with both free and combined chlorine products. The data are, thus, somewhat confused and probably vary because of the different chemical reactions involved in each experiment. The toxicity of chlorine to the larger invertebrates has been reviewed by Becker and Thatcher (1973), Brungs (1973), Evins (1975), Whitehouse (1975), Mattice and Zittel (1976), and by Morgan and Carpenter (1978). Field data are much scarcer than bio-assay or experimental data.

Snails (*Acculosa* sp.) exposed to the effluent from cooling-towers at the Appalachian Carbo Power Plant suffered 50 per cent mortality at 0.04 mg l^{-1} chlorine and 80 mg l^{-1} copper, but the authors made no attempt to distinguish the main causal agent of death (Dickson, *et al.*, 1974).

In surveys on several British rivers, Langford (1967, 1971b) concluded that the fauna downstream of some thermal discharges was reduced in species diversity, but that chlorine or scour may have been more responsible than temperature, though no chemical or physical data were presented. Brown and Aston (1975) noted that the resistance of some freshwater invertebrates to chlorine was fairly high. They found that the leech *Erpobdella octoculata*, the hog-louse *Asellus aquaticus*, the snail *Limnea pereger*, together with Tubifici-dae and Chironomidae, colonized tanks fed by the chlorinated cooling-water system at Ratcliffe-on-Soar. Similarly, Milner and Langford (unpublished) have listed more than twelve invertebrate species colonizing cooling-tower ponds including, in addition to those above, the snail *Potomapyrgus jenkinsi*, the colonial bryozoan *Plumatella repens*, flatworms, and damsel-fly larvae. Langford (1967, 1971b) also listed several organisms found in chlorinated cooling-systems.

Freshwater amphipods exposed to chlorinated thermal plumes were killed if exposed for long periods in the discharge canal of a power plant, but not in the outer plume (Lanza, *et al*., 1975). There was also evidence that at least one estuarine species (*Gammarus daiberi*) could detect chlorinated effluents and would actively avoid high residuals (Ginn and O'Connor, 1976). Oyster (*Ostrea edulis*) and barnacle (*Elminius modestus*) larvae can tolerate short-term exposure to concentrations similar to those found in most British power plants, i.e. about $0.2–0.5$ mg 1^{-1} (Waugh, 1964). The larvae of American oysters, however, have 48 h LC50s of less than 0.005 mg 1^{-1} (Mattice and Zittel, 1976).

Prolonged sub-lethal exposure to chlorine or chloramine was found to affect respiration in the larvae of the lobster *Homarus americanus*, causing marked increases after 30- and 60-min exposures to 1.0 and 0.1 mg 1^{-1} (Capuzzo, *et al*., 1977). Chloramines had a more rapid effect than free chlorine.

Chlorine may also inhibit reproduction in some species. For example, the freshwater amphipod, *Gammarus pseudolimnaeus*, produced fewer young than normal after 15 weeks' exposure at 0.0034 mg 1^{-1}. At 0.035 mg 1^{-1} no young were produced and at 0.22 mg 1^{-1} adults suffered 50 per cent mortality in 4 days (Arthur and Eaton, 1971).

Suppression of reproductive success or its elimination by chlorine, are reported for several macro-invertebrates in fresh- and salt-water and growth of some species may also be restricted (McLean, 1973; Mattice and Zittel, 1976; Capuzzo, 1979).

7.7.3. Fishes

(a) **Mortalities**

There have been a number of reports of fish kills in chlorinated power station effluents in the U.S.A. at concentrations of $0.01–3.05$ mg 1^{-1} in fresh-water (Brungs, 1973; Truchan, 1977). Chlorinated sewage discharges have also caused fish mortalities (E.I.F.A.C., 1973; Bellanca and Bailey, 1977). Mortalities appear to have occurred where fish migrated into power station outfall channels or plumes during non-chlorination periods and were then killed by a rapid increase in chlorine concentration at the beginning of a chlorination period. The numbers of fish involved in mortalities are quite small compared with those at intakes (see chapter 6). Brown and Aston (1975) found that carp (*Cyprinus carpio*) and eels (*Anguilla anguilla*) being cultured in a power station cooling-water circuit survived the normal intermittent chlorination schedule where peaks reached 0·3–0·5 mg l^{-1}, for short periods.

Recently, the data on the effects of chlorine, chlorine derivatives, and their toxicity to fish have accumulated rapidly. Most are summarized and reviewed by Zillich (1972), Brungs (1973), and Morgan and Carpenter (1978), and they deal with both intermittent and continuous dosage in relation to eggs, larvae, juveniles, and adults of various species. Most of the research has been carried out under experimental conditions.

(b) Eggs and larvae

Eggs and larvae of white perch (*Morone americana*), striped bass (*Morone saxatilis*), and blue-black herring (*Alosa aestivalis*) showed age-related tolerance to chlorine (Morgan and Prince, 1977). Older eggs were more tolerant than newly laid eggs, but older larvae were less tolerant than newly hatched larvae. Abnormal larvae hatched from eggs exposed to 0·31–0·38 mg l^{-1}, though the abnormality rate was only about 15 per cent. Other work (Lauer, *et al.*, 1974) showed that eggs of *Morone saxatilis* were tolerant to short exposure to chlorine, though larvae were susceptible to low concentrations. Lanza, *et al.* (1975) found that on exposure of striped bass larvae to simulated chlorine/temperature plume conditions, mortalities did not increase significantly at 0·16 mg l^{-1} residuals. The larvae of plaice (*Pleuronectes platessa*) and soles (*Solea solea*) were also found to be less tolerant of chlorine than eggs (Morgan and Carpenter, 1978).

The larvae of freshwater fishes show similar tendencies, and eggs of salmonids are generally less sensitive than newly hatched larvae (E.I.F.A.C., 1973). Short-term thermal stress (24–34°C) and total residual chlorine (0·05–1·0 mg l^{-1}) applied simultaneously for 6·5–60 min to the eggs, embryos, and larvae of the estuarine mummichog (*Fundulus heteroclitus*), and striped bass (*Morone saxatilis*) showed that synergistic effects occurred on older larvae, but that thermal stress was mainly responsible for egg and embryo mortality (Middaugh, *et al.*, 1977, 1978).

(c) Juveniles and adults

Mattice and Zittel (1976) tabulated relevant data for species of freshwater and salt-water fishes, and Morgan and Carpenter (1978) and Thatcher (1978) have summarized the more recent data. In common with the other organisms, different species of fish have different tolerances to chlorine. In fresh waters, salmonids are generally less tolerant than cyprinids or other 'coarse' (rough) fishes. Similarly, freshwater fishes are more sensitive than marine species to chronic exposure to chlorine, but the reverse is true for acute exposures (Mattice and Zittel, 1976) (see chapter 6).

Lethal concentrations for fishes begin to be effective at about 0·008–0·01 mg l^{-1} for freshwater fishes and about 0·02 mg l^{-1} for salt-water fishes

(E.I.F.A.C., 1973; Mattice and Zittel, 1976), but tolerance may be affected by the presence of other compounds, for example nitrogenous substances (Katz, 1977), and by temperature or the physiological condition of the fish (Morgan and Carpenter, 1978). Katz (1977) concluded that the presence of organic nitrogenous compounds could indirectly decrease the toxicity of chlorine to freshwater fish. Hoss, *et al.* (1975) also showed that chlorine doses of $0\cdot5$ mg l^{-1} reduced survival of fishes in water containing $1\cdot0$ mg l^{-1} of copper. Low oxygen levels may also increase the toxic effects of chlorine on fishes (E.I.F.A.C., 1973), and pH is important as it affects the relative proportions of free chlorine and hypochlorous acid present and thus the potential toxicity (Evins, 1975).

Exposure data for periods of 30–60 min are probably the most relevant to fishes in discharge canals or entrained in plumes (Mattice and Zittel, 1976; Seegert, *et al.*, 1977). In many marine situations immediate plume or canal concentrations may exceed $0\cdot4$ mg l^{-1} during both continuous and intermittent chlorination (Jensen, 1977; Jolley, *et al.*, 1978; Cherry, *et al.*, 1977). Fish trapped and exposed for long periods to these levels would be endangered.

Seegert and Brooks (1978a) exposed perch and rainbow trout to single 30-min and triple 5-min chlorine concentrations varying from $0\cdot7$ to $22\cdot6$ mg l^{-1} in fresh-water. They found that trout could survive over $4\cdot0$ mg l^{-1} for single 5-min doses, but only $2\cdot87$ mg l^{-1} for triple doses at 5-min intervals. They also found that some batches of trout were much more sensitive than others. Thirty-minute exposures produced LC50s of $0\cdot99$ and $0\cdot94$ mg l^{-1} at $10°C$ and $15°C$ respectively. No mortalities occurred below $0\cdot54$ mg l^{-1} while 100 per cent mortality occurred at $1\cdot6$ mg l^{-1}. Juvenile perch were more sensitive with an absolute minimum of $0\cdot48$ mg l^{-1} and maximum of $0\cdot54$ mg l^{-1}. Some studies have shown that fish are more sensitive to chlorine toxicity at higher temperatures, though the increase in sensitivity appears to occur usually when the fish are nearing their temperature-tolerance limits (Stober and Hanson, 1974; Hoss, *et al.*, 1975; Thatcher, *et al.*, 1976; Brooks and Seegert, 1977). In studies where temperatures were well below thermal thresholds little increase in sensitivities were shown with increased temperature (Heath, 1977; Middaugh, *et al.*, 1978).

The form of the halogen in the environment may determine the degree of toxicity. For example, Larsen, *et al.* (1977) showed that chloramines reduced growth of Coho salmon alevins at concentrations of $0\cdot022$–$0\cdot023$ mg l^{-1} and mortalities occurred at $0\cdot047$ mg l^{-1} in fresh water. Comparing the effects of free chlorines and of chloramines, Capuzzo, *et al.* (1977) found that the former was generally more toxic to fishes, though other studies have produced conflicting results (Holland, *et al.*, 1960; Morgan and Carpenter, 1978). Age of fish may also influence toxicity. For example, 32-day-old pink salmon had a 72 h LC50 of $0\cdot084$ mg l^{-1} while 61-day-olds had an LC50 of $0\cdot91$ mg l^{-1} (Holland, *et al.*, 1960). Larval fish and eggs exposed to various total residual chlorine concentrations showed decreased growth- and hatching-rates, together with increased anatomical abnormality rates (Morgan and Prince, 1977).

Fishes also show sub-lethal responses to chlorine residuals including changes in reproductive processes, metabolic activity, and behaviour. Arthur and Eaton (1971) found that chloramine concentrations of 0·085 mg l^{-1} almost eliminated spawning of the fathead minnow *Pimephales promelas*, and that numbers of spawnings per female and numbers of eggs per spawning decreased at 0·043 mg l^{-1}.

Chlorinated white perch (*Morone americana*) showed decreases in blood pH and increases in haematocrit in experiments (Block, *et al.*, 1978), and a breakdown in osmoregulatory processes was also noted. There is evidence, however, that fish exposed for short periods can recover within 24 h (Zeitoun, 1978).

The behaviour of fishes affected by chlorinated effluents is highly significant to effects in the field. Several studies have noted and studied avoidance responses to chlorinated water (E.I.F.A.C., 1973; Morgan and Carpenter, 1978). Thresholds for avoidance vary with species but may be as low as 0·001 mg l^{-1} for trout (Sprague and Drury, 1969; White, 1972).

In estuarine fishes avoidance thresholds range from 0·04 to 0·15 mg l^{-1}, (Middaugh, *et al.*, 1977).

Different acclimatization temperatures produce different chlorine avoidance thresholds with different species of fish (Cherry, *et al.*, 1977). Field studies showed no fish to be present where total chlorine residuals in the discharge channel of the Glen Lyn Power Plant reached 0·46 mg l^{-1} and free-chlorine residual 0·27 mg l^{-1}. In another interesting field study of behaviour, sonic-tagged white bass were tracked and located at points in the chlorinated thermal discharge of the R. L. Hearn Power Station near Toronto. Fish were shown to select certain home areas during non-chlorination periods and then to move into areas of low residual chlorine once chlorination proceeded. They were found to avoid areas with more than 0·035 mg l^{-1} (Grieve, *et al.*, 1978).

Experimental work has shown no consistent relationship with temperature. Cherry, *et al.* (1977) found that the presence of hypochlorous acid (HOCl) was the main stimulant for the avoidance response of two species. It was concluded in a further study in fresh water (Cherry, *et al.*, 1977), that the relative concentrations of the component fractions of the total residual, for example free chlorine, chloramines, hypochlorous acid, hypochlorite ions, etc., may affect toxicity.

(d) Causes of chlorine mortality and mode of death

We have seen that sub-lethal doses of chlorine can affect the metabolic processes of fish. The exact causes and mode of death in lethal concentrations are not yet fully understood.

It appears that chlorine generally causes the gill epithelium to slough off and the production of mucus (Cairns, *et al.*, 1975). At cellular level, enzymes may become altered by the oxidation of sulphydryl ions, and the enzyme activity irreversible destroyed (E.I.F.A.C., 1973).

Histopathological effects of intermittent chlorination on bluegills and rainbow trout showed that gill lesions formed, the epithelium swelled, liver glycogen was depleted, and liver necrosis occurred. The ultimate effect is blockage of the respiratory process and of gas transport across the gill epithelium (Bass, *et al.*, 1977). Marked changes in the blood constituents also occur on acute exposure including haemoconcentration, haemolysis, and disruption of plasma electrolytes (Zeitoun, 1978).

(e) Effects of other antifouling agents

There has been little work done as yet on the effects of other antifouling agents (see chapter 3) and their wider ecological consequences, though it was found that ozone may have much more intensive and rapid deleterious effects on the respiration and osmo-regulation of white perch than chlorine (Block, *et al.*, 1978). Also, bromine chloride was found to be 2 to 4 times less toxic to estuarine organisms than chlorine, with a more rapid decay rate. However, the efficiency as an antifouling agent is no doubt also somewhat lower (Roberts and Gleeson, 1978).

(f) Comparison of chlorine toxicity in fresh and salt waters

The problems involved in measuring low levels of chlorine or total residual oxidants in sea-water cause difficulties in comparing effects of these low concentrations (Carpenter, J., 1977). Davis and Coughlan (1978) illustrated clearly that bacterial activity was suppressed at levels below the chemically detectable limits. The complex chemistry of chlorine in sea-water and even polluted fresh-water also makes biological assessments and establishments of safe criteria a difficult task.

There seems little doubt, however, that chronic toxicity thresholds are lower for freshwater than for salt-water organisms, and that the opposite is true for acute exposures (Mattice and Zittel, 1976; Fig. 6.10).

Safe criteria may be partially site specific, depending upon the water conditions and the fauna. The major requirement for predictions of ecological consequences is an assessment of the extent and decay and dilution of chlorine and its derivatives in any thermal plume, together with a knowledge of the tolerances (time and temperature-based) of the relevant organisms. Some species in the flora and fauna may, of course, be common to many sites in any one region of the world.

To summarize, therefore, chlorine is without doubt an aquatic hazard, though to what extent is difficult to gauge. From the field data it is evident that, as with heat, large areas of water are not drastically denuded of a flora or fauna by discharges, though organisms may be killed at times by chlorinated power plant effluents if trapped or exposed suddenly. The increase in power generation will create a greater need for further antifouling agents and though chlorine may be restricted for ecological reasons, a suitable economic alternative has yet to be found. Heat seems to be one possibility, but the increased heat load into the environment may again raise problems (Stock and Strachan, 1977). The possible long-term effects of accumulations of organo-halogen compounds are as yet speculative with respect to power plant discharges.

7.8. ATMOSPHERIC EMISSIONS

It is well known that aquatic organisms can tolerate limited ranges of pH and that the long-term survival of salmonid fish populations in waters with pH less than 4·0 is doubtful owing to effects on eggs and fry (E.I.F.A.C., 1968a; Carrick, 1979). Table 7.1 summarizes some general data on pH tolerances among fishes. It is also known that many natural lakes and streams in regions with low calcium content in surface soils or bedrocks have low pH and the water has poor buffering capacity (Hutchinson, 1957; Dillon, *et al.*, 1978; Seip and Tollan, 1978; Jeffries, *et al.*, 1979).

The capacity of fish to maintain and regulate their salt budget is seriously impaired by high concentrations of hydrogen ions, i.e. low pH. In some regions of the world, notably Scandinavia, fish deaths have occurred for many years at times of natural snow melt when large amounts of water with very low pH and low salt concentrations have diluted the already ion-poor lakes and rivers (Scriven and Howells, 1977; Schofield, 1976) (see chapter 3).

Field studies in natural streams in Britain have shown clearly that certain groups of invertebrates, notably mayflies, molluscs, and the shrimp *Gammarus pulex* are not present where pH is less than 5·5 (Sutcliffe and Carrick, 1973). Frost (1939) showed that trout in acid streams (pH 4·6–6·8) grew much more slowly than in those with higher pH (i.e. 7·4–8·4), though Brown, M. (1957) suggested that ionic composition of the water was not the sole cause and low general productivity may have been responsible for the poor growth.

Herricks and Cairns (1974–6) acidified part of a stream artificially, reducing the pH from 8·0 to 4·0 for 15 min. Both diversity and density of the fauna were reduced, though full recovery occurred over 19–28 days. This system was, however, much different to that in the regions which are reported as ecologically affected by airborne acidification, i.e. parts of North America and Scandinavia. In these areas the continuous precipitation of acidified rain and deposition of acidic materials, supposedly from long-range and short-range sources (power stations, smelters), has caused a gradual but insidious decline in the pH of already poorly buffered acid waters (Likens, *et al.*, 1972; Jensen and Snekvik, 1972; Brosset, 1973; Likens and Bormann, 1974; Bjor, *et al.*, 1974; Dochinger and Seliga, 1975; Leivestad, *et al.*, 1976; Schofield, 1976; Scriven and Howells, 1977; Dillon, *et al.*, 1978).

While there is little doubt that airborne acidity is partly responsible for the

Table 7.1. *Summary of the effect of pH values on fish*

Range	Effect
3·0–3·5	Unlikely that any fish can survive for more than a few hours in this range although some plants and invertebrates can be found at pH values lower than this.
3·5–4·0	This range is lethal to salmonids. There is evidence that roach, tench, perch, and pike can survive in this range, presumably after a period of acclimatization to slightly higher, non-lethal levels, but the lower end of this range may still be lethal for roach.
4·0–4·5	Likely to be harmful to salmonids, tench, bream, roach, goldfish, and common carp which have not previously been acclimatized to low pH values, although the resistance to this pH range increases with the size and age of the fish. Fish can become acclimatized to these levels, but of perch, bream, roach, and pike, only the last named may be able to breed.
4·5–5·0	Likely to be harmful to the eggs and fry of salmonids and, in the long term, persistence of these values will be detrimental to such fisheries. Can be harmful to common carp.
5·0–6·0	Unlikely to be harmful to any species unless either the concentration of free carbon dioxide is greater than 20 p.p.m., or the water contains iron salts which are precipitated as ferric hydroxide, the precise toxicity of which is not known.
6·0–6·5	Unlikely to be harmful to fish unless free carbon dioxide is present in excess of 100 p.p.m.
6·5–9·0	Harmless to fish, although the toxicity of other poisons may be affected by changes within this range.
9·0–9·5	Likely to be harmful to salmonids and perch if present for a considerable length of time.
9·5–10·0	Lethal to salmonids over a prolonged period of time, but can be withstood for short periods. May be harmful to developmental stages of some species.
10·0–10·5	Can be withstood by roach and salmonids for short periods but lethal over a prolonged period.
10·5–11·0	Rapidly lethal to salmonids. Prolonged exposure to the upper limit of this range is lethal to carp, tench, goldfish, and pike.
11·0–11·5	Rapidly lethal to all species of fish.

Reference is made to different species on the basis of information known to us the absence of a reference indicates only that insufficient data exist. (From E.I.F.A.C. Report, 1968a.)

declining pH in waters in parts of Norway, Sweden, and North America (Schofield, 1976; Dillon, *et al.*, 1978), the actual origins of the acidity and the proportional effects of natural soil-breakdown products and aerial acidity are as yet unknown (Scriven and Howells, 1977; Seip and Tollan, 1978). There have been long-standing problems of acidity in Norwegian streams, and liming has been used to combat low pH and calcium concentrations since at least the 1920s (Henriksen and Johannessen, 1975). The fact that fish may become acclimatized to low pH (Lloyd and Jordan, 1964) and that there may be some genetic selection for tolerance in some rivers has been suggested

(Brown, D., 1978). There would be, therefore, problems involved in establishing fisheries in such rivers or lakes using stocked fish reared in other waters. Changes in forestry, land use, and fishery management may also be contributory factors in the decline of some fisheries (Dochinger and Seliga, 1975; Newman, 1975; Braekke, 1976). In Canada and the U.S.A. the evidence for increased acidification of some lakes by airborne emission is unequivocal (Schofield, 1976; Dillon, *et al.*, 1978), particularly in the regions affected by the Sudbury (Ontario) smelters. The decline of fisheries, including those of lake trout (*Salvelinus namaycush*), lake herring (*Coregonus artedii*), and white-suckers (*Catostomus commersoni*) is well documented.

Ultimately, in this book were are concerned with emissions from power stations and as yet the effects of these are not always separable from those from other sources, though it is possible to identify particles from power station emissions in the atmosphere (Scriven and Howells, 1977). The final ecological decision about Scandinavian waters is economically important as it has been estimated that to remove sulphur from power station emissions in the U.K. alone the capital cost may exceed £100 $\times$ 10^6 sterling. There are obviously many qualifying and complicating factors such as soil chemistry, changes in land and river use, types of fishing, and changes in management. Much of the scientific assessment is hampered by incomplete or absent biological data from the nineteenth century and earlier.

7.9. RADIO-ACTIVITY

Most concern about discharges of radio-activity in the aquatic environment is, as might be expected, focused on the potential long- and short-term effects on the human population, either *directly* through contact with effluents or sediments, or *indirectly* through consumption or organisms. The fact that most aquatic organisms, at all trophic levels, accumulate radio-active elements (radio-nuclides) in their tissues is well documented for many regions of the world (e.g. Rice, 1965; I.A.E.A., 1966, 1969, 1971, 1974a,b, 1975a,b, 1979; Polikarpov, 1966; Nelson, *et al.*, 1972; Nelson and Evans, 1973; Eisenbud, 1973; Rice and Baptist, 1974; Harvey, R., 1974; Roberts, 1977; Pentreath, 1977, 1978a,b; Hewett and Jeffries, 1978). It is this property, together with the magnified accumulation through food chains, which poses the threat to man. In many regions organisms of direct importance in the food supply are constantly monitored, for example fish, crustaceans, mussels, oysters, and seaweeds (see references), and the links from source of the radio-active element to man via one of these organisms is known as a critical pathway. The processes which tend to dilute, concentrate, and disperse radio-active materials in an aquatic environment are shown in Fig. 7.3.

Radio-active substances are toxic to the aquatic organisms themselves (Polikarpov, 1966; Rice and Baptist, 1974). Although the amounts of radio-active elements discharged from nuclear power stations and processing

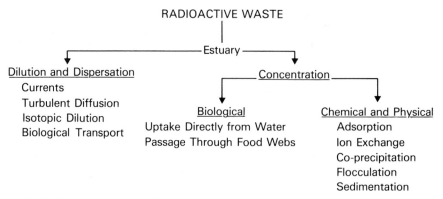

Fig. 7.3. Processes tending to dilute or concentrate radio-active materials added to a marine environment.

plants are very small, the toxicity of such elements is far greater than that of non-radio-active elements (Eisenbud, 1973). As an example, industrial health limits of lead (Pb) in air for breathing may be as high as 100 g cm^{-3}, but the International Commission on Radio-active Protection (I.C.R.P.) maximum permissible limit of *radio-active* lead in air (^{210}Pb) is 4×10^{-4} Ci m^{-3} (0·0148 k Bq m^{-3}) which is equivalent to only 5×10^{-6} g m^{-3}. To put the potential exposure into perspective, however, it is unlikely that more than a few microgrammes of ^{210}Pb would be found in lead processing plants or industry whereas non-radio-active lead is used by the millions of tonnes (Eisenbud, 1973).

Before reading the following section it should be borne in mind that the danger of radio-active isotopes to man and other biological systems has been known for many years and that many of the safety limits for man were set before nuclear power plants were evolved (Eisenbud, 1973). We are also concerned here with the steady low level discharge of waste radio-active materials to the environment and not the results of accidents or nuclear weaponry, which produce massive instantaneous and generally short-lived releases of radio-active material (Dunster, 1978). For a detailed account of the physical and physiological effects of radio-activity on tissues, together with a summary of terminology and types of irradiation, the reader is referred to a report of the United Nations General Assembly (U.N. General Assembly Official Records, 17th Session, Supp. 16. A/5216–1962), and to the I.A.E.A. (1979) report on methods of assessing radiological effects on aquatic ecosystems.

Briefly, the acute effects of ionizing radio-toxins are to produce protein and cell damage through reactions with water and other molecules. Decay of a nuclide incorporated into tissues can also cause similar damage, though often the effects of these are delayed and damage may not become evident until some time after irradiation. Longer-term effects may also be physical, chemical, or physiological, though the most emphasis is placed on danger to genetic material, for example DNA and chromosomes, which may result in hereditary damage to organisms (U.N., 1962; I.A.E.A., 1979).

7.9.1. Lethal doses (acute toxicity)

The doses of ionizing radiation required to kill various types of aquatic organisms may differ by almost three orders of magnitude (Polikarpov, 1966; Rice and Baptist, 1974). Figure 7.4 summarizes the LD50 ranges for the various groups of organisms, irradiated by X-rays or gamma-rays. Bacteria, protozoa, and algae are generally much more tolerant than the higher organisms. Also, the early stages of higher organisms tend to be most sensitive (Polikarpov, 1966; Rice and Baptist, 1974). As an example, the LD50s of various stages in the development of rainbow trout ranges from $12 \cdot 9 \times 10^3$ Ckg for gametes and up to $38 \cdot 7 \times 10^3$ Ckg for adults. Within groups of organisms some species are also much more tolerant than others. An example quoted by Rice and Baptist (1974) is that of two crustaceans, the grass shrimp *Palaemonetes pugio* and the blue crab *Callinectes sapidus*, where the 40-day LD50 for the latter was found to be greater by two orders of magnitude than that for the former. Similar differences are found within other groups (Polikarpov, 1966). Mortality may also depend on the period of exposure at dose rates which are not instantly lethal. Increased temperature also increases sensitivity to radiation (Rice and Baptist, 1974). After a single large dose

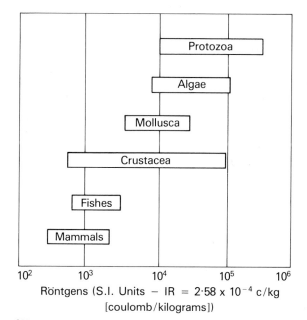

Fig. 7.4. Dose of X-rays or gamma-rays required to kill 50 per cent of organisms, redrawn after Rice and Baptist (1974).

damage may occur immediately or after some long delay. Polikarpov (1966) quotes cases of smaller organisms dying up to 1 month after dosage. Also, adult fishes (*Lebistes* sp.) lost weight, refused food, and developed sores up to 14 days after irradiation. Mortality occurred between 18 and 50 days afterwards. Till (1978) found that although eggs of the carp and fathead minnow absorbed plutonium—238 (IV) there was no evidence of effects on their development under typical environmental exposures.

In the natural environment only one case of radiation mortality has been definitely established among aquatic organisms (Rice and Baptist, 1974). Carnivorous fish in the Marshall Islands received sufficient ^{131}I from weapon fallout to cause thyroid damage and death. They had fed on herbivorous fish which in turn had eaten algae containing large amounts of ^{131}I. Rice and Baptist also state that 'ecological investigations in the Columbia River have yielded no direct evidence of deleterious effects to aquatic communities from the thousands of bequerels of radio-activity discharged for more than 25 years by the Hanford production reactors'. In British waters no evidence of radiation damage to aquatic organisms has ever been recorded (Preston, 1971; Hetherington, 1976; Mitchell, N., 1977a).

7.9.2. Chronic exposure

Nowhere in the world do nuclear power stations produce liquid radio-active discharges with sufficient toxicity to cause mortalities of aquatic organisms. Should any damage develop it would be from extremely long-term exposure to extremely low levels of irradiation, though as yet data on such exposure are scarce (Polikarpov, 1966; Rice and Baptist, 1974; Abbot and Mix, 1979). Irradiation damage may ensue from external exposure or from radionucleides ingested into organisms (Rice and Baptist, 1974).

Various authors disagree about the levels which may cause chronic effects. Rice and Baptist (1974) quote Russian research on exposure of plaice (*Pleuronectes platessa*) eggs to low concentrations of some elements. This low-level irradiation produced larval malformation at 4–5 times the control rate. American research has suggested, however, that higher external doses did not produce such abnormalities. There is no explanation for the different conclusions. The data are reviewed at length by Polikarpov (1966) and summarized by Rice and Baptist (1974).

Long-term exposures of populations of various organisms under experimental conditions have shown interesting results. For example, chronic exposure of the algae *Nitzchsia* sp. to ^{137}Cs showed that the populations were very resistant with no evidence of reproductive inhibition. Similar work on the cladoceran *Daphnia pulex* suggested, however, that production was inhibited by radiation. Chronic irradiation was believed to cause *increased* fecundity in mosquito fish (*Gambusia affinis*) in a contaminated lake, though the numbers of dead embryos and abnormal larvae were greater in the irradiated broods (Blaylock, 1969—cited in Rice and Baptist, 1974).

Temperature and other environmental factors may influence chronic effects. For example, Grayum (1973) found that there was a significant interaction effect on the alga *Chlamydomonas* as a result of thermal shock and gamma radiation. ^{14}C uptake was drastically reduced. Angelovic, *et al.* (1969) attempted to determine the interactions of other factors with chronic gamma radiation. Radiation doses of zero, 0·0083 and 0·0128 Gy h^{-1}, salinities of 10, 20, and 30 per cent and temperatures of 15, 20, and 25°C were used in observations on the morphology of post-larval pinfish (*Lagodon rhomboides*). They found that neither radiation nor salinity affected growth, but temperature did. Low-level radiation also appeared to stimulate growth, resulting in a deeper-bodied, longer fish at 0·0083 Gy h^{-1}. Changes in salinity and temperature and radiation also caused certain morphological changes, altering the size of several characteristic structures. Temperature appeared to be much more effective than either of the other environmental factors in causing growth and morphological changes.

In fouling communities, very low-level irradiation appears to stimulate the development of organisms. Polikarpov (1966) quotes several studies which show that growths of periphyton in aquaria were so stimulated. Similar results were shown for other organisms, for example barnacles, ascidians, and bryozoans. At higher levels of radiation, however, growth and survival may be reduced. Acute exposure to ^{60}Co was found to be suitable for killing and controlling clams (*Corbicula fluminea*) fouling power station heat exchangers (Tilly, *et al.*, 1978).

Williams and Murdoch (1973) exposed sessile marine invertebrates (namely, sponge, coral, oyster, slipper shell, barnacle, and sea-squirt) to different levels of ^{60}Co radiation, over periods from 3 to 7 months. Ten dose rates were used from $8·5 \times 10^6$ to 0·212 Gy h^{-1}. Growth of new sponge tissue was reduced by radiation at 0·0085 Gy h^{-1}. Growth and survival of oysters and coral were reduced slightly by 0·0042 Gy h^{-1}, but slipper shells grew well at 0·085 Gy h^{-1} and sea-squirts at 0·017 Gy h^{-1}. So far, the predicted long-term genetic damage by very low levels of radiation has not been adequately demonstrated in the field even though radionuclides from major sources are detectable some distance away from discharges (see references).

7.9.3. Bio-accumulation

The fate of any radionuclide discharged to the aquatic environment may be complex. It can remain in suspension, precipitate and settle on the bed, or be ingested and accumulated by an organism (see chapter 3, Fig. 7.3). The ratio between the concentration of an element in the water and that in the organism is known as the concentration factor (C.F.). Polikarpov (1966) and Rice and Baptist (1974) list concentration factors for many organisms calculated from data produced in field experimental studies (Tables 7.2 and 7.3). Two points emerge:

(i) There are wide ranges for each group, reflecting species differences.

Table 7.2. *Reported values of concentration factors (C.F.) for various classes of marine organisms*

Element	Group	C.F. range	Mean C.F.
Cs	Plants	17–240	51
	Molluscs	3–28	15
	Crustacea	0·5–26	18
	Fish	5–244	48
Sr	Plants	0·2–82	21
	Molluscs	0·1–10	1·7
	Crustacea	0·1–1·1	0·6
	Fish	0·1–1·5	0·43
Mn	Plants	2,000–20,000	5,230
	Molluscs	170–150,000	22,080
	Crustacea	600–7,500	2,270
	Fish	35–1,800	363
Co	Plants	60–1,400	553
L	Molluscs	1–210	166
	Crustacea	300–4,000	1,700
	Fish	20–5,000	650
Zu	Plants	80–2,500	900
	Molluscs	2,100–330,000	47,000
	Crustacea	1,700–15,000	5,300
	Fish	280–15,500	3,400
Fe	Plants	300–6,000	2,260
	Molluscs	1,000–13,000	7,600
	Crustacea	1,000–4,000	2,000
	Fish	600–3,000	1,800
I	Plants	30–6,800	1,065
	Molluscs	20–20,000	5,010
	Crustacea	20–48	31
	Fish	3–15	10
Ce	Plants	120–4,500	1,610
	Molluscs	100–350	240
	Crustacea	5–220	88
	Fish	0·3–538	99
K	Plants	4–31	13
	Molluscs	3·5–10	8
	Crustacea	8–19	12
	Fish	6·7–34	16
Ca	Plants	1·8–31	10
	Molluscs	0·2–112	16·5
	Crustacea	0·5–250	40
	Fish	0·5–7·6	1·9
Cu	Plants	—	1,000
	Molluscs	—	286
	Fish	0·1–5	2·55
Mo	Plants	12–42	23
	Molluscs	11–27	17
	Crustacea	8·9–17·3	13
	Fish	7·6–23·8	17

Table 7.2. *(Cont.)*

Element	Group	C.F. range	Mean C.F.
Mu	Plants	15–2,000	448
	Molluscs	1–3·6	2·2
	Crustacea	1–100	38
	Fish	0·4–26	6·6
Ar–Nb	Plants	170–2,900	1,119
	Molluscs	8–165	81
	Crustacea	1–100	51
	Fish	0·05–247	86

From Eisenbud, 1973.

Table 7.3. *Reported concentration factors (C.F.) for various classes of freshwater organisms*

Element	Group	C.F. range	Mean C.F.
Cs	Plants	80–4,000	907
	Fish	120–22,000	3,680
Sr	Plants	80–410	200
	Fish	0·85–90	14
Mn	Plants	1,300–600,000	150,000
	Molluscs	1,100–1,600,000	~300,000
	Crustacea	1,700–250,000	125,000
	Fish	0·1–400	81
Co	Plants	300–30,000	6,760
	Molluscs	300–85,000	32,408
	Crustacea	—	—
	Fish	60–3,450	1,615
Zn	Plants	140–15,000	3,155
	Molluscs	30–140,000	33,544
	Crustacea	300–4,000	1,800
	Fish	10–7,600	1,744
Fe	Plants	40–45,000	6,675
	Molluscs	20–80,000	25,170
	Crustacea	60–1,800	930
	Fish	0·1–1,225	191
I	Plants	10–200	69
	Molluscs	60–1,000	320
	Crustacea	—	—
	Fish	0·5–25	9
Ce	Plants	200–35,000	3,180
	Molluscs	400–1,500	1,100
	Crustacea	300–1,000	600
	Fish	2–160	81
K	Fish	340–18,000	4,400
Ca	Plants	64–720	350
	Fish	0·5–470	70

From Eisenbud, 1973, after various authors

(ii) Concentration factors are generally greater in freshwater organisms than in marine organisms. The reason for this is that mineral concentrations generally are lower in fresh-water than in sea-water, including those of the radio-nuclides. The difference between the water and cell concentrations is greater (Eisenbud, 1973). The wide differences between species and groups may also have been due to experimental procedures and timing, as it is known that an organism may show a high C.F. initially, but this declines when the animal's intake and output of minerals equilibrates (Eisenbud, 1973). Temperature may also influence bio-accumulation rates, though algae are less affected than higher organisms and several species showed inverse relationships between temperatures and absorption (Harvey, 1974).

A number of known critical pathways to man have been studied (Preston, 1971; Hetherington, 1976; Mitchell, N., 1977a,b).

Perhaps one of the most well-known and publicized pathways is through the littoral sea-weed *Porphyra umbilicalis*, which is eaten as laverbread and is well established in the diet of groups of people in South Wales (U.K.) (Preston, 1971). A major collection area for this weed was, at one time, close to the discharge from the Windscale Nuclear Fuel Processing Plant in Cumbria, and it was found that the consumers in South Wales were receiving doses of radio-nuclides originating from the laverbread collected at Windscale. Although the doses were only a fraction of the recommended I.C.R.P. limits, monitoring of the weed concentrations and doses to consumers was instigated and still continues, though the laverbread industry has declined (Mitchell, N., 1977b). Consumption of fish and shellfish in the Windscale area is monitored and the most recent estimates show that the average consumer would absorb only 7 per cent of the I.C.R.P. recommended dose limits (Mitchell, N., 1977a).

At Trawsfynydd Nuclear Power Station, which discharges effluent into an enclosed, oligotrophic freshwater lake in Wales, consumption of trout caught from the lake was higher than that of fish at marine sites, but the dose level to consumers was still only 7 per cent of the I.C.R.P. dose limit (Mitchell, N., 1977b). ^{137}Cs levels were higher than those of most other radio-nuclides.

Data from all the other power station discharges in the U.K. show that at all sites traces of radio-active materials are found in some organisms, but at most these are barely detectable. At the marine and estuarine sites where dispersion and dilution are rapid, total exposure to man from consumption of shellfish or crustaceans has rarely exceeded 0·1 per cent of the I.C.R.P. dose limit for radio-caesium and most other elements (Mitchell, N., 1977a,b).

The significance of bio-accumulation of any element in a food organism lies in the tissue in which it accumulates. For example, if a radio-nuclide accumulates in the muscle of a fish it is much more likely to be consumed than if it accumulates in the liver, kidney, or other non-consumable organ. Polikarpov (1966) found that in the crab *Carcinus maenas* (not usually consumed) ^{137}Cs was most concentrated in the muscle, but the total burden of ^{137}Cs was greater in the shell. Clams, oysters, scallops, and certain crabs store

^{90}Sr in shells which are not consumed, but ^{65}Zn and ^{60}Co are accumulated in the soft, edible tissues (Preston, 1971; Eisenbud, 1973; Rice and Baptist, 1974). The freshwater lamellibranch *Anodonta piscinalis* was found to accumulate radio-nuclides in both shell and body tissues, though the concentrations were found to vary over a period of 125 days, probably owing to seasonal and physiological changes (Garder and Skulberg, 1965). Where migratory or highly mobile species of fish absorb radio-nuclides, the area of influence is extended as the fish disperse. Thus, fish containing radio-active materials picked up near Windscale are found dispersed over many kilometres in the Irish Sea (Mitchell, N., 1977a). Similarly, juvenile salmonids in the Columbia may transport any absorbed radio-activity to sea. Experiments with stocked Chinook salmon smolts irradiated daily during hatching and the alevin stage suggested, however, that there were no obvious effects on returning adults. (Donaldson—cited in Rice and Baptist, 1974). Chronic irradiation of Chinook salmon eggs and alevins did not affect their development or adult return to rivers until doses exceeded 2580 C/kg day. Above this, growth was retarded and adult returns lower (Hershberger, *et al.*, 1978).

Most evidence of bio-accumulation has been in the vicinities of nuclear fuel processing plants and industries producing nuclear weapons (Eisenbud, 1973; Rice and Baptist, 1974; Mitchell, N., 1977a,b).

To summarize, therefore, although this section has only given a general outline of the mass of data in existence, the indications are clear, namely that the discharge of radio-active materials from the nuclear power industry is insufficient to: (a) to cause effects on the biota *in situ*, and, (b) even given large potential accumulation factors, to result in levels of radio-activity in edible organisms which will approach the recommended dose limits set by the I.C.R.P. for man.

Evidence is, in fact, that as far as the biota are concerned, temperature and chlorine may be more significant environmental factors. We are, however, left with the remotely possible very long-term effects of elements which will persist in organisms, and which may have genetic consequencies though how these will be measured is a major problem.

7.10. SUMMARY

There is a wide variety of chemicals discharged from power stations though most are in relatively small amounts. In the short-term there appears to be little danger of acute toxicity from most substances, with the exception of chlorine which may have already been responsible for fish mortalities at several sites. The area in which knowledge is lacking is in the chronic long-term effects of a low level of chlorine compounds, heavy metals, and radio-nuclides. Put into perspective against other industries, power station contaminants are of little over-all significance. The awareness of radio-nuclide accumulation and the massive programmes of monitoring are likely to forewarn of possible dangers well in advance.

Chapter 8 Beneficial aspects of ecological changes in waters affected by electricity generation

8.1. INTRODUCTION

Earlier chapters have described the ecological changes which have occurred in natural water-bodies used for power generation. While habitats for some species have been at least partially destroyed, particularly in hydro-electric developments, other species have benefited by the changes. The massive lakes in the tropics, created by damming river reaches have become huge, productive commercial fisheries, even though dominant species have changed (Lowe-McConnell, 1966; Ackerman, et al., 1973; Beadle, 1974; Mordukhai-Boltovskoi, 1979). However, Balon (1978) has questioned the over-all advantage when compared with the disadvantages of the ecological and social disturbance. Impoundment of fast-flowing rivers in North America has opened up large reservoirs for recreational fishing and other water-based activities and made water space available for much larger numbers of people than previously (O.R.R.C., 1962).

In various parts of the world, attempts are being made to use the heat discharged from power stations for domestic heating, horticulture, and aquaculture (Balligand, et al., 1975; Lee and Sengupta, 1977).

8.2. FISHERIES IN HYDRO-ELECTRICITY RESERVOIRS

Many of the reservoirs created for hydro-electricity generation have multiple uses including potable water storage, navigation control, and commercial and recreational fisheries. The merits and demerits of the new fisheries, however, may depend upon the social advantages and disadvantages for the local population. For example, in tropical Africa, riverside tribes may have evolved their annual food and agriculture programmes to match the natural river conditions. A newly created lake may extend and improve the fishery but the reduction of the annual flooding and silt deposition may completely destroy the previous agricultural system. Also, fishermen have to adapt and develop methods and techniques to exploit the different type of fishery in the lake (Hammerton, 1972; Ackermann, et al., 1973; Rzoska, 1976; Mordukhai-Boltovskoi, 1979).

In North America, Britain, and Europe, the development of large lakes has opened up sport fishing, using both residential and stocked fishes, to many less

238

affluent anglers (O.R.R.C., 1962). At the same time, however, migratory fish populations exploited either for sporting or commercial purposes have become seriously depleted owing to the blocking of spawning runs by dams, destruction of spawning gravels, and mortalities in lakes and turbines (see chapter 4).

8.2.1. Commercial fisheries

Probably the best documented studies of commercial fishery development are from man-made lakes of Africa and the U.S.S.R. (see Lowe-McConnell, 1966; Hammerton, 1972; Ackermann, et al., 1973; Mordukhai-Boltovskoi, 1979), though some commercial fishing occurs in North American impoundments (Jester, 1971). In the U.S.S.R., reservoir fishes account for a large proportion of the total consumed. Of the larger African lakes created for hydro-electricity generation, four, namely Kariba, Volta, Kainji, and Aswan, occupy in total over 2,800 km² territory. In these and many others, commercial fisheries have grown up since the early 1960s to provide important food supplies, some of which are exported to other countries (Lowe-McConnell, 1966).

Harding (1966) described the development of fisheries on Lake Kariba over a period of almost 20 years. Before impoundment, most riverine fish species bred when the river was in a rising flood condition. Juveniles utilized the flooded banks for feeding and protection, and in these early stages growth was rapid. However, during low-water periods when the river was reduced to a narrow channel, growth was slower and mortalities high because of over-crowding. The voracious tiger fish *Hydrocyon vittatus* was the main predator, although this was also one of the commercially exploitable species (see Ackermann, et al., 1973). The change from riverine to lacustrine conditions, the clearance of large tracts of land before flooding, and the absence of violent water level fluctuations, benefited the whole fish populations. Predation decreased and a higher survival rate of juveniles resulted. Catches increased markedly in 1959 and 1960, but then gradually fell as commercial fishing developed. On the north bank of Kariba, fish production rose from 500 tonnes in 1960 to 3,500 tonnes in 1963, all caught by Tonga tribesmen. Harding (1966) suggested, however, that yields could decline as the lake equilibrates (see chapters 2, 3, and 4).

The lake fishery was based mainly on three species, *Labeo congoro, L. altivelis,* and *Tilapia mossambica,* and although 7 tonnes of fingerlings of *Tilapia macrochir,* a commercial species in other lakes, were introduced to Kariba in 1962, the success of the species was not obvious. The explosive growth of the water fern *Salvinia auriculata* was a serious hindrance to fishing at first but, once the lake was full, the mats of weed broke up as a result of wind and wave action (see chapter 4).

High productivity in the early stages of a new lake development is very typical of tropical and temperate impoundments (Jackson, 1966) and the populations can take some years to stabilize. However, while the standing crops of lacustrine and static-water fishes such as cyprinids, catfishes, centrarchids, cichlids, and others may increase, migratory species such as eels and salmonids may decline and the potential yield may also be reduced by the normal drawdown periods in hydro-reservoirs (see chapter 4).

In North America, commercial fishing for coarse (rough) fishes is used in impoundments as a means of control, to reduce competition with the game and high-quality sport fishes. Jester (1971) described the effects of commercial operations on the smallmouth buffalo (*Ictiobus bubalus*); river carpsucker (*Carpiodes carpio*); and carp (*Cyprinus carpio*) of Elephant Butte reservoir in New Mexico. In 6 years, 96,674 smallmouth buffalo, weighing 97,764 kg were harvested. The actual reduction of smallmouth buffalo over the last 3 years of this work represented over 14 per cent of the population of nettable fish of that species. Jester's paper does not actually show how the harvested fish were utilized, whether for direct human consumption or for fishmeal processing.

The dramatic changes in the Volga as a result of extensive impoundment also resulted in the expected change from riverine to limnophilous fishes. Unlike Western Europe, the cyprinid and other coarse fishes are heavily utilized as food in the U.S.S.R. Thus, although sturgeon catches in the impounded reaches fell from 27×10^3 tonnes in 1910 to 4–6×10^3 tonnes in 1977, the catches of bream, roach, pike, and other limnophilous species rose sharply by up to 10 times. Stocks of most species are augmented by introductions from farms and hatcheries and new potentially commercial species, such as the 'peled' *Coregonus lavaretus* are introduced and acclimatized. Various papers in Mordukhai-Boltovskoi (1979) describe the lakes, the fish, and fisheries in detail.

There are probably still many man-made lakes which have highly productive fisheries with commercially viable populations but have, for some reason, not been exploited. Riggs (1958) discussed the potentials and showed that as long ago as 1953, 3·4 per cent of the total commercial fish catch in the U.S.A. came from fresh waters, representing 7·8 per cent of the monetary value for the country. Since that year many large reservoirs have been formed by the damming of the larger rivers (Stroud, 1966; Hall, 1971).

Riggs also quotes standing crops for cold-water reservoirs as about 100 kg ha^{-1} and he estimated that at a cropping level of 30 per cent the total potential catch from all reservoirs was over 150 million kg consisting of sport fishes, game fishes, and coarse (rough) fish.

The percentage of the standing crop that can be harvested annually without deleterious effects on the fishery depends on species composition, and the resilience of each species to cropping.

In the Tennessee Valley area, standing crops of 217 kg ha^{-1} have been recorded in storage reservoirs and up to 367 kg ha^{-1} in mainstream impoundments. Commercial food and coarse fish accounted for 27 per cent, forage fish 57 per cent, 'pan' fish 10 per cent, and larger sport fish 6 per cent. Working on figures published in 1962 (O.R.R.C., 1962), the potential commercially harvestable populations of fish in T.V.A. reservoirs alone, could amount to up to 2.2×10^6 kg per annum, and these reservoirs were then only a small proportion of those available now (Stroud, 1966). Introduction of fishes to utilize new lacustrine food resources such as phytoplankton and zooplankton may fill niches and increase total fish production markedly. For example the 4,000 ha Hiwassee Reservoir in North Carolina produced much lower yields than the Norris Reservoir in Tennessee. Introduction of a plankton-feeding pelagic species (*Dorosoma cepedianum*) filled a gap and production rose sharply (Hall, 1971). In Russia, where there is, according to Rzoska (1966) a strong emphasis on exploitable fish production, fertilization has been used to stimulate phytoplankton in reservoirs with falling yields, plankti-verous fishes have been stocked and 'intermediate' food organisms such as mysids, snails, and gammarids have successfully been introduced (Borodich and Havlena, 1973; see Mordukhai-Boltovskoi, 1979). By 1968 reservoir catch in the U.S.S.R. was 44×10^3 tonnes and the aim is for 170×10^3 tonnes eventually (Lowe-McConnell, 1973).

8.2.2. Recreational fisheries

While commercial fisheries and food resources are the main benefits to the generally poorer people of the tropics, in the more affluent regions, for example, North America, Britain, and Europe, impoundments have become a major recreational fishing resource.

In 1960 there were already more surface acres of man-made reservoirs in the U.S.A. (including those of 4.0 ha or less, up to reservoirs of over 200 surface ha) than of natural lakes, excluding the Great Lakes and those in Alaska. Of the 5.2 million ha of man-made lakes, 3.0 million were in large reservoirs and, of these, almost 2.7 million were in reservoirs used for multiple purposes, for example hydro-electricity, water-storage, flood control, recreation, or a combination of some of these. Of all the recreational pursuits on these waters, fishing was by far the most popular (O.R.R.C., 1962).

In 1963, fishing effort totalled 96 million man-days on large reservoirs/impoundments, i.e. over 25 man-days ha^{-1}, comprising 25 per cent of all angling effort in U.S.A. fresh waters. By 1976, an increase of 75 million man-days (85 per cent) was forecast for these reservoirs (Stroud, 1966). In 1960, average angler harvests from large reservoirs were about 16 kg ha^{-1} in warm waters, and 9.0 kg ha^{-1} in cold waters. The catch per angler/day was about 0.6 kg and 0.45 kg respectively. An increase of about 30 per cent in harvest was anticipated as necessary to meet the 1976 demands.

Examples of the benefits to the ordinary angler are illustrated by the increases in angling pressure on reservoirs in the U.S.A. Since Norris Dam was completed in Tennessee during 1935, there has been considerable development of dams in the Tennessee River System, and the O.R.R.C. Report (1962) study states that 'reservoirs administered by the Tennessee Valley Authority, have developed into a sport-fishing playground second only to the Great Lakes, among inland fishing waters'. Between 1935 and 1960, reservoir area in this region increased from 46×10^3 ha to 240×10^3 ha with a combined shoreline of more than 16×10^3 km. Some 6·5 million man-days were spent on T.V.A. reservoirs as long ago as 1960, and considerable increases have occurred since that time. In Oklahoma, as another example, the state Game and Fish Department estimated that there were some 5·5 million man-days of fishing spent on reservoirs of all types in 1960. The total catch from 175×10^3 ha was estimated at $5·4 \times 10^6$ kg and the average catch as 30 kg per surface ha per year, i.e. about 1·1 kg per angler/day. Turner (1971) found that fishing effort increased from 28 man-h day^{-1} to an average of 1,058 man-h day^{-1} when the Rough River, in Kentucky, was impounded. Annual yield also rose from about 410 kg to 28×10^3 kg over a longer season, and the catch per man-h rose from 0·07 to 0·12 kg though this declined slightly some 4 years after impoundment. The reservoir was, however, stocked with fishes over this period. Similar figures may be provided for other areas, and they demonstrate very readily the vast over-all contribution and benefits which have accrued in fisheries formed by impoundments. An analysis of the variables which affected angler-crop in U.S.A. reservoirs showed that reservoir area, length of growing season and age of impoundment had the most significant influence of the ten variables tested. Actual standing crop was, however, related to other factors, including length of shoreline, depth of outlets, and total dissolved solids, none of which were correlated with angler-harvest (Jenkins and Morais, 1971). In mainland Europe, the same kind of trend has occurred.

In Scotland the lakes created upstream of hydro-dams are highly utilized for trout fishing and many are managed on a 'put-and-take' basis. In the Central Electricity Generating Board area in the U.K. the only large-scale fishery upstream of a hydro-electric power station is at Nant-y-moch and Dinas in the Rheidol scheme in mid-Wales. The lake is naturally oligotrophic and only small indigenous trout are present. Stocking with almost catchable sized trout, however, has attracted large numbers of fly-fishermen. Since the lake opened in 1962, the annual total has increased from about 100 to over 5,000 angler/days. More recently, the lower lake in the Ffestiniog pumped-storage scheme has also been developed as a 'put-and-take' fishery, and the lake at Trawsfynydd, used both for cooling the nuclear power station *and* as the water source for the Maentwrog hydro-station, has been a noted trout fishery for many years. In the past 5 years, large populations of perch (*Perca fluviatilis*), rudd (*Scardinius erythrophthalmus*), and minnows (*Phoxinus phoxinus*) have also developed in the lake almost to nuisance proportions. The total productivity of the lake and angler access is, however, greatly increased over that of the small stream which occupied the valley previously.

8.3. TAIL-RACE FISHING

The attraction of fishes to the tail-races of hydro-electric power stations has caused these reaches to be very attractive to anglers. In most British rivers, angling is restricted immediately below hydro-dams owing to the dangers to anglers of sudden rises in water level and velocity. In the larger rivers of the U.S.A., however, this is not always the case, and tail-races are very popular venues (O.R.R.C., 1962; see Hall, 1971). On nine tail-race waters in the T.V.A. region some 724,000 anglers were counted in 1960, an average use of 150 anglers ha year^{-1}. The author has observed tail-race-fishing by boat in the T.V.A. area, below Wilson Dam. Anglers drive upstream in high-powered boats to within a few metres of the outlet grid. A grapnel attached to a line is then flung at the grid and the boat moored in the fast water while the anglers fish. Although this may be illegal in some waters the method can be much more rewarding than fishing from the bank. Many species of fish are found in tail-race populations, usually reservoir species which arrive from upstream via turbine outlets together with upstream migrants attracted to the flow (see chapter 4).

8.4. OTHER BENEFITS

At most British hydro-reservoirs fishing is the major recreation. Sailing, swimming, water-skiing, and other similar pursuits have not been generally encouraged except in selected parts of reservoirs, though the pressure for these pursuits is increasing.

In the U.S.A. and other countries, although fishing is the major recreational pursuit, swimming, boating, canoeing, and water-skiing are all allowed and encouraged on most major impoundments again with restrictions to certain areas in many cases. Because of these activities reservoirs become major holiday attractions, especially in the middle of continents where sea-based water sports are relatively inaccessible.

The depth of man-made lakes facilitates water transport and in large rivers dams often have huge locks for access by shipping. The increased transport facilities also encourage the distribution of the commercial fish catches, often by large refrigerated boats in tropical regions.

8.5. USES OF THERMAL DISCHARGES

8.5.1. Aquaculture

The fact that most aquatic animals will, given enough food, grow better at higher temperatures than at low temperatures is well established (see chapter 5). The cooling-water discharged from power stations is not always of sufficient quality to be used directly for fish culture as it may originate from polluted rivers or polluted tidal waters. Therefore, the selection of any power station site for aquaculture projects can depend very much on the condition of the river, lake, estuary, or coastal water used for a cooling-water source (Sylvester, 1975b). Chlorine or other biocide dosage may also limit scope for direct use. However, there are, at the present time, at least thirty power stations throughout the world at which pilot scale or commercial aquaculture facilities are being developed. Twenty of these use fresh water (Aston and Milner, 1976; P.S.E.G., 1977; Hooper, *et al.*, 1978). Fishes, shellfish, and crustaceans are the major organisms being cultured. As long ago as 1960, Iles surveyed the quality of water used by the C.E.G.B. and some Scottish power stations and recommended several as sites for possible aquaculture projects. At Grove Road Power Station in London, carp (*Cyrpinus carpio*), grass carp (*Ctenopharyngodon idella*), and tilapia (*Tilapia zilli*) were kept in cooling-water ponds and their growth monitored (Iles, 1963). The grass carp and tilapia did not show very encouraging growth-rates and ordinary carp grew best, i.e. from about 65 to 500 g in 16 months. From about 1963 to 1973, little work on fish culture was carried out by the C.E.G.B., though at Hunterston Nuclear Power Station in Scotland, and at Carmarthen Bay in South Wales, marine fishes (mainly flatfishes) were being grown in the cooling-water discharges (Kerr, N., 1976). Nash (1968, 1969, 1970) and Shelbourne (1970) have given details of the experimental farming of plaice (*Pleuronectes platessa*), turbot (*Scophthalmus maximus*), and Dover sole (*Solea solea*), at Hunterston Power Station. In the mid 1960s large-scale trials with plaice and halibut (*Hippoglossus hippoglossus*) together with the other two species were begun. At effluent temperatures of 10–23°C, 9-month-old plaice and sole grew on to marketable size of about 25–27 cm in 12 months. Under natural temperature regimes the 'growing-on' period would have been twice as long. When finally costed in 1972, it was found that neither plaice nor halibut had any economic future for warm-water production. For turbot and dover sole, however, which are both high-priced, high-quality fishes, there were still commercial possibilities.

Turbot were stocked in tanks at a density of about 22 kg m^{-3} with a flow rate through tanks of 0·04 m^3 h for each 1 kg of live fish. Oxygen levels were maintained above 5 p.p.m., though moderate aeration was used. From egg to the marketable size of 500 g, a turbot required about 550 days, and about 91·4 m^3 of water in total (Kerr, N., 1976).

For each species there are, of course, optimal temperature requirements for growth and food conversion. For turbot the *tolerable* range is 6–19°C with a short-term maximum of 23°C. *Optimal* range is 16–18°C. For Dover sole, the tolerable range is 10–21°C, short-term maximum 24°C and desirable range 18–20°C (Kerr, N., 1976). Hunterston 'A' produced a ΔT of 8°C in the cooling water discharge. The newer power station Hunterston 'B', runs at a ΔT of 14°C and Fig. 8.1 shows the period during which the desirable temperature ranges for both turbot and Dover sole are reached under normal operating

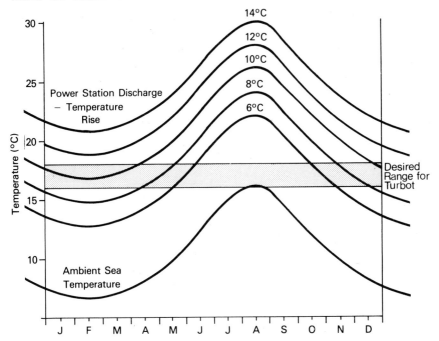

Fig. 8.1. The effect of a power station discharge on the temperature required for optimum growth for turbot (Hunterston Power Station), redrawn after Kerr (1976).

conditions. Optimal temperatures cannot be maintained all the year round using only heated water, and thus in summer abient sea-water is pumped in to cool the rearing tanks.

Chlorine has been a limiting factor in the maintenance of the farming potential at Hunterston (see 8.5.2.b), though vigorous aeration and cascades, together with the natural chlorine-demand of the water, have reduced the 0·5 p.p.m. intake concentration to a residual of about 0·02 p.p.m. in the fish tanks. Once past the juvenile stage, flatfish will tolerate about 0·05 mg l^{-1} continuously and up to 0·2 mg l^{-1} for periods up to 4 h. The very young fish are kept in clean sea-water, and heat is provided by passing the discharge water through a heat-exchanger, though at best a 3–5°C rise is achieved in the rearing tanks. At times when other chemicals are injected into the power station cooling-water system, i.e. after boiler-cleaning, etc., the farm hot-water supply has to be discontinued.

Another hazard at power station fish farms is the sudden shut down at feeder power stations owing to emergency breakdown or a fall in demand. At such times discharge temperatures may fall by 8–14°C. For eurythermal species this may only result in a slowing of growth-rate, but if warm-water species are being cultured heavy mortalities could occur.

Two species have been found to have considerable growth potential at freshwater (inland) sites in Britain (Aston and Brown, 1975; Aston and Milner, 1976). They carried out a series of thorough experiments on the growth of eels (*Anguilla anguilla*) and carp (*Cyprinus carpio*) in tanks fed from cooling-water system at the 2,000 MW Ratcliffe-on-Soar Power Station which uses the polluted River Trent as a cooling-water source. The power station incorporates a closed cycle cooling system (see chapter 2). Water was supplied to the tanks from three sources:

(i) Make-up (river water) at ambient temperature. Range 5–25°C.

(ii) Condenser water, i.e. direct from the condenser outlets. Range 15–35°C.

(iii) Cooling-pond water, i.e. after passing through cooling-towers. Range 10–27°C.

Figure 8.2 shows the mean growth rate of eels in the tanks and Fig. 8.3 shows similar data for carp. By far the fastest growth was found in the hottest water for both species. Food was supplied at about 10 per cent of fish weight per day, and was usually in excess. Although highest mortalities among eels occurred in the condenser water mainly from predation, this group still produced the greatest biomass after 11 or 12 months. Eels grew from a mean weight of about 1 g up to 25 g (mean) over a 12-month period, though the size range in condenser water tanks at the end of the experiment was very wide. Similar ranges were shown by Descamps, *et al.* (1976). To optimize the system, continuous grading would need to be used, sorting eels into larger and smaller size groups. At best, a marketable eel would take about 2–2½ years to grow, in comparison to 10–14 years in the wild.

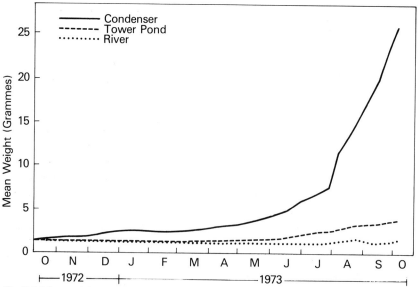

Fig. 8.2. Mean weights of eels in condenser, tower pond, and river tanks from October 1972 to October 1973, Ratcliffe-on-Soar Power Station, U.K., redrawn after Aston and Brown (1975).

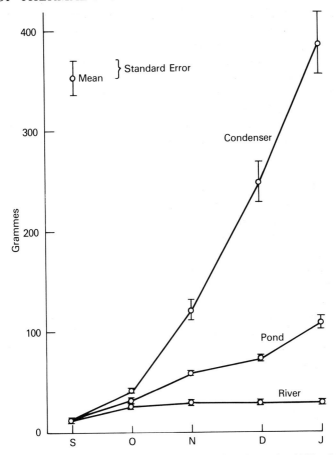

Fig. 8.3. Carp weights in condenser, pond and river-water from September 1974 to 3 January 1975. Ratcliffe-on-Soar Power Station, redrawn after Aston and Milner (1976).

Carp (*Cyprinus carpio*)—also grew extremely well in condenser water, averaging 13 g at the beginning of the work and 958 g (range 268–2,052 g) after 10 months. In river water at more or less ambient temperatures, the fish averaged only 172 g (range 79–325 g) after 10 months (Fig. 8.3). As a result of this work, and studies in other countries, a commercial scale eel farm is now in operation at Drax Power Station in Yorkshire, with a proposed initial production of 200 tonnes.

Recent experiments, using a relatively constant temperature of about 25°C obtained by balancing condenser- and pond-water flows, has shown that both growth and food conversion in carp are enhanced by this modification (Aston and Macqueen, 1978). Little success has been achieved with other freshwater species, though experiments on improved growth and hatching of salmonids have been attempted (Bulleid, 1974). Lawrie (1976) described the first year's trials using a heated-water installation (not from a power station) to rear Atlantic salmon smolts. A maintained temperature of 10°C was supplied to eyed-ova (i.e. about 4–5°C above ambient for mild winters).

On first feeding the temperature was raised to 12–14°C. Once ambients reached 9°C, heating was discontinued. In 1975, 65 per cent of fish reached the 1-year smolt stage, instead of the normal 24 per cent at ambient temperatures. Normally fish are released to the river to migrate at the smolt stage. Both Lawrie (1976) and Bulleid (1974) show advances in egg-hatching, growth, and smolt development using heated water, but in both cases, the 'growth' advance was mainly a result of the earlier hatching, and actual rates of growth once feeding began were not much enhanced. Lawrie estimated the extra cost of heating the water was about 10p (£0·1 sterling) per smolt. Survival in heated water was better than at ambient.

In other countries, substantial progress has been made at some power station sites (Beall, *et al.*, 1977). In Table 8.1 are listed known species being cultured at inland power stations at present, including the freshwater prawn

Table 8.1. *Fish species cultured at inland power stations in various countries*

Country	Power station	Species cultured
U.S.A.	Gallatin	Catfish
	Colorado City Plant	Catfish
	Hanford	Catfish
	Fremont (Nebraska)	Catfish and *Tilapia*
	Lake Trinidad Plant	Catfish
	Hutchinson	Catfish
	Mercer	Trout and *Macrobrachium*
U.K.	Ratcliffe-on-Soar	Carp and Eels
	Trawsfynydd	Trout
	Ironbridge	Carp (farm under construction)
France	Cadarache	Eels
Netherlands	Flevo	Carp, grass carp, and trout
E. Germany	Finkenheerd	Carp
	Rheinberg	Carp, eels, and trout
Poland	Konin	Carp, grass carp, silver carp, and big-head carp
Hungary	Szazamlombatta	Carp, grass carp, silver carp, and big-head carp
Soviet Union	Klasson	Grass carp and carp
	Krivorozh	Grass carp and carp
	Shaktinsk	*Tilapia*
	Kirishi	Carp and trout

Various authors; after Aston, *et al.*, 1976.

Macrobrachium rosenbergii. Korneev (1976) also mentions the culture of sturgeon, a hybrid sturgeon and sterlet (bester), eels, and trout in the U.S.S.R. He quotes a figure of 72×10^3 ha of cooling-ponds in the U.S.S.R., used for thermally enhanced fish culture, and proposed production of 600–800 tonnes per ha as a realistic potential. Carp production of 150–186 kg m^3 of pond/tank, volume is already being achieved, providing marketable fish weighing about 450–500 g. Commercial production of caviare from sturgeon reared in power station cooling-water is also forecast.

Japanese projects include using waste heat for the culture of eels, shrimps, sea-bream, and white fishes (Yee, 1972) and some commercial benefit is claimed for eels and shrimps.

As yet, the commercial benefit of heated water over natural waters has not been universally proved, though it is claimed that commercially viable farms are in operation at Mason Power Station (Maine) and at the Lake Colorado City Plant (Aston, 1976; Beall, *et al.*, 1977).

At Northport, a marine site, an oyster-culture project was begun in the 1960s. A two-phase system is used. First the oysters are reared in tanks from eggs to tiny larvae which are then fed on phytoplankton. Secondly, the tiny oysters are 'planted' out on mesh trays in the effluent lagoon at the power station. They grow quickly for about 6–12 weeks and are then 'transplanted' to open water in Long Island Sound for their final growth. They are cropped after about 2 years. The rapid growth and grading in the warm-water lagoon is claimed to be of considerable advantage.

ΔTs are around $14\cdot5°C$, superimposed on ambients from -1 to $24°C$. In summer, a supply of diluting water at ambient temperature is pumped into the effluent lagoon to avoid exceeding the lethal temperature. Commercial sales amounting to $5 million dollars were achieved in 1977 (Beall, *et al.*, 1977). As in some other U.S.A. sites used for fish culture, antifouling procedures within the power plant are totally mechanical, using brushes and sponge balls under pressure for condenser tube cleaning, so there are no problems with biocides (Aston, 1976).

On the negative side, the growth of American lobsters (*Homarus americanus*), from eggs to 450 g over 3 years in the effluent from the Ecina Power Plant owned by the San Diego Power and Light Co., was considered too slow for commercial farming development.

At the Gallatin Power Plant of the T.V.A. and at the Lake Colorado City Power Plant in Texas, channel catfish (*Ictalurus punctatus*) are being successfully cultured, though again the commercial viability of the system is not fully clarified.

At the Gallatin Plant, raceways, i.e. long-channels with a rapid through-flow, are used to culture the catfish, though raceway culture is probably more expensive than pond or cage/effluent systems, both in capital outlay and maintenance costs.

At Lake Colorado, the fish are reared in cages, floating in the effluent discharge canal, mainly during winter at temperatures of 21–25°C, and reported yields are about 250 tonnes ha^{-1} yr^{-1}, with intensive feeding. A similar system is used to rear brook trout (*Salvelinus fontinalis*) and lake trout (*S. namaycush*) at a Canadian Power Plant (Hooper, *et al.*, 1978). Because different species have different temperature tolerances and optimal growth ranges, systems have been developed which will culture more than one species throughout a year.

Thus, at the Mercer Power Station, New Jersey, a pilot farm is geared to produce the freshwater prawn *Macrobrachium rosenbergii* in the summer months when effluent temperatures are highest, and rainbow trout (*Salmo gairdneri*) during the winter (Eble, *et al.*, 1975). Until the present time most effort has been put into *Macrobrachium* culture. Eggs are hatched and juveniles reared in vessels in two large greenhouses. Juveniles are then stocked either into a raceway or a pond for growing on. In June 1974, juveniles stocked at a density of 11 per m^2 and mean length of 2 cm, increased in length by an average of 1·6 cm per month until October.

Juveniles stocked in May 1975 at a mean length of 1·3 cm and densities of 54 per m^2 and 28 per m^2 increased by an average of 1·3 cm per month. These densities are still well below those required for commercial production, and experimental work on more intensive culture at high densities is in progress (P.S.E.G., 1977).

The Mercer Power Plant is chlorinated intermittently and supplies of effluent water to the pilot farm are cut off automatically during the injection period. At the Flevo Power Station in Holland, a double-cropping system also operates. Trout are reared in discharge water in winter and carp and eels in summer (Koops, 1976).

The major running cost for fast-growing fish and invertebrates is for food, usually of high quality and high protein content. To reduce such costs, polyculture systems may be used in which plankton or invertebrates are grown in high nutrient media such as sewage sludge or liquids. These organisms then provide food for the cultured species or intermediate prey species. In Israel, where polyculture systems are operated at ambient temperatures, yields as high as 8,000 kg ha^{-1} have been recorded (Beall, *et al.*, 1977).

Instead of organic wastes, cold, nutrient-rich sea-water abstracted from oceanic depths and used for cooling in a power station, may be seeded with phytoplankton cultures and tapped off into shallow ponds, where the subsequent algal blooms can be utilized as in other polyculture systems (Gundersen and Bienfang, 1972).

Although no *fully* costed breakdown of the advantages of warm-water aquaculture has yet been published, there are a number of experimental projects in progress in the U.S.A. and Western Europe which show commercial possibilities. In Eastern Europe and in Japan, the commercial viability of thermal-discharge aquaculture is apparently already accepted, usually because different species are acceptable to markets.

In the U.K., main consumer demand has, to date, been for higher-quality marine fishes together with trout and salmon, but with the increase in demand for 'block' fishes, that is, processed frozen fish, fish-fingers, etc., freshwater species such as carp, catfishes, *tilapia*, grass carp, and marine species as bass, whiting, flounders, and others, may become acceptable and may be generally easier to culture on an intensive, large scale.

In other countries where carps, catfishes, and 'lower-quality' marine species are already acceptable, marketable food-fishes and the commercial potentials are more easily established. In the U.K. present consumer demand for carp and other cyprinids is mainly by anglers for stocking fishing pools and lakes, though carp are imported for consumption by certain ethnic or religious groups.

The eel, however, shows much better and immediate commercial potential. Aston and Milner (1976) assessed world consumption at about 50,000 tonnes per year, and increasing. 'Farming,' they state, 'is the only way to increase supplies, as wild stocks are being rapidly depleted by over-exploitation'. There is a vigorous international eel trade, which can be expected to increase.

8.5.2. Problems with aquaculture at power stations

(a) Temperature

One of the main problems with using cooling-water for other purposes, including aquaculture, is that it generally provides low-grade heat.

Thus, if the water has to be transported before use, either directly or in heat-exchanger systems, it may have cooled below required temperatures (Hugenin, 1976). Also, heat-exchangers do not work at all effectively when transferring heat into flowing systems. For instance, with a ΔT of 8°C, the ratio of water flow to be heated to that supplying heat has to be very low to produce a ΔT of even 4–5°C. In static systems, heat-exchangers may be more efficient.

To maintain temperatures over 25°C at direct-cooled sites in temperate regions, therefore, even in summer, farms need to be sited very near to power stations. At stations with cooling-towers, temperature maintenance may be much simpler although, as Aston and Milner (1976) have shown, there can be fluctuations of up to 10°C or more daily. There are, at such sites, three sources of water available (see 8.2), combinations of which may be used to help maintain more consistent temperature ranges throughout the year (Aston and Macqueen, 1978). The fluctuations shown in the intake temperature at Ratcliffe-on-Soar, i.e. up to 6°C daily (Aston and Milner, 1976), are unusual for a large British river, and are mainly a result of the warm-water discharge from Castle Donington Power Station, some 10 km upstream of the Ratcliffe intake. In a normal large river, ambients vary very little, almost always less than 2°C on any day (see chapter 2). While it is true that power stations discharge 'low-grade' heat, there are of course still problems with potentially 'lethal' temperatures in warmer regions, if effluents are used directly. Most fish and invertebrate species worth culturing will not tolerate long periods in captivity at temperatures over 35–36°C though they may survive for short periods (see chapter 6). Strawn (1970) found that only one species, viz. *Lagodon rhomboides* out of nine which were kept in cages in the effluent at a Texas power station showed good survival all year round. However, the 'double-cropping' or 'seasonal' culture systems may well avoid using species which may be at or near their thermal death point in summer. The use of intake water to cool discharge water, a procedure already used for environmental reasons at some power stations, may also partly answer the excess temperature problems in aquaculture systems. Apart from problems with the discharge temperature a power station may shut down either for emergency breakdown repairs, normal maintenance or because of demand reduction. Some shut-downs may be predicted but, whatever the cause, there is as yet no valid economic reason for keeping a power station running solely to maintain a fish farm. In the case of emergency shut-downs there is often little warning. Operations may cease and effluent temperatures fall in a few hours or less. As has already been shown (chapter 5), this sudden temperature shock may lead to fish mortalities in effluent channels in winter (Block, 1974; Cox and Coutant, 1976; Wolters and Coutant, 1976). In an enclosed fish farm the results may even be more drastic as no fish would be able to escape. Thus, for absolute safety the fish farm would need either an auxiliary heating system *or* the cultured species would need to be rapidly adaptable to temperature changes.

Carp and eels appear to fall into this latter category though trout do not.

(b) Chlorine

We have already seen that chlorine is used widely as a biocide to prevent culvert and condenser fouling, and that chlorine residuals are extremely toxic to many species of fishes and invertebrates (chapters 6 and 7).

At power station fish farms, chlorine residuals in the effluent may be critical. usually there is a need for either some method of de-chlorination or a chlorine detection system which cuts off the water supply. In marine systems the problems involved in detecting low levels of chlorine render automatic detection difficult (Carpenter, J., 1977).

Neither chlorine alarm systems nor operators are infallable, and any fish farm at a power station is faced with the possibility of a chlorine overdose caused by mechanical or human failure. Whatever the cause, mass mortalities of cultured organisms can result.

The toxicity of chlorine or its derivatives depends very much on other factors such as pH, temperature, time of exposure, and synergisms (see chapter 6). Although some species of fish and invertebrates may tolerate fairly high total halogen residuals, the most desirable power stations are where mechanical or physical antifouling procedures are used (see chapter 3).

At power stations with fish farms, various systems are used to prevent chlorine kills. At Hunterston (U.K.) and at the Mercer Pilot Farm (New Jersey, U.S.A.) automatic detection apparatus indicates chlorine levels and the flow from the power station is shut-off either manually or automatically (Nash, 1974; Aston, 1976). Page-Jones (1971) described the automatic alarm at Hunterston and noted the alarm threshold was set at about 0.3 mg l^{-1} though the instrument could detect as little as 0.03 mg l^{-1}.

Nash (1974) suggested that effects of residuals of $0.02–0.10$ mg l^{-1} may be negligible in fish tanks if the water exchange (flow-rate) is limited to less than five volume changes per day. He also stated that there still is an obvious need for basic research on methods of chlorine removal, including the use of de-chlorinating chemicals. For example, sodium thiosulphate injected at a $1:1$ ratio of thiosulphate to chlorine is a noted method though the effects of the thiosulphate on fishes are not yet known. Sulphur dioxide (SO_2) also reacts freely and instantaneously with both free and combined chlorine as follows:

1. $SO_2 + H_2O \rightarrow H_2SO_3 + HCl$
2. $NH_4Cl + H_2SO_3 + H_2O \rightarrow NH_4SO_4 + HCl$

The possibility of de-chlorination using power station flue-gases which contain a high percentage of SO_2 requires further investigation.

Seegert and Brooks (1978b) compared de-chlorination methods using activated carbon, ultra-violet light, and sodium sulphite. They eventually developed a method using a combination of carbon and sodium sulphite which they showed to be 'safe, effective, and reliable'.

Mechanical condenser cleaning methods involving sponge balls or brushes are used at several power stations supplying water to fish farms. At Northport, in the U.S.A., the 'Mann' system using rotating brushes is also used.

Aston (1976) states that at Northport, plant operation has evolved to the stage where the 'threat of any danger to the oysters from cleaning processes has been virtually eliminated'.

Even when chlorine residuals are low enough in themselves, there remains concern that synergisms will occur at high temperatures and/or with other contaminants such as heavy metals. It has already been noted that with pink and Chinook salmon (*Onchorhyncus gorbuscha* and *O. tshawystcha*), the toxicity of chlorine under experimental conditions increased as temperatures rose from 0 to $10.0°C$ (Stober and Hanson, 1974). At 0.1 p.p.m. residual chlorine, the time to death of Chinook salmon decreased from 10^5 min at $25°C$ to 10^2 min at $30°C$. Similarly that for pink salmon decreased from 8×10^3 min at $11°C$ to 1.1×10^2 min at $22°C$.

It has been suggested (Evins, 1975), that long-term exposure to 0.01 mg l^{-1} of free chlorine in fresh-water could be dangerous to salmonids. In freshwater fish farms at power stations exposure to highest levels of chlorine will generally be of short duration three times each day (Brown and Aston, 1975). The peak in fact may be sustained for only a few minutes.

(c) Heavy metals

The other major contaminants to which cultured organisms may be exposed in cooling-water are heavy metals, particularly copper, zinc, cadmium, iron, and nickel. The toxicity of heavy metals is discussed briefly in chapter 7 but, generally, levels discharged from power stations are not within critical toxic ranges for most fishes. However, as we have seen already, toxicity can be increased by both low dissolved oxygen levels and high temperatures. In Britain, Romeril (1974) and Davis, M. (1977), have published data on trace metal levels in organisms and sediments in the vicinity of power station cooling-water outfalls (see chapter 7). Romeril and Davis (1976) and Davis (personal communication) have also analysed tissues of eels and carp grown in the closed circuit cooling-system at Ratcliffe-on-Soar for trace metals (see also chapter 7). The River Trent, from which the cooling-water from Ratcliffe Power Station is abstracted contains high concentrations of some heavy metals (Trent River Authority, 1964–74). In the power station cooling-system soluble copper showed a threefold increase in concentration from that of the river water (Table 8.2). Cadmium and nickel also increased slightly, but iron showed a marked decrease. In the particulate phase, all metals showed increased levels, but this may have been due to an over-all increase of suspended solids in the cooling-water system. There was no evidence of concentration of metals in muscle tissues of the eels in the cooling-water (Table 8.2). The highest levels of zinc and nickel were found in eels kept in unheated Trent river water. Copper concentrations were lowest in eels reared in cooling-pond water. There was no evidence of bio-accumulation, even in this rather polluted system, and in fact the authors conclude that metal accumulation was slower than tissue growth, so that in fast growing eels, the metals are 'diluted'. Total 'body-burden' (i.e. the total weight of metal ion in the whole fish) may however have been high in some specimens. In this system,

Table 8.2. *Trace metal levels in condenser, cooling pond, and river water at Ratcliffe-on-Soar power station, 1975*

Date	Water sample	Soluble phase (μg l^{-1})					Particulate*				
		Cd	Cu	Fe	Ni	Zn	Cd	Cu	Fe	Ni	Zn
24 October	River	0·40	5·06	62·9	17·8	27·8	1·88	15·0	0·75	7·03	1·63
	Pond	0·82	16·3	27·5	25·1	31·4	4·88	43·9	2·72	33·1	4·42
	Condenser	0·78	14·9	28·5	26·5	39·4	5·10	33·2	2·30	27·8	4·10
25 October	River†	0·55	6·45	46·8	18·7	32·6	4·49	18·3	0·77	9·52	2·63
	River	0·29	5·20	59·1	19·2	27·4	3·20	13·0	0·76	9·10	2·80
	Pond	1·55	15·3	31·0	25·6	23·0	6·40	39·2	2·32	27·9	6·76
	Condenser	0·54	14·8	38·0	26·4	16·0	7·71	37·1	2·36	32·8	6·74

* Expressed in μg l^{-1} for Cd, Cu, and Ni and in mg l^{-1} for Fe and Zn.
† Taken from the river near the intake. All other samples collected at experimental rig.
From Romeril and Davis, 1976.

as in most culture systems, eels were fed on artificial diets and there is thus less chance of accumulation of contaminants through other organisms in the food-chain.

None of the cadmium concentrations in the muscle of the Ratcliffe eels approached levels 'likely to constitute a public health hazard' and comparison with eels from other localities showed that they were much lower than in some other 'wild' stock. A recent analysis on carp (*Cyprinus carpio*) grown in the Ratcliffe cooling system also showed that there was little evidence of accumulation of copper in muscle tissues though levels in the water were at the time quite high (Davis, personal communication). Molluscs are generally more efficient than fishes at accumulating heavy metals and many species show selective accumulation of metal ions in tissues as we have already seen (see chapter 7). Thus the culture of shellfish in general needs to be in uncontaminated water, though it has been shown that if the animals are allowed to purge themselves in clean water for at least 24 h, some metal levels may decrease significantly (Davis, 1977). Metals leaching from equipment used in a culture unit may be more important than those in the relevant power station cooling water (Boyden and Romeril, 1974).

(d) Dissolved gases
In some of the fishes and shellfishes cultured in thermal discharge canals or heated ponds in the U.S.A., symptoms of gas bubble disease have been noted resulting from supersaturation with oxygen (Goldberg, 1978). Usually, high temperature and high oxygen concentrations lead to the gas coming out of solution in the fishes' body fluids, causing embolisms and death in the same way as in gas bubble disease found below hydro-installations. In such situations some form of alleviation of high dissolved gas levels is necessary either through artificial turbulence or cascades (see chapter 4).

(e) Radio-activity
Oysters (*Crassostrea virginica*) cultured in the heated effluent for 26 months at the Maine Yankee Nuclear Power Plant were found to grow faster, but also to demonstrate an increased rate of accumulation of gamma-emitting radio-nuclides (Price, *et al.*, 1976). The influence of temperature on radio-nuclide accumulation is noted in chapter 7. Molluscs cultured in such effluents would need to be monitored before being marketed and there would doubtless be market resistance.

8.6. OTHER USES OF WASTE HEAT
There have been many discussions on the uses of waste heat from power stations in horticulture, agriculture, and domestic heating. In some colder countries, for example, Finland, U.S.S.R., and Iceland, dual purpose plants are built which produce both electricity and hot water or steam for domestic heating (Belter, 1975; Beall, *et al.*, 1977). However, as these plants are often

designed for the dual-purpose role, the heat is not in the strict sense 'waste', unlike that from a remote-sited single-purpose electricity generating plant. Fenton and Norris (1972) reviewed the economics of heat/power and total energy installations and concluded that a change in economic attitudes was necessary before these could be globally acceptable.

In Britain, an example of waste heat use for domestic heating is at Battersea (London), where cooling-water is piped to nearby flats (apartments). However, auxiliary heating for the water is necessary at times to maintain required temperatures.

There are a number of schemes in progress for using cooling-water (waste heat) in agriculture, though as with aquaculture the economics are not yet fully justifiable in most cases (Beall, et al., 1977; Lee and Sengupta, 1977). In the German Democratic Republic, large hot-houses ($0.2 \times 10^6 \text{ m}^2$) are heated by cooling-water from power stations, producing crops all year round (Mitzinger, 1974). The usage is considered to be of great efficiency. Also, the first batch of tomatoes has recently been produced from a greenhouse in the U.K. heated by cooling-water from a power station.* In Oregon (U.S.A.), cooling-water has been used experimentally on a 68 ha plot, as a source of spray-irrigation to raise high-value crops (Thompson, 1971). Crops germinated earlier and were eventually marketed 2 weeks before the normally reared crops. In France, fish culture and greenhouse heating are being tried (Balligand, et al., 1975) and similar work is in progress in Czechoslovakia (Hatle and Lampar, 1975; Granby, 1978). Other uses include under-soil heating by means of pipes, to reduce soil frosts.

Although the concept of using waste heat in such schemes may be highly acceptable, in practice, the economics will depend very much on the climate of the country, the locality and siting of power stations, water-quality and demand for the finished products (Beall, et al., 1977).

The development of agricultural methods, and new strains of fast-growing, cold-tolerant species have reduced the economic attractions of capital-intensive installations using low-grade waste heat. On the other hand the increasing costs of oil and other fuels may enhance their prospects particularly if, as is happening in many places, power stations are planned with possible waste-heat usage in mind.

One problem remains, however, and that is to whatever use the cooling-water is put, there are such massive amounts available that only a small proportion can ever be utilized. The heat disposal problem will therefore, not disappear and dispersal into the environment will still be necessary perhaps even more so as larger power stations come into use.

---

* There is now a fully productive commercial system growing lettuces and other viable crops operating in the U.K.

8.7. ANGLING IN THERMAL DISCHARGE CANALS

It has, in the author's experience, long been believed by anglers that thermal discharge canals are the best place to fish, particularly in autumn, winter, and spring. In chapter 5 it has been shown clearly that fishes appear to congregate

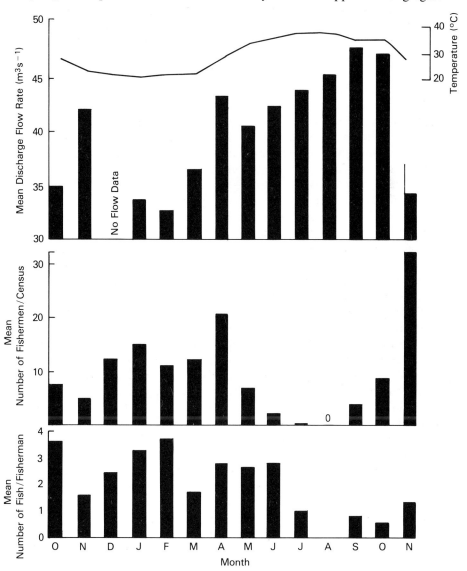

Fig. 8.4. Monthly mean number of fish per fisherman, number of fishermen per census, discharge flow rate (m³ s⁻¹) and water temperature (°C) recorded during creel censuses at the P. H. Robinson Generating Station discharge canal from October 1968 to November 1969. Redrawn after Landry and Strawn (1973).

in such canals in winter. In an analysis of angler catches in the River Trent (U.K.) it was noted that anglers actually complained of poor fishing coinciding with lack of thermal discharges from certain power stations (Whiting, *et al.*, 1975). At several power stations in the U.S.A., censuses have shown that discharge canals are favoured angling venues (Fig. 8.4), particularly during October to April (Landry and Strawn, 1973; Moore, *et al.*, 1973; Marcy, 1976a). At the Chalk Point Power Station on the Patuxent, 50 per cent of the annual angler trips in that area of the river were to the thermal discharge canal. During October to April this proportion rose to 90 per cent.

However, it is notable that in summer these fisheries were hardly visited at all. The increase in fishing success in such canals is probably a result of increased activity at higher temperatures causing increasing demand for food, with some localized accumulation.

In many regions, however, these locations have created special venues for winter fishermen which did not exist prior to power station developments.

8.8. SUMMARY

Both hydro-developments and thermal power stations produce changes in aquatic habitats which may benefit fisheries and may enhance the production of fish though these advantages may be offset by other alterations. Large man-made lakes have increased the availability of angling and other water-based sports for the larger part of the human population. Thermal discharge canals have also provided new winter fishing for many anglers. The economic viability of aquaculture in power station cooling-waters remains to be generally verified, except in limited regions of the world, though many studies are in progress. Winter growth of plants may be enhanced by waste heat and the efficiency of horticulture improved. The problem of waste-heat disposal will be little altered by these relatively small-scale uses, especially in view of other agricultural progress and expansion of electricity generation.

Chapter 9 Legal aspects of water use in electricity generation in relation to ecological effects

9.1. INTRODUCTION

The complexities of the law and the constraints relating to water use and pollution control in relation to electricity generation in different countries are such that it would be impractical here to attempt even to summarize all the parts which are relevant. Basically, there is little over-all consistency from country to country in the approach to regulating such aspects as minimum compensation flow and fish passage at dams, or the nature, composition, and temperature of thermal power station effluents. In smaller countries with 'self-contained' drainage-basins such as Britain, the regulation of water use and effluent discharge is generally a much simpler process than for those countries having massive rivers such as the Rhine, Danube, or the Columbia which cross international boundaries (O'Sullivan, 1974). Further, countries which have separate federal and state or provincial legislation, for example the U.S.A., Canada, South Africa, and West Germany also have two or three tier systems of legal controls, and federal standards may differ from state or provincial standards for various reasons.

The basic principles on which the legal constraints are developed and the relevance of these constraints to the industry and the observed ecological effects are the main concern of this chapter.

9.2. FISHERY PROTECTION AT HYDRO-ELECTRICITY SCHEMES

The fact that dams and weirs may obstruct the passage of migratory fishes such as salmon and eels has been recognized by European law for many hundreds of years (Netboy, 1968; Gregory, 1974). However, even as late as the 1960s a large dam, i.e. the Iron Gate Dam on the Danube was completed without fish passes, cutting off hundreds of miles of available spawning grounds for migratory shad, salmon, and sturgeons. Developments at present under construction in northern Canada include dams and diversion schemes which will impede, delay or divert migratory salmonids and in some of which no provision is made for their free passage (Efford, 1975; Power, 1978). Controversy still also appears in Britain and recent arguments have occurred over delay of salmon at weirs and at fish passes in hydro-electricity dams in Scotland (*Sunday Times*, 1978).

About the time that hydro-electricity was beginning to develop in Britain a group of Commissioners for English Fisheries was appointed and although they appeared to be concerned more about fishing weirs, fixed nets or catching devices, and fishing-mill dams, it was established by 1873 that all new dam works in England and Wales would have fish passes. This, of course, covered the new, higher dams of the proliferating hydro-electric schemes. Perhaps the major legislation on fisheries in Britain was the Salmon and Freshwater Fisheries Act (1923), which is still largely incorporated into current legislation. Part II of the 1923 Act laid down strict rules to ensure that fish are not unduly impeded in their migrations, though the concern was only for salmon and migratory trout. The Act ensured not only that no new fixed catching devices could be built, but that no structures 'for detaining or obstructing' the passage of migratory trout and salmon could be 'erected or used in any inland or tidal water except by river authorities which have been given power to do so'. The 1923 Act also ensured that owners and operators of dams or weirs constructed fish passes, made proper provision for flows in these passes, and protected the fish during passage.

It was obviously realized that migratory fish tended to congregate below or above artificial, as well as natural obstructions. Thus provisions were made to prevent the taking of fish within set distances (for example 50 m above and 100 m below) a dam unless by rod and line. Further, it was decreed that where water was diverted for use, provision should be made to prevent migratory fish entering such diversions or pipes if the water was to be used for consumption or in a canal. The appropriate authorities (Fishery Boards, now Regional Water Authorities) were also given powers to prevent ingress of fish where necessary, where this was not done satisfactorily by owners or occupiers. Thus as a piece of comprehensive fishery legislation to protect salmonids it is still in the main satisfactory, at least as far as the construction of dams is concerned.

In Scotland, laws intended to regulate the construction of dams obstructing salmonids were in existence in the seventeenth century, and during the development of the large, hydro-electricity schemes a great deal of research and consideration was applied to easing the passage of fish both upstream and downstream around these obstructions (Scottish Home Department, 1957). The first recorded fish pass was constructed on the River Teith around 1830 (Berry, 1955). To allow the proper development of integrated hydro-schemes and thermal generation in Scotland, Acts of Parliament were necessary (e.g. Hydro-Electric Development Act (Scotland) 1943; Electricity Re-organization (Scotland) Act 1954) because the schemes would contravene parts of other Acts relating to rivers and fisheries. Also, because of fishery and land legislation, the construction of each major hydro-electricity scheme in North Wales was preceded by an Act of Parliament (North Wales Hydro-electric Power Acts, 1952, 1955, 1973).

Under the 1943 Act, the North of Scotland Hydro-electricity Board was given wide powers to construct schemes (with approval and subject to various provisions), but significantly the Act provided for the appointment of Amenity and Fisheries Committees, which would advise and be consulted on ecological implications of schemes.

Similarly, the terms of the 1957 Act which created the Central Electricity Generating Board to develop power generation in England and Wales, required the Board to 'develop and maintain an efficient co-ordinated and economical supply of electricity in bulk for all parts of England and Wales', and 'also in formulating or considering any proposals relating to the function' (of the industry, etc.) shall each (Board, Area Board, Council) take into account any effect which the proposals would have on the natural beauty of the countryside or on any such flora, fauna, features (of the countryside), buildings or objects, of historical or architectural interest'. Both of these are statutory obligations regarded as of equal importance (Leason, 1974).

In all the Acts covering the building of dams by the C.E.G.B. since 1952, provision has been made for improvement or protection of the fisheries which might be affected. For many years major developments in the U.K. have been preceded by legislation, public discussion, and consultation with interested bodies such as local authorities, amenity groups, and conservationists. Problems still arise, however, mainly through lack of detailed ecological data by any party.

Britain has been reasonably fortunate in that some concern about the 'flora and fauna' in relation to electricity developments has been written into the Acts of Parliament for many years even though the actual ecological research was not given high priority until the last two decades. In other countries, particularly the developing countries such as those of the African and South American continents or those with rapidly expanding economies such as the U.S.A. and Canada, the need for electricity was (and still is in some places), given overriding consideration and many of the hydro-electricity schemes were licensed with only cursory consideration for aquatic or terrestrial ecology. In fact, even the social aspects were largely ignored in many African schemes and whole villages were moved without proper planning (Lowe-McConnell, 1966; Hammerton, 1972; Ackermann, et al., 1973).

In the U.S.A. the rapid expansion of hydro-electricity schemes has caused many thousands of miles of salmon spawning rivers to be blocked off (Schweibert, 1977). For example in the Columbia, the series of dams, many originally without fish passage facilities, led to a serious decline in some salmon stocks. Retrospective fitting of passage facilities, smolt rearing and transport, and the building of salmon ladders in later dams has involved about $500 million in capital costs, and $10 million in annual maintenance and operating costs (see Mains, in Schweibert, 1977).

Although many of the more recent dams and some of the older ones incorporated fish passes and were legally bound to maintain passes and minimum flow conditions, not all dams have complied. Problems with gas supersaturation and gas bubble disease in the Columbia have now forced power companies to fit tail-race deflectors to try to alleviate the effects of pressure changes (Schweibert, 1977).

Today, there is not the same ready acceptance of hydro-electricity schemes in North America. In Canada the massive La Grande hydro-electricity project in Quebec has been delayed for some years by objections, and eventually by an injunction requested by ethnic groups such as the Cree Indians and the Innuit (Eskimos) seeking protection of their hunting and fishing rights (Penn, 1975; Power, 1978). This project has been somewhat of a legal, social, political, and ecological landmark in Canada and led to the establishment of a full 'Environmental Impact Assessment' for the scheme (Penn, 1975). Legal constraints on the development of hydro-schemes have not been particularly stringent in many countries until recent years, though it is doubtful if schemes in the more democratic countries will escape stringent constraints in future.

In the U.S.S.R. there is, currently, massive development of some river systems, for example the Volga, Dnieper, Kama, and others (Yourinov, *et al.*, 1974) and emphasis is placed on the passage of migratory fishes at most dams.

Where this is not possible, facilities for rearing salmonids and sturgeon are developed (Lowe-McConnell, 1973; see also Mordukhai-Boltovskoi, 1979). The decline of Atlantic salmon in rivers is well documented by Netboy (1968). In both Europe and the U.S.A., dams, water-abstraction and diversion, hydro-electricity generation, and pollution have decimated stocks, often leading to the extinction of natural indigenous fish from some rivers. In countries such as Sweden and some parts of Eastern Europe, stocks are only maintained by artificial propogation and transporting. It is significant that in Sweden, the Salmon Research Trust is heavily financed by the electricity industry. The legal administration of hydro-schemes on multi-national rivers is difficult, probably almost impossible, as the economic, political, and ecological priorities may vary widely in importance from country to country.

9.3. MINIMUM FLOW CRITERIA

As we have seen in chapter 2 the flow of rivers fluctuates widely under natural conditions. Control by dams and operation of hydro-schemes may either even out flow patterns or, if badly planned and operated, cause even more violent fluctuations than in the natural state.

The flow requirements for each fish or invertebrate species may differ markedly. Further, species may require different flow regimes to complete various life-history stages. For example, adult salmonids require fairly high river flows to enable them to ascend to spawning grounds. In any river the minimum flow must always provide enough depth of water over shallows or fish passes to enable large fish to traverse these obstacles during upstream migration, though the actual discharge may not be as important here as the way it is channelled. After spawning, and during the growth period of the young fish, high flows may displace eggs or individuals from territories and inhibit feeding. Periods of stable, moderate flows thus provide optimum conditions for this stage. In non-salmonid rivers, particularly in lowland regions, excellent coarse fisheries exist where inflows may be almost nil at times (Langford, 1963, 1966, 1971b). These rivers often have 'U'-shaped channels and, as flows lessen, there is no excessive exposure of the available substrate like that found in the shallower stony streams (Baxter, 1961).

Various ways of using impounded waters for enhancing the migration of salmonidae in rivers have been propounded (Banks, 1969). Baxter (1961) suggested that the quantity of compensation water necessary to maintain all phases of salmonid life-history in a river over a whole year would be 20 per cent of the average dry-weather flow (A.D.F.). Wesche (1976) reported that in U.S.A. streams, 10 per cent of the average annual discharge (A.A.D.) was necessary for short-term survival of salmonids and 30 per cent of the A.A.D. was a 'satisfactory' flow to maintain a fishery. Wesche proposed as a 'general rule of thumb', that a minimum of 30 per cent of the A.A.D. is needed to maintain Wyoming trout stream fisheries. It is possible that the river discharge only determines the rate at which fish will ascend or descend rivers when the topography of the bed is such that there is not sufficient depth of water for fish to pass over a riffle or an obstacle. Thus, where low flows occur some channelling of the bed in shallow areas may facilitate fish passage (Langford, et al., 1977). The flow criteria and constraints for any river may not necessarily be wholly determined by the discharge, but also by the way in which the topography of the river allows the water to flow. The amount of compensation water needed to maintain a spawning area may vary with season, and criteria should include where possible some allowance for seasonally fluctuating requirements. There is also a need for mixing of surface, mid-depth, and hypolimnial compensation water in reservoirs or rivers likely to stratify. This will provide water of the correct temperature and dissolved gas concentrations downstream. Modelling studies should allow prediction of these parameters in developments, provided time is allowed for adequate pre-planning research (Krenkel and Parker, 1969a). Compensation water flow and the type of discharge are often decreed by law or at least specified approximately in modern licensed hydro- or water-storage schemes, though in many early schemes throughout the world legal requirements were not fixed in the original consents and flows recommended by ecologists for fishery protection were not met.

Discharges from turbines should also be made on a gradual increasing and decreasing pattern rather than sudden starts and stops. Sudden start-up may not only cause massive scour, displace fish downstream and cause turbine mortalities but may be dangerous to anglers fishing in the river downstream. Also, sudden shut-down may cause fish attracted upstream during the operating period to be stranded as the river bed dries out rapidly. As hydro-electricity plant becomes older and as thermal power generation increases, the operating pattern at hydro-installations may change from base-load with steady water use, to peak-load with widely fluctuating use, and this change may have drastic effects on the river regime and the ecosystem (see chapters 2 and 4). There can be no really generalized criteria for either operating or design features of hydro-electricity schemes which will protect fisheries in every situation. Each scheme must be studied in detail, mainly with regard to the hydrography of the channel; water discharge patterns, temperature, and chemistry. Ecological predictions can be made from such data together with the basic biological information, and the scheme should be tailored as far as is economically and ecologically practical to suit the requirements of maintaining the particular fishery.

9.4. LEGAL ASPECTS OF WATER USE AT THERMAL POWER STATIONS

9.4.1. Effluent constraints

Different legal constraints and negotiating methods are used in various countries for the control of effluents from thermal power stations (E.P.A., 1972; McLoughlin, 1976; Johnson, 1979). In England and Wales an Act of Parliament (Control of Pollution Act, 1974) allows the appropriate pollution control authorities (Regional Water Authorities) to grant consents for water abstraction and discharge by the C.E.G.B. for cooling-water and other effluents from power stations. Prior to organized water pollution control in Britain legal action had been taken under Common Law against the discharge of water at other than normal temperatures to a stream (Wisdom, 1956).

Although Leason (1974) argued that cooling-water is not strictly a 'polluting discharge' it is covered by successive Prevention of Pollution Acts, and heat was noted as a potential hazard as long ago as 1919 (Electricity Supply Act, 1919). On ecological grounds it does constitute an alteration in the environment, though the biological data do not generally suggest dramatic adverse effects. Section 7 of the 1951 Rivers (Prevention of Pollution) Act (U.K.) now incorporated in the Control of Pollution Act, 1974, specified conditions as to 'the nature, and composition, temperature, volume, or rate of discharge of effluent' which may be imposed by the Regional Water Authority. Conditions have been to date, therefore, set by the Water or River Authority, not only on the temperature limit of a discharge, but also on its chemical composition and volume. Discharges of non-thermal wastes, for example, sewage, ash-effluents, and surface drainage, are all subject to separate consent conditions. In addition, the Water Authority also grants a licence to abstract set volumes of water for cooling and places limitations on amounts evaporated from cooling-towers. The system allows for variation in consents so that where necessary for the protection of specific river qualities mores stringent criteria than usual may be required.

Consent conditions on power station discharges in Britain may vary from site to site and river to river, but for cooling-waters in England and Wales most are set at a maximim temperature of around 30°C with ΔTs of about 8°C, sometimes 5–8°C in summer and 9–10°C in winter. Chemical consents vary slightly with the design of the site and the quality of the receiving water. As examples, Table 9.1 shows the consent conditions for two major modern power stations in England, namely Ironbridge 'B' sited on a clean, salmonid river, the Severn; and Ratcliffe-on-Soar sited on the polluted reaches of the Trent, which has a mainly coarse (cyprinid) fishery. Both stations used closed-circuit tower-cooling with similar evaporation-rates, though different total amounts. Consents for discharge to tidal waters have not generally been widely applied in Western Europe though under the terms of the Control of Pollution Act, 1974, the British Water Authorities were intended to have jurisdiction over 'those tidal waters and parts of the sea adjoining the coast of the Water Authority Area to which any of the provisions of the Rivers (Prevention of Pollution) Act 1951 apply'. The implementation of the second part of the Control of Pollution Act 1974 relating to tidal waters was expected in 1979 but was delayed. Recent legislation by the E.E.C. will however affect the discharge of power station effluents, both in fresh and tidal waters (Johnson, 1979).

The application of some of the 'blanket'-type criteria proposed by E.E.C. directives is generally regarded in Britain as impracticable (House of Lords, 1977). For example, the restriction of thermal discharges such that water temperatures in summer should never exceed 25°C in cyprinid waters would close down most river-cooled British power stations as they operate at

Table 9.1. *Consent conditions for the composition of liquid discharges from two U.K. power stations*

	A	B
Power Station:	Ironbridge B	Ratcliffe-on-Soar
River:	River Severn—clean reaches	River Trent—polluted, recovering
Size:	1,000 MW	2,000 MW
Cooling:	Closed circuit—4 towers	Closed circuit—8 towers
Cooling-water discharge:	*Max. temp.* 30°C	30°C
	Suspended solids Max. 1·5 × intake level	Max. 1·5 × intake level
	pH 6–9	No constraint
Ash lagoon outflow:	*Suspended solids* Max. 50 p.p.m.	Max. 100 p.p.m.
	pH No constraint	5–10
	Toxic metals Max. 0·5 mg l^{-1} (each element)	No constraint

Published by permission of the C.E.G.B.

present. In the hot, dry summer of 1976 for example *natural* river temperatures in Britain exceeded the 20°C and 25°C limits (salmonid and cyprinid rivers) set by the E.E.C. with records of 25–26°C in both salmonid and cyprinid reaches (Langford and Howells, 1976). The evidence from the research is that coarse fisheries are not affected by water temperatures of 28–30°C provided that these are not exceeded for long periods and provided there is some horizontal or vertical stratification. Salmonid fisheries should not, however exceed 24–25°C for long periods. Avoidance data (chapter 5) also suggests that fish would move away from areas of adverse temperature, provided cooler areas exist. The full implementation of a 2°C ΔT limit on marine thermal discharges into shellfish waters would close down at least one modern British power station. The specification of such 'blanket' limits ignores the fact that most marine thermal discharges stratify, particularly if the outfall is offshore, and the warm chlorinated water rarely comes into contact with benthic organisms (see chapters 2 and 5). Blanket-type criteria may also be a retrograde step in Britain where specific situations may be controlled by specific constraints.

Any chlorine limits set must also take into account the difficulty of measurement at low levels in sea-water.

In most European countries temperature limits are set which are not to be exceeded at the point of discharge and these are usually 30°C for fresh-water and 35°C for the sea. Mixed temperatures may not usually exceed 22–28°C depending on the country and the water quality of the river. ΔTs are also fixed, and vary from 3°C to 15°C depending upon season and river category. For example, in Class I rivers in Poland (drinking water, salmonid fisheries) mixed temperatures may not exceed 22°C. In Class III rivers (industrial water supply, irrigation only) the maximum allowable is 26°C. Some countries also specify levels of oxygen and where necessary request that thermal discharges should be aerated (U.N.I.P.E.D.E. Study Doc. 20/D4). Only the U.K. does not publish such set national criteria and consents are negotiated on individual merits taking into account local water conditions and uses.

The situation in the U.S.A. is complicated by the existence of state and federal systems which may set different criteria for discharges (Federal Power Commission Report, 1969). The National Environmental Policy Act, 1969, and subsequently the Federal Water Pollution Control Act Amendments of 1972, implemented the need for permits for every point source of discharge including waste heat and these permits may be issued by the state if the state meets federal legal requirements. If this is not so, the Federal government, represented by the Environment Protection Agency (E.P.A.) retains the authority. The Federal Water Pollution Control Act Amendments of 1972 (F.W.P.C.A.) PL 92–500, set two national targets:

(i) To eliminate the discharge of pollutants into navigable waters by 1985.

(ii) To achieve wherever attainable an interim goal of water quality which provides for the protection and propagation of fish, shellfish, and wildlife, and provides for recreation in and on the water by 1 July 1983.

To these ends every discharge requires a permit to operate.

For power stations the E.P.A. effluent and intake guidelines require 'the best technology' available to minimize ecological impact. Cooling-towers (closed-cycle cooling) are regarded as the best technology.

When licences are granted for building a power station, the state or E.P.A. set discharge guidelines, but under Section 316 (a) of the F.W.P.C.A. (1972) the owner or operator of a steam electric (thermal) power station has the opportunity to demonstrate that the thermal effluent limitations are more stringent than is necessary to: 'assure the protection and propagation of a balanced, indigenous population of shellfish, fish and wildlife in and on the body of water into which the discharge is to be made'.

The demonstration usually takes the form of a research or survey programme under operating conditions or of predictive studies which indicate the extent of ecological consequences. If successful the Administrator or the E.P.A. or the state may impose alternate, or less stringent limitations when granting a permit to operate.

The problem with this type of legislation, which is a 'play-safe' philosophy, is that operators and E.P.A. faced with a multimillion dollar capital investment of power plant which does not meet effluent standards and which is not licensed to operate though finished, may enter protracted negotiations which may result in huge financial losses, probably passed on to the consumer. Similarly, if cooling-towers are built, at huge expense where they are obviously not ecologically necessary, the consumer must foot the bill. The vast waste in research money and effort which ensues from the repeated duplication of very similar Section 316 (a) demonstrations at new power plants is little short of ridiculous, though many private consultants and universities may benefit from the funds available.

Effluent temperature criteria recommended by the E.P.A. for thermal discharges are, for example, maximum of 90°F (33°C) with maximum permissible rises above ambients of 5°F (2·8°C) in streams and 3°F (1·7°C) in lakes.

In estuaries and coastal waters, recommended ΔTs are 4°F (2·2°C) from June through to August. Tarzwell (1970) suggested that even these may be too high and that requirements may need to become more restrictive, though his paper was published before much of the recent research was completed.

An important consideration in determining temperature criteria in U.S.A. waters was the 'allowable mixing zone'. This area of effluent dispersion and heat dissipation was required to be limited so as to allow passage to migratory fish or drifting organisms, though as explained in chapter 5 little evidence really exists of the 'thermal barrier' theory.

Mount (1971) has reservations about the regulation and enforcement of mixing zones and it is evident that rather complex monitoring procedures would need to be established either remotely or manually to ensure compliance with a too tightly delimited mixing zone. Mixing zones fell out of fashion after the N.E.P.A. (1969) and some state legislation specified 'no mixing zone' as a constraint, though this a very difficult constraint to design and operate. E.P.A. criteria suggest that mixing zones for all effluents should allow maintenance of 75 per cent of the cross-sectional area for passage of organisms. This is impossible to achieve in small rivers such as those in Britain, nor is there much evidence to show them to be necessary (see chapter 5). However, for sensitive migratory species in warmer regions it is a reasonable safety measure.

9.4.2. The relevance of temperature criteria

Whilst it is realized that there must be controls on aquatic discharges if pollution is to be controlled and if waters are to retain fisheries, nevertheless it should be evident by now that almost none of the scientific data produced so far indicate the need for some of the rigid, inflexible (and unnecessarily stringent in most cases) temperature criteria used in the U.S.A. and to some extent in Europe. Effluent limits as low as 25°C in temperate waters and ΔTs of 0·5–3·0°C are ecologically unnecessary in most situations, apart from the fact that they may be lower than the natural ambients and fluctuations in many waters (see chapter 2). In fact, most of the evidence is that allowable increases of up to 10–12°C in winter and 6–8°C in summer, may not be harmful at more than a tiny handful of sites on temperate waters and even this is conservative. Discharges with ΔTs of up to 20°C may well be harmless provided heat dispersal and dilution are adequate. Most of the detectable biological effects encountered have been found either in effluent channels or in restricted environments such as cooling-ponds or small, warm-water rivers. There is little evidence at all that in an open environment such as a large lake, river or tidal areas, freely dispersed thermal effluents cause adverse effects which can be attributed to temperature. Scour and chlorine and flow changes could be much more important, though in areas of rapid dispersal these are again of little ecological significance, except in the immediate vicinity of an outfall.

The most stringent limits probably need to be in tropical waters, though even here ΔTs can be up to 5–6°C or more in some seasons, without ecological hazards.

The blanket E.E.C. criteria, which seem to be based very much on those recommended by the E.P.A. for the U.S.A. are unrealistic for most European waters and when implemented will prove expensive and ecologically unnecessary. Criteria should be tailored to each individual situation, more or less stringent as this demands for protection of the biota.

9.4.3. Constraints on intakes

In Europe and the U.K. there are as yet no legal constraints on intakes except on the total amounts of water which may be abstracted. In the U.S.A., Section 316 (b) of the F.W.P.C.A. (1972) requires 'that the location, design, construction and capacity of cooling-water intake structures reflect the best technology available for minimizing adverse environmental impact'. This basically refers to impingement, entrainment, and entrapment of organisms (see chapter 6). Demonstrations of the effects of the intake on the populations of organisms in the water body have to be made before a licence to operate is granted. Again this has led to a multitude of repetitious site-specific research and survey programmes, most of which are probably mainly inconclusive, though population dynamics models have been produced for a number of species (Van Winkle, 1977). The constraint has also led to many attempts at retrospective fitting of fish and fish larval rescue and deflection schemes in the U.S.A. at considerable cost.

So far deflection schemes for fish have been generally unsuccessful and although massive impingement mortalities have been recorded, adverse effects of these on populations have not been demonstrated (see chapter 6). Current velocities of about $0.15–0.2$ m s^{-1} do seem to produce less impingement than higher velocities but they are sometimes expensive to achieve as the size of intake structures and pumps have to be increased significantly over normal designs. Also they may not be effective when fish are seasonally present in huge numbers. Some recommendations for screen siting and design criteria are given in chapter 6.

9.5. CONTROL OF RADIO-ACTIVE DISCHARGES

The International Commission on Radiological Protection was formed as long ago as 1928 and since that time the Commission, consisting now of about fifty eminent scientists in the field, have been instrumental in setting safety standards for permissible doses to man and discharges to the environment. Other international bodies involved are the International Atomic Energy Agency (I.A.E.A.), the World Health Organization, and the Food and Agricultural Organization of the United Nations (I.A.E.A., 1974a,b, 1979). In the U.S.A. the Atomic Energy Commission licenses nuclear power plants but the E.P.A. is now charged with recommending safe liquid discharge concentrations of radio-nuclides. In Britain, the U.K.A.E.A., C.E.G.B., and the other electricity undertakings (N.S.H.E.B. and S.S.E.B.) are responsible for the development and safety of nuclear power stations. A constant monitoring programme in aquatic environments is operated by the Fisheries Radiobiological Laboratory, part of the Ministry of Agriculture, Fisheries and Food (Hetherington, 1976, Mitchell, N., 1977a,b). The C.E.G.B. also monitors power station discharges in conjunction with the government laboratories. The permissible discharges of different radio-nuclides vary with area, with potential exposure limits and pathways involved, but care is taken when calculating maximum permissible doses to leave large safety margins, often of two or three orders of magnitude. The problem is knowing precisely which nuclides to monitor in every situation.

In Britain, a standard known as the Derived Working Limit (D.W.L.) is

Table 9.2. Principal exposure pathways from the discharge of liquid radio-active wastes

Site	Critical material	Critical exposure category	Principal exposed group
UNITED KINGDOM ATOMIC ENERGY AUTHORITY			
Winfrith	Fish and shellfish	Beta dose to GI tract*	Local fishermen and families
Harwell	Drinking water	Beta/gamma dose to whole body (somatic and genetic hazard)	General public (Greater London)
Dounreay	Detritus associated with fishing gear	Beta dose to hands	Local fishermen
	Beach sludge	Gamma dose to whole body	Local fishermen and others
THE RADIOCHEMICAL CENTRE LIMITED			
Amersham	Drinking water	Beta/gamma dose to whole body (somatic and genetic hazard)	General public (Greater London)
CENTRAL ELECTRICITY GENERATING BOARD AND SOUTH OF SCOTLAND ELECTRICITY BOARD			
Berkeley	Estuarine sediment	Gamma dose to whole body	Salmon fishermen/river authority workers
	Shrimp and salmon flesh	Beta/gamma dose to whole body	Local fishermen and families
Bradwell	Fish flesh	Beta/gamma dose to whole body	Local fishermen and families
Dungeness	Fish flesh	Beta/gamma dose to whole body	Local fishermen and families
	Beach sediment	Gamma dose to whole body	Bait diggers

	Fish and shrimp flesh	Beta/gamma dose to whole body	Local fishermen and families
Hinkley Point	Beach sediment	Gamma dose to whole body	Local fishermen
Oldbury	Estuarine sediment	Gamma dose to whole body	Salmon fishermen/river authority workers
	Shrimp and salmon flesh	Beta/gamma dose to whole body	Local fishermen and families
Sizewell	Fish and shellfish flesh	Beta/gamma dose to whole body	Local fishermen and families
	Beach sediment	Gamma dose to whole body	Local fishermen
Trawsfynydd	Trout flesh	Beta/gamma dose to whole body	Local fishermen and families
Hunterston	Fish flesh	Beta/gamma dose to whole body	Local fishermen and families
	Beach sediment	Gamma dose to whole body	Shellfish collectors

DEFENCE ESTABLISHMENTS

Aldermaston	Drinking water	Beta/gamma dose to whole body (somatic and genetic hazard)	General public (Greater London)
Chatham	Estuarine sediment	Gamma dose to whole body	General public (houseboat dwellers)
Devonport	Estuarine sediment	Gamma dose to whole body	General public
Faslane	Foreshore sediment	Gamma dose to whole body	Boatyard workers
Holy Loch	Estuarine sediment	Gamma dose to whole body	General public
Rosyth	Estuarine sediment	Gamma dose to whole body	Dredgermen

* Gastro-intestinal tract.
From Mitchell, N., 1977a.

Table 9.3. *Estimates of public radiation exposure from liquid radio-active waste disposal in the U.K., 1976*

Site	Pathway	Maximum exposure* of an individual (percentage of I.C.R.P.-recommended dose limit)
UNITED KINGDOM ATOMIC ENERGY AUTHORITY		
Winfrith	Shellfish	< 1
Harwell*	Drinking water	< 1
Dounreay†	External dose (foreshore)	< 1
	Beta dose (fishermen)	< 1
	Shellfish	< 1
CENTRAL ELECTRICITY GENERATING BOARD		
Berkeley/Oldbury	External dose	< 0·1
	Fish/shellfish	< 0·1
Bradwell	Oysters	0·05
	Fish	0·2
Dungeness	External dose	< 0·1
	Fish	< 0·1
Hinkley Point	External dose	< 0·1
	Fish/shellfish	< 0·1
Sizewell	External dose	< 0·1
	Fish/shellfish	< 0·1
Trawsfynydd	Lake fish	21
Wylfa	Fish/shellfish	< 0·1
	External dose	< 0·1
SOUTH OF SCOTLAND ELECTRICITY BOARD		
Hunterston	External dose	< 0·1
	Fish/shellfish	< 0·1
DEFENCE ESTABLISHMENTS		
Chatham	External dose	< 0·1
Devonport	External dose	< 0·1
Faslane	External dose	< 0·1
Holy Loch	External dose	< 0·1
Rosyth	External dose	< 0·1
BRITISH NUCLEAR FUELS LIMITED		
Windscale	Fish/shellfish	44
	External dose	8
	Porphyra/laverbread	0·2
Springfields	External dose	< 1
Chapelcross	External dose	< 1
	Shellfish	< 1

From Mitchell, N., 1977a,b.

used a great deal in monitoring radio-active exposure (Preston, 1971; I.A.E.A., 1979). This expresses the estimated maximum rate of radio-active material release to the environment which will not exceed the maximum permissible dose to nearby inhabitants or those most exposed. This D.W.L. takes into account the likely pathways which might be involved, for example, fish, shellfish, nets, laverbread, etc. Some of these pathways for the major British sources of radio-active waste are shown in Table 9.2. Table 9.3. shows the amounts of discharge authorized by licensing authorities (usually government) and the percentage actually discharged to surface and coastal waters in 1974 (Hetherington, 1976). As can be seen, there is still some safety margin for many substances, even though the authorized discharges are set well within calculated safety limits.

The E.P.A. limits for some radio-nuclides in water in 1972 were:

Gross radiation: 1,000 pCi l^{-1} (27 × 10^3 Bq l^{-1})

Radium 226: 3 pCi l^{-1} (80·1 Bq l^{-1}).

There are a number of criteria developed for radiological protection using different parameters and these are described in detail by Eisenbud (1973), and in the I.A.E.A. 1979 report.

To summarize, however, it is obvious that discharges of radio-activity from nuclear power plants are small and unlikely to be acutely lethal but there is yet much research to do on their chronic long-term effects. As Eisenbud (1973) states 'there may not be an absolutely safe dose of ionizing radiation'. Of course, if that is true then exposure of man to other sources than nuclear power stations, for example luminous paints, television, general industrial activities, and natural background radiation should receive equal research investigation.

9.6. THE ENVIRONMENTAL IMPACT STATEMENT

In the U.S.A. the National Environmental Policy Act, 1969, initiated a process whereby all the potential ecological, social, and economic effects of a new development, including power stations would be assessed and reported in a document known as an Environmental Impact Statement. Since then thousands of such statements have been written (Baker, et al., 1977; Ward, 1978; Landy, 1979). Unfortunately, state environmental acts set up similar and separate systems and these have also spawned a huge number of such statements. Not all, of course, relate to power stations. Because of the massive proliferation of paper from the statements, the usefulness of the E.I.S. has been called into question and severely criticized as a decision-making tool (Bendix and Graham, 1978; Wandesford-Smith, 1978).

Tucker (1978) reviewed the process of environmental assessment and concluded that it has in many ways become counterproductive in the U.S.A., both for the developer and for the conservationists, mainly because of bureaucracy. Each new E.I.S. has been larger than the last. In 1970, an

Environmental Report filed by the Florida Power and Light Company, relating to a nuclear site, contained about 250 pages. The most recent report consisted of 8 three-inch thick volumes. The cost rose from about 100,000 dollars to over 6,000,000 dollars. While this type of operation has employed large numbers of students, research institutions, power-company staff, and private consultants for several years, it is doubtful if more than a small fraction of the papers will ever be read and used.

Further, it is evident from reports of the many ecological studies carried out for E.I.S.s that the quality of research work ranges from excellent to extremely poor, the latter usually where an institution has tried to short-cut or failed to plan properly. It is essential that such studies should always be of good academic standard, both for their accuracy of prediction and for their credibility. So far, the formal 'statement' is not widely used in Britain and other European countries, though such systems are proposed. Without adequate control this too could be counterproductive. The system does, however, have advantages in that many areas of potential environmental problems not foreseen by planners and engineers can be thoroughly discussed and investigated before decisions are made to site and construct. The problem with less widely discussed and reported systems is that ecological and operational environmental hazards may have to be investigated after the final siting decisions have been taken on economic and engineering grounds.

However, although the system of E.I.A. is probably to be used in the countries of the E.E.C. the personal view of the author is that the E.E.C. should learn from the experiences of the U.S.A. and not emulate the system without severe criticism and strict controls. From the analysis of the data produced so far and reviewed, albeit with inadequacies, in this book, it should be evident that long-term studies of every new site and proposed development ought not to be necessary, though obviously some basic data are needed on which to base predictions. As it is administered in the U.S.A., the E.I.A. system is grossly wasteful in research effort though highly profitable to universities, consultancies, and the legal profession. The political, public relations, and financial necessities are, of course, recognized (CEQ, EDA, 1972) and should be clearly stated, though they are outside the scope of this work.

9.7. SOME NOTABLE LEGAL ACTIONS INVOLVING POWER STATIONS

There are few published legal actions of any significance concerning aquatic effects of power stations. Perhaps one of the most notable in the U.K. was that involving the Pride of Derby Angling Association and a group of industries on the River Derwent (All England Law Reports, 1952). Here the Spondon Power Station was accused with several others of causing a reach of the Derwent to become fishless over a period of years.

In this case, known as the 'Pride of Derby' case, outfalls from a sewage works, Spondon Power Station, and an artificial-textile factory, were sited on the reach which deteriorated, over a period of years from a well-known productive fishery to being fishless. The Angling Association together with another body sued all three installations for damages in 1952, claiming that the increased temperature from the power station effluent had enhanced the effects of the other, chemical discharges. Estimates at the time suggested that the discharges from the textile factory were responsible for about 85 per cent of the pollution load and the sewage works 15 per cent, with the effects of the power station being negligible although water temperatures in the discharge reached over 100°F (38°C). This was not accepted by the court and when damages of many thousands of pounds were apportioned the textile firm had to pay 50 per cent and the other installations 25 per cent each of the damages. Ironically, when the cooling-system of the power station was drained for alterations to be made, the channels which had been heated to over 35°C at regular intervals over the years contained large populations of healthy roach (*Rutilus rutilus*) (Ross, personal communication).

The judgement against the Electricity Authority was based on the fact that 'water heated to 100°F or over must be injurious to any fish native to English waters' plus the secondary fact that chlorine was present in the effluent. At that time, little was known about effects of effluent stratification, acclimatization, or oxygenation by cooling-towers. However, the case was probably a landmark as it, together with the Hobday Report (Ministry of Health, 1949), stimulated 'thermal discharge' research in the U.K. and possibly in the U.S.A.

A notable legal action in the U.S.A., on another aspect of cooling-water systems was that against the Indian Point No. 2 Power Station on the Hudson River. On 11 January and 26 February 1972, an estimated 163,000 tiny striped bass (*Roccus saxatilis*) were reported killed by impingement on the intake screens at Indian Point No. 2. As a result a fine was levied, after a law-suit, by the State of New York on Consolidated Edison who operated the power plant. The fine totalled $1,638,000, based on a $500 Civil Penalty plus $10 per fish. Amazingly each individual fish weighed only 5 g (0·17 oz) and the total weight of fish killed was about 728 kg (1,600 lb). The fish was valued therefore at something like $1,000 per pound (2·2 kg), and the fine was based on the assumption that each of the tiny, juvenile bass was a potential adult sport fish. It is, however, a well-established biological fact that the natural mortality from juvenile to adult is extremely high in fishes, probably between 75 and 90 per cent in the first year and over 99·9 per cent before maturity (Van Winkle, 1977).

Subsequent modelling studies by Englert *et al.* (1976) demonstrated that anglers 'crop' 16 per cent of the one million adult striped bass in the Hudson and these have a spawning potential of $1·6 \times 10^{11}$ eggs. Some eggs and larvae would be at risk from entrainment or impingement, but the total adult stock production of eggs (10^{12} per annum) is naturally depleted by 70 per cent by natural predation and other factors in the first year. It is evident that in fact the normal angler crops more sport-fish by taking a few adults than the power station did by killing 163,000 juveniles.

Although this is by no means intended to condone fish mortalities by impingement, it is necessary to put such effects into perspective and set them against the assets gained by the human population from an electricity supply. In economic and ecological terms the cost (ultimately to the consumer), of the fine and legal action was out of all proportion to the actual ecological damage.

Another landmark in the impact assessment of power stations was the decision by the U.S. Court of Appeals for the District of Columbia (Calvert Cliffs Co-ordinating Committee, *et al. v.* U.S. Atomic Energy Commission, 1971) over the Calvert Cliffs Nuclear Power Station in Maryland. This mandated a complete and thorough review of the thermal effects of nuclear power plant discharges and also extended the 'environmental-impact' approach (Rickhus, 1977).

In the field of hydro-electricity developments, the La Grande–James Bay injunction in Canada produced a major revision of the approach to the assessment of biological impacts of hydro-schemes. At one stage the judge hearing the case (Judge Malouf) granted an injunction against the developers and ordered them to:

(*a*) immediately cease, desist and refrain from carrying out works operations and projects in the territory, etc., and (*b*) to cease, desist and refrain from interfering and from causing damage to the environment and natural resources in the territory.

Within 2 weeks, however, the decision was reversed and the project was resumed, though legal actions, discussions, and objections are still continuing (Power, 1978).

9.8. FUTURE GENERATION DEVELOPMENTS AND RESEARCH NEEDS

There have been many speculations about new and potential sources of energy for power generation and, as we have seen in chapter 1, research on such sources is already in progress. Leason (1975) and England (1978) summarized the potential of certain sources such as wave power, tidal power, and wind power. Of the three, England suggested that only wave power would have any possible economic advantage without potential environmental objections, though this is perhaps an optimistic viewpoint. Glendenning (1977) also reviewed the prospects for wave power, tidal power, and ocean thermal energy conversion (O.T.E.C.). This last prospect would operate on the temperature difference between naturally heated surface water and colder water from below the thermocline in tropical seas. As with wave power, the work on O.T.E.C. is still on a very experimental scale, and although Glendenning suggests that any natural source of energy such as wave power or O.T.E.C.

will not have environmental problems, he is also suffering the optimism of the non-ecologist. Already, there have been suggestions that the extra shelter given by large installation of wave power units in a selected site off the north-west coast of Scotland may produce changes on the coastline and in the communities of organisms acclimatized to the action of waves. Also, there are bound to be problems associated with fouling on large marine structures which would entail the use of biocides or antifouling treatments. These in turn may have effects on the natural flora and fauna. The area of solar power cells required to provide 2,000 MW of electricity in the U.K. for example would be about 130 km^2 (50 mile^2) and this would need back-up from the power stations on a dull day (England, 1978). Similarly 4,000 windmills would be needed to supply power on a windy day equivalent to one or two modern power stations. On a windless day the power would not be available. However, all the renewable energy systems have been used in specific sites and in certain situations. Ironically, the major objections to large-scale use of alternative renewable sources of energy are most likely to be on the grounds of ecological damage, either visual, terrestrial, or aquatic.

With an excess of base-load generation at present and for the next few years, the electricity demand in the U.K. is certain to be for peak-topping generation from gas-turbines or pumped-storage schemes like Dinorwic. Pumped-storage schemes will require a great deal of ecological research which may well be done partly at sites already operating. In the U.S.A. and other countries pumped-storage is also likely to be further developed, but data on the ecology of pumped-storage reservoirs are still very scarce. In this area perhaps lie the most practical and immediate research needs. The ecological effects of the changing proportion of thermal power to hydro-power in countries such as Canada and the U.S.A. will create new ecological problems as base-load and steady water use gives way to peak load output and widely fluctuating daily water use (see chapters 2 and 4). This may also apply to countries such as Switzerland, Spain, and Sweden where nuclear power is expanding.

As far as thermal power generation is concerned, the expanding nuclear programme in many countries will add to water demand and waste heat disposal, though it should be evident from this book that this latter is only likely to cause ecological problems in specific or very localized areas. The area of effect at any one site needs however to be assessed and put into perspective. The effect of entrainment and impingement mortalities on localized and more widespread populations of fish, invertebrates, and other organisms is another area where more research is needed. There is a requirement to quantify effects, and establish whether expenditure on fish rescues, screening and exclusion of fishes is *really* necessary to conserve species. The development of valid biological models is a high priority.

The uses of waste heat are being studied, though there is much more to do. Biocide problems really need to be overcome generally before large-scale aquaculture is a safe operation. The economics of waste-heat use also require classification. Alternative antifouling systems may also have economic advantages if developed.

The research and monitoring of radio-active wastes must continue, mainly because at the present time this subject is still very much exposed to public and conservationist fears. Although our knowledge is vast, it is still full of gaps, particularly about long-term effects of low-level doses to both aquatic organisms and to man. With the possible proliferation of nuclear power, the public will need the reassurance of sound monitoring programmes and research, though there needs to be more effort to provide the public with real digestible data rather than emotive argument from both sides. Apart from the general issues, there will always be specific problems at specific sites, some answers to which can be provided from the literature and some from properly planned long-term research programmes. Ultimately, however, studies at the site may well have to be made, perhaps of specific organisms.

9.9. IN CONCLUSION

It is perhaps inadvisable to try to summarize the mass of information presented in this book. There is, however, one major point which has, more than any, emerged. It is that there is no doubt at all that as far as the electricity industry is concerned that major alterations in the ecology of natural water bodies have come from the development of river dams, diversions, and lake control schemes, rather than from nuclear, coal, or oil-fired generating stations.

Hydro-electric power is not 'clean power' in the ecological sense even though there are few additions of contaminants. It has been responsible for massive and dramatic ecological changes in most rivers of the world. It also is responsible for extensive chemical changes in rivers, especially below hypolimnial discharges.

Whether any hydro-electricity scheme is totally harmful or not depends a great deal on the actual situation.

There is no doubt at all, however, that very high dams such as that proposed in the Fraser River in Canada, would need a great deal of expensive modification to allow fish passage and conserve fisheries. On the other hand the lower dams on oligotrophic rivers such as those in Scotland, provided that they have fish passes and ladders, and multi-level outlets probably have had much less effect on migratory fish than originally predicted, though the river invertebrate faunas upstream have been altered. In many cases positive benefits have accrued, apart from electricity, though not all authorities would agree as to the extent of extra benefits (Balon, 1978).

In comparison, in spite of predictions of dire consequences, the measurable biological effects of thermal power plants have so far been small, even allowing for public concern, financial outlay and welter of scientific publications.

The actual areas of detectable *biological* effect caused by hydro-schemes and thermal power stations in water, if compared, show clearly the extent of their respective impact. It is doubtful if true thermal effects could be detected in biological terms over more than a few thousand hectares in the world, at a cost in research, legislation, alleviation and monitoring, of millions of dollars per hectare. In contrast, lacustrinization alone at hydro-stations has altered the ecology of some millions of hectares of river upstream, apart from the downstream effects of changes in sediment transport, gases, and flows.

What of future research approaches therefore? How is it possible in countries such as the U.S.A. or in Europe to prevent unnecessary constraints being placed on industries (electricity or otherwise), as a result of speculation or extrapolation from irrelevant and inapplicable data? At least one way would be co-ordination of ecological research by the various organizations involved and continuous resistance by industries to hasty constraints, until valid data are available. Rational approaches by politicians and government departments will also be necessary, to resist minority opinion often unsupported by scientific data.

The second approach is to have expert active research ecologists directly employed by industries, as is done already in many countries using their experience, data and contact with other ecologists and the specialist literature, to forecast possible areas of future concern and plan research programmes accordingly. In this way data are available should pressures arise. One of the most common complaints by ecologists concerned with power and other developments is, that although planners know of the proposed schemes many years in advance, often the ecologist is asked for his predictions only when the developments are already planned in detail (Efford, 1975).

Often ecologists are not an integral part of planning teams and are presented with a *fait accompli* scheme which they are then almost expected to support with data and prediction. In some cases, it may be that the hesitancy of ecologists to make specific predictions is alien to engineers and planners. However, there are many situations in which these latter have had much 'leisure in which to repent their haste'.

Many of the yieldings to the conservationist pressures by industry and others in the past has been because of political or economic considerations together with a lack of good relevant ecological data and the expertise to interpret them. Often it is easier to yield to political pressures and pass on the costs of possibly useless but good 'public relations conservation' measures to the consumer. However, the costs of pollution control and environmental constraints may today be considerable.

Thus the aim of the ecologist, whether employed in industry or not, should be to analyse data objectively and never to subject industry or any other organizations to constraints which are economically and ecologically unnecessary, and which cannot be *fully* justified on a scientific basis. Another danger today is that some research may tend to be self-perpetuating because of the institutions set-up to do it and in many cases this leads to overcomplications and the search for effects so subtle that no method could detect them. Also in many instances ecologists have not attempted to interpret their data and communicate them to the planner or engineer in a comprehensible form.

As far as research programmes themselves are concerned, there is not only a scientific necessity but also a moral requirement to have much closer co-ordination between industry and research institutions and, for example, to reduce, co-ordinate, and rationalize the numbers of repetitive and replicated research, monitoring, and 'demonstration' studies of the type found in the U.S.A. at present. Although each site may have some unique problem, the basic ecological effects of any power plant may be partly evaluated by selective 'generic' research with wide application.

Throughout the literature read, annotated, and digested for this book, there is considerable evidence of extrapolation and prediction from often quite irrelevant and inappropriate data. There is also much evidence of the 'bandwagon' research and legislative approach—for example in the shifts of emphasis from thermal effects, to entrainment, to impingement and to chlorine studies—in spite of the fact that power station operating methods have not changed though of course the industry has expanded. It is, also, obvious that, in the case of impoundment effects, there was considerable underestimate of the ecological damage which could be caused by impounding eutrophic water.

One problem for the future is that the organizations which have depended on environmental issues and impact assessment at power stations in the U.S.A. will either need to discover new and even more subtle areas for possible impacts or will find themselves fundless as the electricity industry and government realize how wasteful and expensive the system has become. One can hope that before the next 'bandwagon' becomes mobile, the scientists and industrialists involved in electricity generation throughout the world will critically review the conclusions so far and decide on a rational and co-ordinated and perhaps much more international approach to environmental problems in natural waters.

Bibliography

1. AASS, P. (1957). 'Fiskeriundersokelsene i Palsbufjord og Tunnhovdfjord. 1949–1956', op. cit. in ref. 2.
2. —— (1971). 'The winter migrations of char (*Salvelinus alpinus L.*) in the hydro-electric reservoirs Tunhovdfjord and Palsbufjord', *Rep. Inst. Freshw. Res. Drottingholm*, **50**, 6–43.
3. —— (1973). 'Some effects of lake impoundment salmonids in Norwegian hydro-electric reservoirs', *Acta Univ. Upsalensis Diss. Sci.*, **234**, 1–4.
4. ABBOTT, D. T., and MIX, M. C. (1979). 'Radiation effects of tritiated seawater on development of the goose barnacle *Pollicipes polymerus*', *Health Physics*, **36**(3), 283–7.
5. ACKERMANN, W. C., *et al.* (eds.) (1973). 'Man-made lakes: their problems and environmental effects', *Geophys. Monogr.*, 17. Washington D.C.: Amer. Geophys. Union.
6. ADAIR, W. D., and DEMONT, D. J. (1971). 'Environmental responses to thermal discharges from Marshall Steam Station, Lake Norman, North Carolina', *Cooling Water Studies*. Edison Electric Co., Baltimore, Maryland: Johns Hopkins University.
7. ——, and HAINS, J. J. (1974). 'Saturation values of dissolved gases associated with the occurrence of gas-bubble disease in fish in a heated effluent', pp. 59–78 in ref. 417.
8. ADAMS, J. R. (1969). 'Ecological investigations related to thermal discharges', Pacif. Coast Elec. Ass. Eng. Operations Section. Ann. Meeting, Los Angeles.
9. AGGUS, L. R. (1971). 'Summer benthos in newly flooded areas of Beaver Reservoir during the second and third years of filling 1965–1966', pp. 139–52 in ref. 461.
10. AITKEN, P. L., *et al.* (1966). 'Fish passes and screens at water-power works', *Proc. Inst. Civ. Engrs*, **35**, 29–57.
11. ALABASTER, J. S. (1958). 'The behaviour of roach (*Rutilus rutilus*) in temperature gradients in large outdoor tanks', *Proc. Indo-Pacific Fisheries Council*, **3**, 49–55.
12. —— (1962). 'The effect of heated effluents on fish', *Int. J. Air and Water Pollut.*, **7**, 541–63.
13. —— (1969). Effects of heated discharges on freshwater fish in Britain', pp. 354–74 in ref. 624.
14. —— (1970). 'River flow and upstream movement and catch of migratory salmonids', *J. Fish Biol.*, **2**, 1–13.
15. ——, and DOWNING, A. L. H. (1966). 'A field and laboratory investigation of the effect of heated effluents on fish', *Fishery Invest. Lond. Ser.*, *1*. **6**(4).
16. ALDERDICE, D. F., and VELSEN, F. P. J. (1978). 'Relation between temperature and incubation time of eggs of chinook salmon (*Oncorhynchus tshawytscha*), *J. Fish. Res. Board Can.*, **35**, 69–75.

281

17. *All England Law Reports* (1952). 1, 1326. Pride of Derby and Derbyshire Angling Association Ltd and Another v. British Celanese Ltd and others.

18. AMES, L. J., *et al.* (1979). 'Amounts of asymmetry in Centrarchid fish inhabiting heated and non-heated reservoirs', *Trans. Amer. Fish. Soc.*, **108,** 489–95.

19. ANDERSON, R. R. (1969). 'Temperature and rooted aquatic plants', *Ches. Sci.*, **10**(3, 4), 157–9.

20. ANDERSON, T. P., and LENAT, D. R. (1978). 'Effect of power plant operation on the zoo-plankton community of Belews Lake, North Carolina', pp. 618–41 in ref. 1067.

21. ANDERSON, N. H., and LEMKUHL, D. M. (1968). 'Catastrophic drift of insects in a woodland stream', *Ecology*, **49,** 189–200.

22. ANDREW, F. J., and GEEN, G. H. (1960). 'Sockeye and pink salmon production in relation to proposed dams in the Fraser River System', *Bull. Internat. Pacif. Salm. Fish. Commission*, 11.

23. ANGELOVIC, J. W., *et al.* (1973). 'Interactions of ionizing radiation, salinity, and temperature on the estuarine fish (*Fundulus heteroclitus*)', pp. 131–41 in ref. 815.

24. —— (1977). *Hydrobiological study of cooling water in the Netherlands. Activities up to 1 June 1977.* U.N.I.P.E.D.E. Commission for Cooling Water Standards. September 1977. Kema, Arnhem.

25. ANSELL, A. D. (1963). 'The biology of *Venus mercenaria* in British waters, and in relation to generating station effluents', *Rep. Challenger Soc.*, **3**(xv).

26. ——, *et al.* (1964). 'Studies on the hard-shell clam. *Venus mercenaria*, in British waters. 1. Growth and reproduction in natural and experimental colonies', *J. Applied Ecol.*, **1,** 63–82.

27. APPOURCHAUX, M. (1952). 'Effets de la temperature de l'eau, sur la Faune et la Flore aquatiques', *L'eau*, **8,** 377.

28. ARMITAGE, P. D. (1977a). 'Invertebrate drift in the regulated River Tees, and an unregulated tributary Maize Beck, below Cow Green dam', *Freshwat. Biol.*, **7,** 167–83.

29. —— (1977b). 'Development of the macro-invertebrate fauna of Cow Green reservoir (Upper Teesdale) in the first five years of its existence', ibid., **7,** 441–54.

30. ARNOLD, G. P. (1974). 'Rheotropism in fishes', *Biol. Revs.*, **49,** 515–76.

31. —— (1978). 'Methods of tracking, deflecting and counting fish in North America', *Fisheries Research Tech. Rep. No. 44.* Min. of Ag. Fish. Food Directorate. Fisheries Research (Lowestoft).

32. ARNDT, H. E. (1968). 'Effect of heated water on a littoral community in Maine', in U.S. Senate Public Wks. Committee on Thermal Pollution. 90th Congress. 2nd Session. *Hearings before subcommittee on Air and Water.*

33. ARTHUR, J. W., and EATON, J. G. (1971). 'Chloramine toxicity to the amphipod *Gammarus pseudolimnaeus*, and the fathead minnow (*Pimphales promelas*), *J. Fish. Res. Board Can.*, **28,** 1841–5.

34. ARTHUR, J. R., *et al.* (1976). 'Parasites of fishes of Aishihik and Stevens lakes, Yukon Territory, and potential consequences of their interlake transfer through a proposed water diversion for hydro-electrical purposes', *J. Fish. Res. Board Can.*, **33,** 2489–99.

35. ASTON, R. J. (1968). 'The effect of temperature on the life cycle, growth and fecundity of *Branchiura sowerbyi* (Oligochaeta: Tubificidae), *J. Zool. (Lond.)*, **154,** 29–40.

36. —— (1973). 'Field and experimental studies on the effects of power station effluents on Tubificids (Oligochaeta, Annelida)', *Hydrobiologia*, **42**(2–3), 225–42.

37. —— (1976). *Aquaculture at power stations in the United States—a summary of information collected during a visit.* C.E.R.L. Lab. Memorandum LM/BIOL/003, Leatherhead, Surrey.

38. ——, and BROWN, D. J. A. (1973). *Local and seasonal variation in populations of the leech* Erpobdella octoculata *(L) in a polluted river warmed by condenser effluents.* C.E.R.L. Report RD/L/R/1845, Leatherhead, Surrey.

39. ——, —— (1975). 'Carp and eel culture in power station cooling-water', *Proc. 7th Coarse Fish. Conf. Liverpool University* (March), 72–80.

40. ——, and MACQUEEN, J. F. (1978). *Temperature control and its advantages in carp culture using power station cooling water.* C.E.R.L. Lab. Note RD/L/N/65/77, Leatherhead, Surrey.

41. ——, and MILNER, A. G. P. (1976). 'Heated water farms at inland power stations., *Fish Farming Int.*, June 1976, 41–44.

42. ASTON, R. J., and MILNER, A. G. P. (1978). *A comparison of populations of* Asellus aquaticus *(L) below power stations in organically polluted reaches of the River Trent.* C.E.R.L. Lab. Report RD/L/R 1977, Leatherhead, Surrey.

43. BAILEY, R. G. (1966). 'Observations on the nature and importance of organic drift in a Devon river'. *Hydrobiologia*, **27**(3–4), 353–67.

44. BAINBRIDGE, R. (1960). 'Speed and stamina in three fish', *J. Exp. Biol.*, **37**(1), 129–53.

45. —— (1975). 'The response of fish to shearing surfaces in the water', pp. 529–40 in ref. 137.

46. BAKER, M. S., *et al.* (1977). *Environmental Impact Statements: A Guide to Preparation and Review.* Practising Law Institute, New York.

47. BALLIGAND, P., *et al.* (1975). 'Utilisation des calories de rejects thermiques pour l'agriculture, la pisciculture et le chauffauge des locaux', pp. 717–30 in ref. 551.

48. BALON, E. K. (1975). 'The eels of Lake Kariba: distribution, taxonomic status, age, growth and density', *J. Fish. Biol.*, **7**, 797–815.

49. —— (1978). 'Kariba: The dubious benefits of large dams', *Ambio.*, **VII**(2), 40–48.

50. BAMBER, R. N. (1978). *The effects of dumped pulverised fuel ash on the benthic fauna of the Northumberland Coast.* PhD Thesis, University of Newcastle-upon-Tyne, U.K.

51. BANKS, J. W. (1969). 'A review of the literature on the upstream migration of adult salmonids'. *J. Fish. Biol.*, **1**, 85–136.

52. BANUS, M. D., and KOLEHMAINEN, S. E. (1976). 'Rooting and growth of red mangrove seedlings from thermally stressed trees', pp. 46–53, in ref. 368.

53. BARBER, Y. M. (1972). Statement as presented before the Fourth Session of the Lake Michigan Enforcement Conference, Sherman House, Chicago, Illinois, September, 19–21.

54. BARDACH, J. E. (1964). *Downstream: A Natural History of the River*, New York: Harper and Row.

55. ——, and DUSSART, B. (1973). 'Effects of man-made lakes on ecosystems', pp. 811–17 in ref. 5.

56. BARNES, R. S. K., and COUGHLAN, J. (1970). *Bradwell biological investigations: the environment in the vicinity of the barrier wall.* C.E.R.L. Lab. Report RD/L/R 1711, Leatherhead, Surrey.

57. BARNETT, P. R. O. (1971). 'Some changes in intertidal sand communities due to thermal pollution'. *Proc. Roy. Soc. Lond. B.*, **177**, 353–64.

58. —— (1972). 'Effects of warm water effluents from power stations on marine life'. *Ibid.*, **180**, 497–509.

59. ——, and HARDY, B. L. S. (1969). 'The effect of temperature on the benthos near Hunterston Generating Station, Scotland., *Ches. Sci.*, **10**, 255–6.

60. BASS, M. L., *et al.* (1977). 'Histopathological effects of intermittent chlorine exposure on bluegill (*Lepomis macrochirus*) and rainbow trout (*Salmo gairdneri*)', *Water Research*, **11**, 731–5.
61. BATES, D. W., and VINSONHALER, R. (1957). 'Use of louvers for guiding fish', *Trans. Amer. Fish. Soc. Proc. 86th A.G.M.*, September 1956, 38–57.
62. BAUER, L., and STERK, G. (1974). 'The energy resources of Austria and their utilization from 1966–1972', paper No. 1. 2–21, *Proc. 9th World Energy Conference, Detroit*. London: W.E.C.
63. BAXTER, G. (1961). 'River utilisation and the preservation of migratory fish life', *Minutes Proc. Instn. Civil Engrs*, **18**, 225–44.
64. BEADLE, L. C. (1974). *The Inland Waters of Tropical Africa. An Introduction to Tropical Limnology*, London: Longman Group.
65. BEALL, S. E., *et al.* (1977). 'Energy from cooling-water'. *Industrial Water Engng.*, **16**(6), 8–14.
66. BEAMISH, F. W. H. (1966). 'Muscular fatigue and mortality in haddock, *Melanogrammus aeglefinus*, caught by Otter trawl'. *J. Fish. Res. Board Can.*, **23**(10), 1507–9.
67. BEAUCHAMP, R. S. A. (1967). 'The effect of biological factors on the design and operation of power stations', pp. 43–51 in Brook, M. (ed.), *Biology and the Manufacturing Industries*. London: Inst. Biol. and Academic Press.
68. —— (1969). 'The use of chlorine in the cooling-water systems of coastal power stations', *Ches. Sci.*, **10**(3–4), 280–1.
69. BECK, A. D. (1978). 'Cumulative effects—A field assessment', pp. 189–210 in ref. 970.
70. BECKER, C. D. (1973). 'Columbia River thermal effects study: reactor effluent problems', *J. Water Pollut. Contr. Fed.*, **45**, 850–69.
71. ——, and THATCHER, T. O. (1973). 'Toxicity of power plant chemicals to aquatic life, *U.S. Atomic Energy Commission Report No. Wash. 1249*. Washington D.C.: U.S. Government Printing Office.
72. ——, *et al.* (1979). 'Synthesis and analysis of ecological information from cooling impoundments', *Report EA-1054*. Palo Alto, Electric Power Research Institute, Ca., U.S.A.
73. BECKER, C. D., and GENOWAY, R. G. (1979). 'Evaluation of the critical thermal maximum for determining thermal tolerance of freshwater fish', *Environ. Biol. Fish.*, **4**(3), 245–6.
74. BEER, L. O., and PIPES, W. O. (1969). *The Effects of Discharge of Condenser Water into the Illinois River*. Northbrook, Illinois, U.S.A.: Industrial Bio-test Labs Inc.
75. BEGG, G. W. (1970). 'Limnological observations on Lake Kariba during 1967 with emphasis on some special features', *Limnol. Oceanogr.*, **15**, 776–88.
76. BEITINGER, T. L., and MAGNUSSON, J. J. (1975). 'Influence of social rank and size on thermoselection behaviour of bluegill (*Lepomis macrochirus*)', *J. Fish. Res. Board Can.*, **32**(11), 2133–6.
77. BELL, F. V. M. (1949). *Timber structures, with particular reference to the maintenance of oil-loading jetties at Queen's Dock, Swansea*. Instn. Civil Engrs. (Marine and Waterways Division), Sess. 1948–9, 3–11.
78. BELLANCA, M. A., and BAILEY, D. S. (1977). 'Effects of chlorinated effluents on aquatic ecosystem in the lower James River', *J. Water Pollut. Contr. Fed.* (April) 639–45.
79. BELTER, W. G. (1975). 'Management of waste heat at nuclear power stations, its possible impact on the environment and possibilities of economic use', pp. 3–23 in ref. 551.

80. BENDER, M. E., *et al.* (1977). 'Effects of residual chlorine on estuarine organisms', pp. 101–8 in ref. 567.
81. BENDIX, S., and GRAHAM, H. R. (eds.) (1978). *Environmental Assessment: Approaching Maturity.* Mich., U.S.A.: Ann Arbor Science Publishers.
82. BENINGEN, K. T., and EBEL, W. F. (1970). 'Effects of the John Day Dam on dissolved nitrogen concentrations and salmon in the Columbia River. 1968', *Trans. Amer. Fish. Soc.,* **99,** 664–71.
83. BENNETT, D. H. (1971). 'Preliminary examination of body temperatures of largemouth bass (*Micropterus salmoides*) from an artificially heated reservoir', *Arch. Hydrobiol.,* **68**(3), 376–82.
84. ——, and GIBBONS, J. W. (1972). 'Food of largemouth bass (*Micropterus salmoides*) from a South Carolina reservoir receiving heated effluent', *Trans. Amer. Fish. Soc.,* **101**(4), 650–4.
85. BENSON, N. G. (ed.) (1970). *A century of fisheries in North America.* Spec. Pub. No. 7. Amer. Fish. Soc.
86. BENSON, W. G., and HUDSON, P. L. (1975). 'Effects of reduced fall drawdown on benthos abundance in Lake Francis Case', *Trans. Amer. Fish. Soc.,* **104**(3), 526–8.
87. BENTLEY, W. W., and RAYMOND, H. L. (1976). 'Delayed migrations of yearling Chinook salmon since completion of Lower Monumental and Little Goose Dams on the Snake River', ibid., **105**(3), 422–4.
88. BEREZAY, G., and GEE, J. H. (1978). 'Buoyancy response to changes in water velocity and its function in Creek Chub (*Semotilus atromaculatus*)', *J. Fish. Res. Board Can.,* **35,** 255–9.
89. BERGSTROM, R. N. (1968). 'Hydrothermal effects of power stations'. Paper to ASCE. Water Resources Conference, May 1968, Chattanooga, Tennessee.
90. BERNHARDT, H. (1967). 'Aeration of Wahbach Reservoir without changing the temperature profile', *J. Amer. Water Works Assoc.,* **59**(8), 943–64.
91. BERRY, J. D. (1955). 'Hydro-electric development and nature conservation in Scotland', *Proc. Roy. Phil. Soc. Glasgow,* **77**(3), 23–36.
92. BEVERTON, R. J. H., and HOLT, S. H. (1957). 'On the dynamics of exploited fish populations', *Fishery Invest. Lond. Ser.,* **2**(19).
93. BEYERS, R. J. (1974). 'Ecological studies in a cooling reservoir in the Southeastern United States', pp. 39–46 in ref. 401.
94. BIRKETT, D. G. (1979). 'Review of potential hydro-electric development in the Scottish Highlands', *Electronics and Power,* May, 339–46.
95. BJOR, K., *et al.* (1976). 'Distribution and chemical enrichment of precipitation in a southern Norway forest stand', July–Dec. 1972, in ref. 124.
96. BLAKE, B. F. (1977a). 'The effect of the impoundment of Lake Kainji, Nigeria, on the indigenous species of mormyrid fishes', *Freshwat. Biol.,* **7,** 37–42.
97. —— (1977b). Lake Kainji, Nigeria. 'A summary of the changes within the fish population since the impoundment of the Niger in 1968', *Hydrobiologia,* **53**(2), 131–7.
98. BLAKE, N. J., *et al.* (1976). 'The macrobenthic community of a thermally altered area of Tampa Bay, Florida', pp. 296–301 in ref. 368.
99. BLAXTER, J. H. S. (1969). 'Swimming speeds of fish', *Fish. Rep. F.A.O.,* **62**(2), 69–100.
100. BLENCH, T. (1972). 'Morphometric changes', pp. 287–308 in ref. 843.
101. BLOCK, R. M. (1974). 'Effects of acute cold shock on the channel catfish', pp. 109–18 in ref. 417.
102. ——, and BURTON, D. T., *et al.* (1978). 'Respiratory and osmoregulatory responses of white perch (*Morone americana*) exposed to chlorine and ozone in estuarine waters', pp. 351–60 in ref. 579.

103. ——, and HELZ, G. R. (eds) (1976). *Proceedings of the Chlorination Workshop.* University of Maryland, Chesapeake Biological Laboratory, 15–18 March.

104. BLUM, J. L. (1957). 'An ecological study of the algae of the Saline River, Michigan', *Hydrobiologia*, **9**, 361–408.

105. BOARD, P. A. (1967). 'Condenser fouling by hydroid flotsam'. C.E.R.L. Lab. Note RD/L/N 116/67, Leatherhead, Surrey.

106. —— (1975). 'Whiteweed fouling of Tilbury "A" Power Station', pp. 83–88 in ref. 205.

107. ——, and COLLINS, T. M. (1965). 'Physical forces in marine fouling', *Discovery*, **26**, 14–18.

108. BOND, W. J., *et al.* (1978). 'The limnology of Cabora Bassa, Moçambique during its first year', *Freshwat. Biol.*, **8**, 433–47.

109. BORGSTROM, E. (1973). 'The effect of increased water level fluctuation upon the brown trout population of Marvann, a Norwegian reservoir', *Norw. J. Zool.*, **4**, 101–12.

110. BORODICH, N. D., and HAVLENA, F. K. (1973). 'The biology of mysids acclimatized in the reservoirs of the Volga River', *Hydrobiologia*, **42**(4), 527–39.

111. BOTT, T. L. (1975). 'Bacterial growth rates and temperature optima in a stream with a fluctuating thermal regime', *Limnol. Oceanogr.*, **20**(2), 191–7.

112. BOUQUET, H. (1977). 'Grass carp in the Netherlands', pp. 108–12, in *Proc. 8th Brit. Coarse Fish. Conf.* London: Janssen Services.

113. BOURQUE, J. E., and ESCH, G. W. (1974). 'Population ecology of parasites in turtles from thermally altered and natural aquatic communities', pp. 551–61 in ref. 417.

114. BOWEN, M. (1976). 'Effects of a thermal effluent on the Ostracods of Par Pond, South Carolina', pp. 219–26 in ref. 368.

115. BOWERS, A. B., and NAYLOR, E. (1964). 'Occurrence of *Atherina boyeri*, Risso.', *Nature (Lond.)*, **202**, 318.

116. BOYAR, H. C. (1961). 'Swimming speed of immature Atlantic herring with reference to the Passamaquoddy tidal project', *Trans. Amer. Fish. Soc.*, **90**(1), 21–6.

117. BOYD, C. E. (1971). 'The limnological role of aquatic macrophytes and their relationship to reservoir management', pp. 153–66 in ref. 461.

118. BOYDEN, C. R., and ROMERIL, M. G. (1974). 'A trace metal problem in pond oyster culture', *Mar. Pollut. Bull.*, **5**(5), 74–8.

119. BOYER, P. B. (1973). 'Gas supersaturation problem in the Columbia River', pp. 701–4 in ref. 5.

120. BOYLEN, C. W., and BROCK T. D. (1973). 'Effects of thermal additions from the Yellowstone geyser basins on the benthic algae of the Firehole River', *Ecology*, **54**, 1283–91.

121. BOITSOV, M. P. (1974). 'Morphology of fingerlings in the Ivankovo Reservoir affected by hot water discharge from the Konakovo Hydroelectric Power Station', *Vopr. Ikhtiol* (U.S.S.R.), **14**, 1046.

122. BRACKETT, C. E. (1967). 'Availability, quality and present utilization of fly-ash', *J. Water Pollut. Contr. Fed.*, **48**(9), 2163.

123. BRADY, D. K., *et al.* (1969). 'Surface heat exchange at power plant cooling lakes', *Cooling Water Studies for Edison Electric Inst: Research Project RP-49. Report no. 5.* Edison Electric Inst., Pub. no. 69–901.

124. BRAEKKE, F. H. (ed.) (1976). *S.N.S.F. Project. FR6/76.* Norway. C.E.G.B., Translation 6579, London.

125. BRAUER, G. A., *et al.* (1974). 'Effects of a power plant on zooplankton distribution and abundance near plant's effluent', *Water Research*, **8**, 485–9.

126. BRAY, E. S. (1971). 'Observations on the reproductive cycle of the roach (*Rutilus rutilis*) with particular reference to the effects of heated effluents'. *Proc. 5th Brit. Coarse Fish. Conf.* University of Liverpool. London: Janssen Services.
127. BREGMAN, J. (1969). Keynote address pp. 3–14 in ref. 624.
128. BRETT, J. R. (1956). 'Some principles in the thermal requirements of fishes', *Qu. Rev. Biol.*, **31**(2), 75–87.
129. —— (1960). 'Thermal requirements of fish', *Robert A. Taft Sanit. Engrg. Cent. Tech. Rep. W60-3.*, 110–17.
130. —— (1964). 'The respiratory metabolism and swimming performance of young sockeye salmon', *J. Fish. Res. Board Can.*, **21**, 1183–1226.
131. —— (1970). 'Fishes, functional responses', chap. 3. (Temperature), pp. 515–60 in ref. 602.
132. ——, and MACKINNON, D. (1954). 'Some observations on olfactory perception in migrating adult coho and spring salmon', *J. Fish. Res. Board Can.*, **11**, 310–18.
133. BRIAND, F. J. P. (1975). 'Effects of power plant cooling-systems on marine phytoplankton', *Marine Biology*, **33**, 135–46.
134. BRISBIN, I. L. (1974). 'Abundance and diversity of waterfowl inhabiting heated and unheated portions of a reactor cooling reservoir', pp. 579–93 in ref. 417.
135. BROCK, T. D. (1970). 'High temperature systems', *Ann. Rev. Ecol. Systematics*, **1**, 191–220.
136. —— (1975). 'Predicting the ecological consequences of thermal pollution from observations on geothermal habitats', pp. 599–621 in ref. 551.
137. BROKAW, C. J., and BRENNEN, C. (1975). *Swimming and flying in Nature.* Vol. 2. New York and London: Plenum Press.
138. BROOK, A. J., and RZÒSKA, J. (1954). 'The influence of the Gebel Aulia Dam on the development of Nile plankton', *J. Anim. Ecol.*, **23**, 101–14.
139. BROOKE, L. T. (1975). 'Effect of different constant incubation temperatures on egg survival and embryonic development in Lake Whitefish (*Coregonus clupeaformis*)', *Trans. Amer. Fish. Soc.*, **104**, 555–9.
140. BROOKER, M. P., and HEMSWORTH, R. J. (1978). 'The effect of the release of an artificial discharge of water on invertebrate drift in the River Wye, Wales', *Hydrobiologia*, **59**(3), 155–63.
141. BROOKS, A. S., and SEEGERT, G. L. (1977). 'The effects of intermittent chlorination on rainbow trout and yellow perch'. *Trans. Amer. Fish. Soc.*, **106**(3), 278–86.
142. BROSSET, C. (1973). 'Air-borne acid', *Ambio*, **2**(1–2), 2–9.
143. BROWN, A. W. A., and DEOM, J. O. (1973). 'Summary: Health aspects of man-made lakes', pp. 755–64 in ref. 5.
144. BROWN, D. J. A. (1977a). *Preliminary observations on the effect of power station cooling water discharges on the size of coarse-fish fry in rivers.* C.E.R.L. Lab. Note RD/L/N 7/73, Leatherhead, Surrey.
145. —— (1973). 'The effect of power station cooling-water discharges on the growth of coarse-fish fry', *Proc. 6th Brit. Coarse Fish Conf.* Liverpool, pp. 191–202. London: Janssen Services.
146. —— (1976). *The effects of power station cooling-water discharges on the growth and distribution of coarse-fish fry. A further year's study.* C.E.R.L. Lab. Note RD/L/N 247/75, Leatherhead, Surrey.
147. —— (1978). *The effects of various cations on the survival of brown trout* (Salmo trutta) *at low pHs'.* C.E.R.L. Lab. Note RD/L/N 155/7, Leatherhead, Surrey.
148. BROWN, D. J. A., and ASTON, R. J. (1972). *A further trial on the use of rotenone for fish recovery before infilling with P.F.A.* C.E.R.L. Lab. Note RD/L/N 271/72, Leatherhead, Surrey.

149. BROWN, D. J. A., and ASTON, R. J. (1975). *Chlorine and chloramine concentrations in the cooling-water of Ratcliffe-on-Soar power station and the potential toxicity to fish.* C.E.R.L. Lab. Note RD/L/N 134/75, Leatherhead, Surrey.

150. BROWN, G. W. (1970). 'Predicting the effect of clearcutting on stream temperature', *J. Water and Soil Conservat.*, **25**, 11–13.

151. ——, and KRYGIER, J. T. (1967). 'Changing water temperatures in small mountain streams', ibid., **22**, 242–4.

152. ——, —— (1970). 'Effects of clear cutting on stream temperature', *Water Resources Research*, **6**, 1133–39.

153. ——, and BRAZIER, J. R. (1972). *Controlling thermal pollution in small streams. Report no. E.P.A. R2-72 083.* Washington. Office of Research and Monitoring. U.S. Environmental Protection Agency.

154. BROWN, J., *et al.* (1976). 'The disposal of pulverised fuel ash in water supply catchment areas', *Water Research*, **10**, 1115–21.

155. BROWN, M.E. (1957). 'Experimental studies on growth', in Brown, M. E. (ed.) *The Physiology of Fishes*, Vol. 1, Metabolism. New York: Academic Press.

156. BROWN, V. M. (1968). 'The calculation of the acute toxicity of mixtures of poisons to Rainbow trout', *Water Research*, **2**, 723–33.

157. BRUNGS, W. A. (1973). 'Effects of residual chlorine on aquatic life', *J. Water Pollut. Contr. Fed.*, **45**, 2180–93.

158. BRYAN, G. W. (1971). 'The effects of heavy metals (other than mercury) on marine and estuarine organisms', *Proc. Roy. Soc.* (Lond. B), **177**, 389–410.

159. ——, and HUMMERSTONE, C. G. (1973). 'Brown seaweed as an indicator of heavy metals in estuaries in south-west England', *J. Mar. Biol. Ass. U.K.*, **53**, 705–20.

160. BUDENHOLZER, R. J., *et al.* (1971). 'Selecting heat rejection systems for future steam electric power plants'. *Westinghouse Eng.*, Nov, 168–75.

161. BULL, H. O. (1937). 'Studies on conditioned responses in fishes. Part VII. Temperature perception in teleosts', *J. Mar. Biol. Ass. U.K.*, **21**(1), 1–27.

162. BULLEID, M. J. (1974). 'A preliminary report on the effect of water temperature on the early growth of Rainbow trout (*Salmo gairdneri*)', *J. Inst. Fish. Mgmt.*, **5**(1), 16–22.

163. BUNTING, D. L. (1974). 'Zooplankton. Thermal regulation and stress', pp. 50–55 in ref. 401.

164. BURNETT, J. M., *et al.* (1974). 'The cooling of power stations', Paper No. 2.2.5. *Proc 9th World Energy Conf., Detroit.* London: W.E.C.

165. BURTON D. T., *et al.* (1976). 'Effects of low ΔT power plant temperatures on estuarine invertebrates', *J. Water Pollut. Contr. Fed.*, **48**, 2259—72.

166. ——, and LIDEN, L. H. (1978). 'Biofouling control alternatives to chlorine as a control agent for cooling water treatment', pp. 717–34 in ref. 579.

167. BUSH, R. M., *et al.* (1974). 'Potential effects of thermal discharges on aquatic systems', *Environ. Sci. and Technol.*, **8**(6), 561–68.

168. BUTCHER, R. W., *et al.* (1930). 'Variations in composition of river waters', *Int. Rev. ges. Hydrobiol. Hydrogr.*, **24**, 47–80.

169. BUTCHER, R. W. (1947). 'Studies on the ecology of rivers. VII. The algae of organically enriched waters', *J. Ecol.*, **33**, 268–83.

170. CADA, G. F., *et al.* (1979). 'A biological evaluation of devices used for reducing entrainment and impingement losses at thermal power plants', pp. 181–214 in *Environmental effects of hydraulic engineering works.* Proceedings of an international symposium. T.V.A. Knoxville, Tenn., 12–14 Sept. 1978.

171. CAIRNS, J. (1956a). 'The effects of heat on fish', *Industr. Wastes.*, May–June, 180–83.

172. —— (1956b). 'The effects of increased temperatures on aquatic organisms', *Proc. 10th Ind. Waste. Conference. Purdue University Eng. Bull.*, **40**(1), 346.

173. —— (1969). 'Ecological management problems caused by heated waste water discharges into the aquatic environment', *Proc. Governors' Conf. on Thermal Pollution*, 18 July 1969. Michigan, Traverse City (mimeo).

174. —— (ed.) (1977). *The Structure and Function of Freshwater Microbial Communities*. Amer. Microscop. Soc.

175. ——, *et al.* (1978). *Effects of temperature on aquatic organism sensitivity to selected chemicals.* Virgina Water Resources Research Center. Bulletin 106. Virginia, U.S.A.: Blacksburg.

176. ——, *et al.* (1970). 'The biological recovery of the Clinch River following a fly-ash pond spill', *Proc. 25th Purdue Ind. Waste Conference*, Purdue University, 5–7 May.

177. ——, *et al.* (1972). 'The response of aquatic communities to spills of hazardous materials., *Proc. Nat. Conf. of Hazardous Material Spills.* 179–97.

178. ——, LANZA, G. R., and PARKER, B. C. (1972). Pollution related to structural and functional changes in aquatic communities with emphasis on freshwater algae and protozoa', *Proc. Acad. of Nat. Sciences, Philadelphia*, **124**(5), 79–127.

179. ——, and PLAFKIN, J. L. (1975). 'Response of protozoan communities exposed to chlorine stress', *Archiv. Für Protistenkunde*, **117**, 47–53.

180. ——, *et al.* (1976). 'Invertebrate response to thermal shock following exposure to acutely sublethal concentrations of chemicals', *Arch. Hydrobiol.*, **77**, 164–75.

181. ——, KAESLER, R. L., *et al.* (1976). 'The influence of natural perturbation on protozoan communities inhabiting artificial substrates', *Trans. Amer. Microscop. Soc.*, **95**(4), 646–53.

182. ——, and MESSENGER, D. (1974). 'An interim report on the effects of prior exposure to sublethal concentrations of toxicants upon the tolerance of snails to thermal shock', *Arch. Hydrobiol.*, **74**(4), 441–7.

183. ——, *et al.* (1975). 'Temperature influence on chemical toxicity to aquatic organisms', *J. Water Pollut. Contr. Fed.*, **47**(2), 267–80.

184. ——, *et al.* (1973). 'The protozoan colonization of polyurethane foam units anchored in the benthic area of Douglas Lake, Michigan', *Trans. Amer. Microscop. Soc.*, **92**(4), 648–56.

185. CALOW, P. (1975). 'The respiratory strategies of two species of freshwater gastropods (*Ancylus fluviatilis* Mull and *Planorbis contortus*. Linn.) in relation to temperature, oxygen concentration, body size and season', *Physiol. Zool.*, **48**, 114–24.

186. Calvert Cliffs' Co-ordinating Committee, Inc., *et al.* v. United States Atomic Energy Commission and United States of America. no. 24,839. D.C. Cir. 23 July 1971.

187. CAPPER, C. B. (1974). 'The protection of open recirculating cooling systems', *Effluent and Water Treatment Jnl.*, **125**, 577–83.

188. —— (1979). 'The effects of halogen toxicants on survival, feeding and egg production of the rotifer, *Brachionus plicatilis*', *Estuarine and Coastal Mar. Sci.*, **8**(4), 307–16.

189. CAPUZZO, J. M., *et al.* (1977). 'The different effects of free and combined chlorine on juvenile marine fish', *Estuarine and Coastal Mar. Sci.*, **5**, 733–41.

190. CARPENTER, E. J. (1973). 'Brackish-water phytoplankton response to temperature elevation', *ibid.*, **1**, 37–44.

191. ——, *et al.* (1972). 'Cooling-water chlorination and productivity of entrained phytoplankton', *Marine Biology*, **16**, 37–40.

192. ——, *et al.* (1974). 'Survival of copepods passing through a nuclear power station on North-eastern Long Island Sound, U.S.A.', ibid., **24**, 49–55.

193. CARPENTER, J. H. (1977). 'Problems in measuring residuals in chlorinated seawater', *Ches. Sci.*, **18**(1), 112.

194. CARR, W. E. S., and GIESEL, J. T. (1975). 'Impact of thermal effluent from a steam electric station on a marshland nursery area during the hot season', *Fishery Bull.*, **73**(1), 67–80.

195. CARRICK, T. R. (1979). 'The effect of acid water on the hatching of salmonid eggs', *J. Fish. Biol.*, **14**, 165–72.

196. CARRIER, R. F., and HANNON, E. H. (1979). *Entrainment: an annotated bibliography.* Palo Alto, Electric Power Research Institute. EA-1049, Interim Report, April 1979.

197. CARSTENS, T. (1975). 'Trapping of heat in sill fjords', pp. 99–111 in ref. 551.

198. CARTHY, J. D., and ARTHUR, D. R. (eds) (1968). *The Biological Effects of Oil Pollution on Littoral Communities.* Supplement to vol. 2 of *Field Studies.* Field Studies Council.

199. C.E.G.B. (1963). *Glossary of Power Station Terms.* Instructions Booklet no. 7. Personnel Department., C.E.G.B., Newgate Street, London.

200. —— (1976). *Proceedings of the Seminar on Aquatic Environmental Studies.* Dinorwic Pumped Storage Scheme, Generat. Construction and Design Division, C.E.G.B., Llanberis, North Wales.

201. —— S.W. Region. (1975). *Uskmouth Power Station. Report on smolt migration 1975.* Bristol. C.E.G.B. S.W. Region.

202. CEMBER, H., *et al.* (1978). 'Mercury bioconcentration in fish: temperature and concentration effects', *Environ. Pollut.*, **17**(4), 311–9.

203. C.E.Q./E.P.A. (1972). *The economic impact of pollution control: A summary of recent studies, Council on Environmental Quality.* Department of Commerce and Environmental Protection Agency, March, U.S. Government Printing Office.

204. C.E.R.L. (1971). Symposium on Freshwater Biology and Electrical Power Generation, C.E.G.B. Research Division. C.E.R.L. Lab. Memo. RD/L/M 312. Parts I and II, Leatherhead, Surrey.

205. —— (1975). Symposium on marine science and electricity generation in Southampton Water, 27 June 1973. Fawley, C.E.R.L. Lab. Memo LM/Biol/001, Leatherhead, Surrey.

206. CHERRY, D. S., *et al.* (1974). 'Temperature influence on bacterial populations in three aquatic systems', *Water Research*, **8**, 149–55.

207. ——, *et al.* (1975). 'Temperatures selected and avoided by fish at various acclimation temperatures', *J. Fish. Res. Board Can.*, **32**(4), 485–91.

208. ——, *et al.* (1976). 'Responses of Mosquito fish (*Gasmbusia affinis*) to ash effluent and thermal stress', *Trans. Amer. Fish Soc.*, **105**(6), 686–94.

209. ——, *et al.* (1977). 'Field-laboratory determined avoidances of the spotfin shiner and the bluntnose minnow to chlorinated discharges', *Water Research Bull.*, **13**(5), 1047–55.

210. ——, LARRICK, S. R., *et al.* (1977). 'Response of eurythermal and stenothermal fish species to chlorinated discharges', pp. 413–18 in Hemphill, D. D. (ed.), *Trace Substances in Environmental Health XL.* A Symposium. Columbia, Missouri: University of Missouri.

211. ——, ——, HOEHN, R. C., and CAIRNS, J. (1977). 'Significance of hypochlorous acid in free residual chlorine to the avoidance response of spotted bass

(*Micropterus punctulatus*) and Rosyface shiner (*Notropis rubellus*)', *J. Fish. Res. Board Can.*, **34**(9), 1365–72.

212. ——, ——, *et al.* (1979). 'Recovery of invertebrate and vertebrate populations in coal ash stressed drainage systems', *J. Fish. Res. Board Can.*, **36**, 1089–96.

213. CHRISTY, E. J., *et al.* (1974). 'Enhanced growth and increased body size of turtles living in thermal and post-termal aquatic systems', pp. 277–84 in ref. 417.

214. CHU, T. Y. J., and OLEM, H. (1980). 'Power industry wastes', *J. Water Pollut. Contr. Fed.*, **52**(6), 1433–1445.

215. CHURCHILL, M. A., and WOJTALIK, T. A. (1969). 'Effects of heated discharges on aquatic environment. The T.V.A. Experience', *American Power Conference Chicago, Illinois.*

216. CHUTTER, F. M. (1963). 'Hydrobiological studies on the Vaal River in the Vereenigung area. Part 1. Introduction, water chemistry, and biological studies on the fauna of habitats other than muddy bottom sediments', *Hydrobiologia*, **21**, 1–65.

217. —— (1972). 'A re-appraisal of Needham and Usinger's data on the variability of a stream fauna when sampled with a Surber sampler'. *Limnol. Oceanogr.*, **17**, 139–41.

218. CLARIDGE, P. N. (1976). *Studies on the fish fauna of the Severn Estuary based on power station sampling.* University of Bath/Report London (mimeo).

219. ——, and GARDENER, D. C. (1978). 'Growth and movements of the twaite shad (*Alosa fallax*) (Lacépède) in the Severn Estuary', *J. Fish Biol.*, **2**, 203–11.

220. CLAY, C. H. (1961). *Design of Fishways and other Fish Facilities.* Ottawa: Queen's Printer.

221. CLEAVER, F. (1977). 'Role of hatcheries in the management of Columbia River salmon', pp. 89–92 in ref. 973.

222. CLEMENT, L. J. (1972). 'The increase in the oxygen content of water in the River Maas, Amer and Donge due to the presence of Amer Power Station (Holland)', *Elektrotechniek*, **50**(21), 838–42. *C.E.G.B. Trans.* 6192.

223. COKE, M. (1968). 'Depth distribution of fish on a bush-cleared area of Lake Kariba, Central Africa', *Trans. Amer. Fish. Soc.*, **97**(4), 460–65.

224. COLE, R. A., and KELLY, J. E. (1978). 'Zoobenthos in thermal discharge to Western Lake Erie', *J. Water Pollut. Contr. Fed.*, **50**(11), 2509–21.

225. —— (1977). 'Chlorination for the control of biofouling in thermal power plant cooling water systems', pp. 29–37 in ref. 567.

226. COLLINS, G. B., *et al.* (1962). 'Ability of Salmonids to ascend high fishways', *Trans Amer. Fish. Soc.*, **91**(1), 1–7.

227. COLLINS, T. M. (1976). *A simple classification of debris in cooling water.* C.E.R.L. Lab. Note RD/L/N66/76, Leatherhead, Surrey.

228. *Concise Oxford Dictionary* (1976). Oxford: Oxford University Press.

229. COOTNER, P. H., and LOF, G. O. (1965). *Water Demand for Stream Electric Generation.* Resources for the Future Inc., Baltimore: Johns Hopkins Press.

230. CORY, R. L., and NAUMAN, J. W. (1969). 'Epifauna and thermal additions', *Ches. Sci.*, **10**, 210.

231. COTILLON, J. (1974). 'La Rance: six years of operating tidal power plant in France', *Water Power*, Oct., 314–32.

232. COUGHLAN, J. (1977). 'Marine wood borers in Southampton Water. 1951–1975', *Proc. Hants Field Club Archaeol. Soc.*, **33**, 5–15.

233. ——, and FLEMING J. M. (1978). 'A versatile pump-sampler for live zooplankton', *Estuaries*, **1**(2), 132–135.

234. ——, and HOLMES, N. J. (1972). *Established communities of fouling organisms in*

Southampton water. C.E.R.L. Lab, Report RD/L/R 1813, Leatherhead, Surrey.

235. ——, and WHITEHOUSE, J. W. (1977). 'Aspects of chlorine utilisation in the United Kingdom', *Ches. Sci.*, **18**(1), 102–11.

236. COULSON, H. J. W., and FORBES, V. A. (1952). *The Law of Waters and Land Drainage.* Hobday S. R. (ed.), (6th ed). London: Sweet and Maxwell.

237. COUTANT, C. C. (1962). 'The effect of a heated water effluent upon the macroinvertebrate riffle fauna of the Delaware River', *Proc. Amer. Acad. Sci.*, **36**, 58–71.

238. —— (1963). 'Stream plankton above and below Green Lane Reservoir', *Proc. Pa. Acad. Sci.*, **36**, 58–71.

239. —— (1967). 'Thermal pollution. Biological effects', *J. Water Pollut. Contr. Fed. Ann. Review.*

240. —— (1968a). 'Thermal pollution—biological effects. A review of the literature of 1967 on wastewater and water pollution control', *J. Water Pollut. Contr. Fed.*, **40**, 1047–52.

241. —— (1968b). 'Effect of temperature on the development rate of bottom organisms', pp. 11–12, in *Biological Effects of Thermal Discharges.* Ann. Rep. Pacif. N.W. Lab. U.S. Atomic Energy Commission. Division of Biological Medicine.

242. —— (1969). 'Thermal pollution—biological effects. A review of the literature of 1968 on wastewater and water pollution control', *J. Water Pollut. Contr. Fed.*, **41**, 1036–53.

243. —— (1970a). 'Thermal pollution—biological effects. A review of the literature of 1969 on wastewater and water pollution control', *J. Water Pollut. Contr. Fed.*, **42**, 1025–57.

244. —— (1970b). 'Biological aspects of thermal pollution—I. Entrainment and discharge canal effects', *C.R.C. Critical Review in Environmental Control*, **1**, 341–81.

245. —— (1971a). 'Thermal pollution—biological effects. A review of the literature of 1970 on water pollution control', *J. Water Pollut. Contr. Fed.*, **43**, 1292–1334.

246. —— (1971b). 'Effects of organisms of entrainment in cooling water: steps towards predictability', *Nuclear Safety*, **12**, 600–7.

247. —— (1973). 'Effects of thermal shock on the vulnerability of juvenile salmonids to predation', *J. Fish. Res. Board Can.*, **31**, 351–4.

248. —— (1975). 'Temperature selection by fish. A factor in power plant impact assessments', pp. 575–95 in ref. 551.

249. —— (1976). 'How to put waste heat to work', *Environ. Sci. and Technol.*, **10**(9), 868–71.

250. ——, *et al.* (1976). 'Further studies of cold-shock effects on susceptibility of young channel catfish to predation', pp. 154–8 in ref. 368.

251. ——, and GENOWAY, R. G. (1968). *Final report on an exploratory study of interaction of increased temperature and nitrogen supersaturation on mortality of adult salmonids.* Battelle Memorial Inst. Pacif. Northwest Labs., Richland, Washington.

252. ——, and GOODYEAR, C. P. (1972). 'Thermal effects', *J. Water Pollut. Contr. Fed.*, **44**, 1250–94.

253. ——, and PFUDERER, H. A. (1973). 'Thermal effects', ibid., **45**, 1331–2593.

254. ——, —— (1974). 'Thermal effects', ibid., **46**, 1476–1540.

255. ——, and TALMAGE, S. S. (1975). 'Thermal effects', ibid., **47**, 1656–1710.

256. ——, —— (1976). 'Thermal effects', ibid., **48**(6), 1487–1544.

257. ——, —— (1977). 'Thermal effects', ibid. (June), 1369–1425.

258. COUTANT, C. C., *et al.* (1978). 'Chemistry and biological hazard of a coal ash seepage stream', *J. Water Pollut. Contr. Fed.*, **50**, 747–52.

259. COVICH, A. P., *et al.* (1978). 'Effects of fluctuating flow rates and water levels on chironomids: Direct and indirect alterations of habitat stability', pp. 141–56 in ref. 1067.

260. COWELL, B. C. (1970). 'The influence of plankton discharges from an upstream reservoir on standing crops in a Missouri River reservoir', *Limnol. Oceanogr.*, **15**(3), 427–41.

261. COWELL, E. B. (ed.) (1971). *The Ecological Effects of Oil Pollution on Littoral Communities.* Institute of Petroleum, Great Britain: Applied Science Publishers Ltd.

262. COX, D. K., and COUTANT, C. C. (1976). 'Acute cold-shock resistance of gizzard shad', pp. 159–161 in ref. 369.

263. CRAGG-HINE, D. (1967). *A preliminary survey of dissolved oxygen content and temperature of the warm water discharge from Peterborough Power Station.* C.E.R.L. Lab. Note RD/L/N62/67, Leatherhead, Surrey.

264. —— (1968). *The use of rotenone (derris extract) for the recovery of live fish.* C.E.R.L. Lab. Note RD/L/N149/67, Leatherhead, Surrey.

265. —— (1969a). *The reproductive cycle of fishes in the effluent channel of Peterborough Power Station. Leatherhead, Surrey.* C.E.R.L. Lab. Note RD/L/N157/69, Leatherhead, Surrey.

266. —— (1969b). *The feeding habits of fish in the effluent channel of Peterborough Power Station.* C.E.R.L. Lab. Report RD/L/R-1556, Leatherhead, Surrey.

267. —— (1971). 'Coarse fish populations in the electricity cut, Peterborough', *Proc. 5th Brit. Coarse Fish. Conf. Liverpool*, 19–28.

268. CRAMER, S. P., and McINTYRE, J. D. (1975). 'Heritable resistance to gas-bubble disease in fall Chinook salmon (*Oncorhynchus tshawytscha*)', *Trans. Amer. Fish. Soc.*, **104**, 934–8.

269. CRAWSHAW, L. I. (1977). 'Physiological and behavioural reactions of fishes to temperature change', *J. Fish. Res. Can.*, **34**(5), 730–4.

270. CRISP, D. J. (ed.) (1964). 'The effects of the severe winter of 1962–1963 on marine life in Britain', *J. Anim. Ecol.*, **33**, 165–210.

271. —— (1965). 'The ecology of marine fouling', pp. 99–117 in Goodman, G. T., Edwards, R. W., and Lambert, J. M. (eds.). *Ecology and the Industrial Society. Brit. Ecol. Soc. Symp. no. 5.* Oxford: Blackwell.

272. ——, and MOLESWORTH, A. H. N. (1951). 'Habitat of *Balanus amphitrite* var. *denticulata* in Britain', *Nature* (Lond.), **167**, 489.

273. CRISP, D. T., and LE CREN, E. D. (1970). 'The temperature of three different small streams in Northwest England', *Hydrobiologia*, **35**(2), 305–23.

274. ——, *et al.* (1978). 'The effects of impoundment and regulation upon the stomach contents of fish at Cow Green, Upper Teesdale', *J. Fish. Biol.*, **12**, 287–301.

275. CUMMINGS, N. W., and RICHARDSON, B. (1927). 'Evaporation from lakes', *Physical Review*, **30**, 527–34.

276. CUMMINS, K. W. (1972). 'What is a river—zoological description', pp. 33–55 in ref. 843.

277. CUSHING, D. H. (1974). 'The natural regulation of fish populations', in Jones, H. (ed.), *Sea Fisheries Research.* London: Paul Elek Ltd.

278. DAHLBERG, M. D., and CONVERS, J. C. (1974). 'Winter fauna in a thermal discharge with observations on a macrobenthos sampler', pp. 414–22 in ref. 417.

279. DALE, H. M., and GILLESPIE, T. (1976). 'The influence of floating vascular

plants on the diurnal fluctuations of temperature near the water surface in early spring', *Hydrobiologia*, **49**(3), 246–56.

280. D'ARCY, B. J., and PUGH-THOMAS, M. (1978). 'The occurrence and numbers of fish in screenings from a cooling tower intake in the Manchester Ship Canal', *Estuarine and Brackish-Water Sci. Assoc.* Bulletin No. 20.

281. DARE, P. J. (1976). 'Settlement, growth and production of the mussel, *Mytilus edulis* L. in Morecambe Bay, England', *Min. of Ag. Fish Food, Fishery Invest.* Series 11, **28**(1). London: H.M.S.O.

282. DAVIES, B. R. (1975). 'Cabora Bassa hazards', *Nature* (Lond.), **254**, 477–8.

283. DAVIES, I. (1966). 'Chemical changes in cooling water towers', *Int. J. Air and Water Pollut.*, **10**, 853–63.

284. DAVIES, R. M., and JENSEN, L. D. (1974). 'Effects of entrainment of zooplankton at three mid-Atlantic power plants', *Cooling-water Disch. Res. Project Rept.* no. 10. Elect. Power Res. Inst. Publn. no 74-049-00-1. Palo Alto, California, U.S.A.

285. ——, —— (1975). 'Zooplankton entrainment at three mid-Atlantic power plants', *J. Water Pollut. Contr. Fed.*, **47**(8), 2130–42.

286. ——, *et al.* (1976). 'Entrainment of estuarine zooplankton into a mid-Atlantic power plant. Delayed effects', pp. 349–57 in ref. 368.

287. DAVIS, D. S. (1967). 'The marine fauna of the Blackwater estuary and adjacent waters', *Essex Naturalist*, **32**(1), 1–12.

288. DAVIS, M. H. (1977). 'Heavy metals in the common cockle, *Cerastoderma edulis* (L) from Southampton Water', M.Sc. Thesis, Brunel University.

289. —— (in press). *The response of entrained phytoplankton to chlorination at a coastal power station.* C.E.R.L., Leatherhead, Surrey.

290. ——, and COUGHLAN, J. (1978). 'Response of entrained plankton to low-level chlorination at a coastal power station', pp. 369–76 in ref. 579.

291. DAVIS, J. M., and COLLINGWOOD, R. W. (1978). 'Destratification in reservoirs', *Water Services*, **82**(990), 487–90.

292. DEACUTIS, C. F. (1978). 'Effect of thermal shock on predator avoidance by larvae of two fish species', *Trans. Amer. Fish. Soc.*, **107**(4), 632–5.

293. DE-JONCKHEERE, J., *et al.* (1975). 'The effect of thermal pollution on the distribution of *Naegleria fowleri*', *J. Hyg. Camb.*, **75**, 7–13.

294. DEMONT, D. J., and MILLER, R. W. (1971). 'First reported incident of gas-bubble disease in the heated effluent of a steam generating station', *Proc. 25th Ann. Conf. S.E. Assoc. Game and Fish Comm.* 392–9, South Carolina, Charleston.

295. DE PALMA, J. R. (1971). *An annotated bibliography of marine fouling for marine scientists and engineers* (mimeo).

296. DESCAMPS, B., *et al.* (1976). 'Étude comparée de la croissance des anguilles en fonction de la temperature dans deux bassins en circuit ouvert'. C.I.E.M./ C.E.C.P.I. Symposium sur la recherche et l'exploitation rationelle des anguilles. Paper No. 6. Helsinki, 9–11 June.

297. DE SYLVA, D. P. (1969). 'Theoretical considerations of the effects of heated effluents on marine fishes', pp. 229–93 in ref. 624.

298. DE TURVILLE, C. M., and JARMAN, R. T. (1965). 'The mixing of warm water from the Uskmouth Power Stations in the estuary of the River Usk', *Int. J. Air and Water Pollut.*, **9**, 239–51.

299. DICKSON, I. W. (1975). 'Hydro-electric development of the Nelson River system in Northern Manitoba', *J. Fish. Res. Board Can.*, **32**, 106–16.

300. DICKSON, K. L., *et al.* (1974). 'Effects of intermittently chlorinated cooling tower blowdown on fish and invertebrates', *Amer. Chem. Soc.*, **8**(9), 845–9.

301. ——, et al. (1977). 'Effects of intermittent chlorination on aquatic organisms and communities', *J. Water Pollut. Contr. Fed.*, **49**, 35–44.
302. DILLON, P. J., et al. (1978). Acidic precipitation in South-Central Ontario: Recent observations', *J. Fish. Res. Board Can.*, **35**, 809–15.
303. DIZON, A. E., et al. (1977). 'Rapid temperature compensation of volitional swimming speeds and lethal temperatures in tropical tunas (Scombridae)', *Environ. Biol. Fish.*, **2**(1), 83–92.
304. DOCHINGER, L. S., and SELIGA, T. A. (1975). 'Acid preicipitation and the forest ecosystem', *J. Air. Pollut. Contr. Assoc.*, **25**(11), 1103–5.
305. DOMINY, C. L. (1973). 'Recent changes in Atlantic Salmon (*Salmo salar*) runs in the light of environmental changes in the Saint John River, New Brunswick, Canada', *Biol. Conservat.*, **5**(2), 105–13.
306. DOUDOROFF, P., and KATZ, M. (1950). 'Critical review of literature on the toxicity of industrial wastes and their components to fish. 1. Alkalis, acids and inorganic gases', *Sewage and Indust. Wastes*, **22**(11), 1432–58.
307. DOUGLAS, B. (1958). 'The ecology of the attached diatoms and other algae in a stony stream', *J. Ecol.*, **46**, 295–322.
308. DOWNS, D. I., and MEDDCOCK, K. R. (1974). 'Design of fish conserving intake system'. *Proc. of the A.S.C.E. Journal of the Power Division*, 11008 (PO2), 191–205.
309. DRALEY, J. E. (1977). 'Biofouling control in cooling towers and closed cycle systems', pp. 23–28 in ref. 567.
310. DREESEN, D. R., et al. (1977). 'Comparison of levels of trace elements extracted from fly-ash and levels found in effluent waters from a coal-fired plant', *Environ. Sci. Technol.*, **11**, 1017–19.
311. DRESSEL, D. M., et al. (1972). 'Vital staining to sort live and dead copepods', *Ches. Sci.*, **13**, 156–9.
312. DRYER, W., and BENSON, N. G. (1957). 'Observations on the influence of the New Johnsonville Steam Plant on fish and plankton populations', *Proc. Ann. Conf. S.E. Assoc. Game and Fish. Comm.*, **10**, 85–91.
313. D.S.I.R. (1964). 'Effects of polluting discharges on the Thames Estuary', *Water Pollution Research. Tech. Paper*, 11, Department of Scientific and Industrial Research. London: H.M.S.O.
314. DUDLEY, R. G. (1974). 'Growth of Tilapia of the Kafue flood plain, Zambia: Predicted effects of the Kafue Gorge Dam', *Trans. Amer. Fish Soc.*, **103**(2), 281–3.
315. —— (1979). 'Changes in growth and size-distribution of *Satherodon macrochir* and *Satherodon andersoni* from the Kafue flood plain, Zambia, since construction of the Kafue Gorge Dam', *J. Fish. Biol.*, **14**, 205–23.
316. DUNSTER, H. J. (1978). 'Pollution resulting from the release of radioactive waste materials to the sea', *Mar. Pollut. Bull.*, **9**, 118–22.
317. DURRETT, C. W. (1972). 'Density, distribution, production and drift of benthic fauna in a reservoir receiving thermal discharges from a steam electric generating station', M.S. Thesis, North Texas State University, Denton, Texas.
318. ——, and PEARSON, W. D. (1975). 'Drift of macroinvertebrates in a channel carrying heated water from a power plant', *Hydrobiologia*, **46**(1), 33–43.
319. DUSSART, B. (1955). 'La temperature des lacs et ses causes de variations', *Verh. int. Ver. Limnol.*, **12**, 78–96.
320. DUTHIE, H. C., and OSTROFSKY, M. L. (1975). 'Environmental impact of the Churchill Falls (Labrador) hydro-electric project: a preliminary assessment', *J. Fish. Res. Board Can.*, **32**(1), 117–25.
321. DUTKIEWITZ, R. K. (1974). 'Pollution and the power industry in South

Africa'. Paper 2.6.9, *Proc. 9th World Energy Conference, Detroit*. London: John Wiley and Sons Ltd.

322. DYER, K. R. (1973). *Estuaries: a physical introduction*. London: Wiley-Interscience.

323. EBEL, W. J. (1969). 'Supersaturation of nitrogen in the Columbia River and its effect on salmon and steelhead trout', *Fish. Bull.*, **68**, 1–11.

324. EBLE, A. F., *et al.* (1975). 'The use of thermal effluents of an electric generating station in New Jersey in aquaculture of the Great Malaysian prawn, *Macrobrachium rosenbergii* and the Rainbow trout, *Salmo gairdneri*', *Proc. Power Plant Waste Heat Utilization in Aquaculture Workshop*, November (6–7), Trenton, N.J.

325. E.D.F. (1977). 'Influence des rejets thermiques sur le milieu vivant en mer et en estuaire', *Journees de la Thermo-ecologique, Electricite de France*.

326. EDINGER, J. E., and GEYER, J. C. (1965). 'Heat Exchange in the Environment'. *Cooling Water Studies for E.E.I. Research Project R.P. 49*. Edison Electrical Institute Pub. No. 65–902.

327. ——, *et al.* (1968). 'The variation of water temperatures due to steam electric cooling operations', *J. Water Pollut. Contr. Fed.*, **40**(9), 1632–9.

328. ——, DUTTWEILER, D. W., *et al.* (1968). 'The response of water temperatures to meteorological conditions', *Water Research*, **4**(5), 1137–43.

329. EDINGTON, J. M. (1966). 'Some observations on stream temperature', *Oikos*, **15**(11), 265–73.

330. EDSALL, T. A., and YOCUM, T. G. (1972). *Review of recent technical information concerning the adverse effects of once-through cooling on Lake Michigan*. U.S. Fish and Wildlife Service. Bureau of Sport Fishing and Wildlife, Ann Arbor, Michigan (mimeo).

331. EDWARDS, D. J. (1974). 'Weed preference and growth of young grass carp in New Zealand', Fisheries Research Publication no. 223. *New Zealand J. Mar. Freshwater Res.*, **8**, 341–5.

332. EDWARDS, T. J., *et al.* (1976). 'An evaluation of the impingement of fishes at four Duke Power Company steam-generating facilities', pp. 373–80 in ref. 368.

333. EDWARDS, R. J. (1978). The effect of hypolimnion reservoir releases on fish distribution and species diversity, *Trans. Amer. Fish. Soc.*, **107**(1), 71–7.

334. E.E.I. (1969). *Environmental projects related to thermal discharges*. New York: Edison Electric Institute.

335. EFFER, W. R., and BRYCE, J. B. (1975). 'Thermal discharge studies on the Great Lakes—the Canadian experience', pp. 371–87 in ref. 551.

336. EFFORD, I. E. (1975). Environmental impact assessment and hydro-electric projects—Hindsight and foresight in Canada', *J. Fish. Res. Board Can.*, **32**(1), 98–209.

337. EGBORGE, A. B. M. (1979). 'The effect of impoundment on the water chemistry of Lake Asejire, Nigeria', *Freshwat. Biol.*, **9**, 403–12.

338. EHRLICH, K. F., HOOD, J. M., MUSZYNSKI, G., and McGOWEN, G. E. (1979). 'Thermal behaviour responses of selected California littoral fishes', *Fish. Bull.*, **76**(4), 837–49.

339. EICHER, G. J. (1970). 'Fish passage', pp. 163–71 in ref. 111.

340. E.I.F.A.C. (1965). 'Water quality for European freshwater fish. Report on finely dissolved solids and inland fisheries', *Int. J. Air and Water Pollut.*, **9**, 151–68.

341. —— (1968a). 'Water quality criteria for European freshwater fish. Report on extreme pH values and inland fisheries', *European Inland Fisheries Advisory Commission Tech. Paper*, no. 4, Rome.

342. —— (1968b). 'Water quality criteria for freshwater fish.—Report on water temperature and inland fisheries', ibid., no. 6, Rome.

343. —— (1970). 'Water quality criteria for European freshwater fish. Report on ammonia and inland fisheries', ibid., no. 11, Rome.

344. —— (1973). 'Water quality criteria for European freshwater fish. Report on chlorine and freshwater fish', ibid., no. 20, Rome.

345. EILER, H. O., and DELFINO, J. T. (1975). 'Comparative effects of two modes of cooling water discharge on Mississippi River biota and environmental ecosystems', pp. 685–92 in ref. 551.

346. EINSTEIN, H. A. (1972). Sedimentation (suspended solids), pp. 309–18 in ref. 843.

347. EISENBUD, M. (1973). *Environmental Radioactivity* (2nd edn), New York, San Francisco, and London: Academic Press.

348. ELECTRICITY COUNCIL REPORT (1973). *Electricity Supply in Great Britain. A Chronology: Organisation and Development* (2 vols). London: Electricity Council.

349. ELLIOT, R. A. (1973). 'The T.V.A. Experience 1933–1971', pp. 251–5 in ref. 5.

350. ELLIOT, J. M. (1965). 'Daily fluctuations of drift invertebrates in a Dartmoor stream', *Nature* (Lond.), **205**, 1127–9.

351. —— (1967). 'Invertebrate drift in a Dartmoor stream', *Arch. Hydrobiol.*, **63**, 202–37.

352. —— (1971). 'Some methods for the statistical analysis of samples of benthic invertebrates', *Sci. Publ. Freshwater Biol. Assn.*, no. 25. Ambleside, England.

353. —— (1972). 'Effect of temperature on the time of hatching in *Baëtis rhodani* (Ephemeroptera: Baëtidae)', *Oecologia*, **9**, 47–51.

354. —— (1973). 'The food of brown and rainbow trout (*Salmo trutta* and *S. gairdneri*) in relation to the abundance of drifting invertebrates in a mountain stream', ibid., **12**, 329–47.

355. —— (1976). 'The energetics of feeding, metabolism and growth of brown trout (*Salmo trutta* L.) in relation to body weight, water temperature and ration size', *J. Anim. Ecol.*, **45**, 923.

356. —— (1978). 'Effect of temperature on the hatching time of eggs of *Ephemerella ignita* (Poda) (*Ephemeroptera-Ephemerellidae*)', *Freshwat. Biol.*, **8**, 51–8.

357. ——, and TULLETT, P. A. (1978). 'A Bibliography of Samplers for Benthic Invertebrates', *Freshwat. Biol. Assn. Occasional Publ.*, no. 4.

358. ELROD, J. H., and HASSLER, T. J. (1971). 'Vital statistics of seven fish species in Lake Sharpe, South Dakota 1964–69', pp. 27–40 in ref. 461.

359. ENGLAND, G. (1978). *Renewable sources of energy—The prospects for electricity*. London: C.E.G.B.

360. ENGLERT, T. L., *et al.* (1976). 'A model of striped bass population dynamics in the Hudson River', pp. 137–50 in Wiley, M. (ed.), *Estuarine Processes*, vol. 1, *Uses, Stresses and Adaptation to the Estuary*. London and N.Y.: Academic Press.

361. E.P.A. (1972). *Water Quality Standards Criteria Digest*. Environmental Protection Agency (August 1972). Washington D.C.

362. EPPLEY, R. W. (1972). 'Temperature and phytoplankton growth in the sea', *Fish. Bull.*, **70**(4), 1063–85.

363. ——, *et al.* (1976). 'Chlorine reactions with sea-water constituents and the inhibition of photosynthesis of natural marine phytoplankton', *Estuarine and Coastal Mar. Sci.*, **4**, 147–61.

364. ERICKSON, J., and FREEMAN, A. J. (1978). 'Toxicity screening of fifteen chlorinated and brominated compounds using four species of marine phytoplankton', pp. 307–10 in ref. 579.

365. ERMAN, D. C. (1973). 'Upstream changes in fish populations following the impoundment of Sagehen Creek, California, *Trans. Amer. Fish. Soc.*, **102**(3), 626–8.

366. ——, and LEIDY, G. R. (1975). 'Downstream movement of rainbow trout fry in a tributary of Sagehen Creek under permanent and intermittent flow', ibid., **104**(3), 467–73.

367. ERTL, M., and TAMAJKA, J. (1973). 'Primary production of the periphyton in the littoral of the Danube', *Hydrobiologia*, **42**(4), 429–44.

368. ESCH, G. W., and McFARLANE, R. W. (eds.) (1976). *Thermal Ecology, II.* Technical Information Centre, Energy Research and Development Administration, *E.R.D.A. Symposium Series* (Conf. 750425), Springfield, Virginia.

369. ESTES, R. D. (1971). 'The effects of the Smith Mountain Pump Storage Project on the fishery of the Lower Reservoir, Leesville, Virginia'. Ph.D. Thesis. Co-operative Fishery Unit, Viriginia Polytechnic Institute and State University, Blacksburg, Virginia.

370. EURE, H. E., and ESCH, G. W. (1974). 'Effects of thermal effluent on the population dynanics of helminth parasites in largemouth bass', pp. 207–15 in ref. 417.

371. EVINS, C. (1975). 'The toxicity of chlorine to some freshwater organisms', *Water Research Centre, Tech. Rept.*, T.R.8, Stevenage, Herts.

372. FAIRBANKS, R. B., *et al.* (1968). *Biological Investigations in the Cape Cod Canal prior to the Operation of a Steam Generating Plant.* Boston, Mass.: Massachusetts Department of Natural Resources.

373. FARREL, J., and ROSE, A. H. (1967). 'Temperature effects on microorganisms', in Rose, A. H. (ed.), *Thermobiology*, pp. 147–215. London: Academic Press.

374. FAST, A. W. (1971). 'Effects of artificial destratification and zooplankton depth distribution', *Trans. Amer. Fish. Soc.*, **100**(2), 355–7.

375. —— (1973). 'Effects of artificial hypolimnion aeration on rainbow trout (*Salmo gairdneri* Richardson) depth distribution', ibid., **4**, 715–22.

376. ——, *et al.* (1975). 'A submerged hypolimnetic aerator', *Water Resources Research*, **11**(2), 287–93.

377. FEDERAL POWER COMMISSION (1969). *Problems in disposal of waste heat from steam electric power plants.* Washington D.C.: Bureau of Power Report.

378. FEDORENKO, A. Y. (1975). 'Feeding characteristics and predation impact of *Chaoborus* (Diptera, Chaoboridae) larvae in a small lake', *Limnol. Oceanogr.*, **20**, 250–5.

379. FENTON, P. J., and NORRIS, T. E. (1972). 'The economics and future of total energy and combined heat/power installations', *Proc. Thermo-Fluids Conference—Thermal Discharges.—Engineering and Ecology.* 4–7 December. Institute Engineering, Sydney, Australia.

380. FERRARIS, C., and WILHM, J. (1977). 'Distribution of benthic macroinvertebrates in an artificially destratified reservoir', *Hydrobiologia*, **54**(2), 169–76.

381. FEY, VON J. M. (1977). 'The heating of a mountain stream and the effects on the zoocoenosis demonstrated on the Lenne, Sauerland', *Arch. Hydrobiol.* (*Suppl.* 53), **3**, 307–63.

382. FICKEISEN, D. H., and SCHNEIDER, M. J. (eds.) (1974). 'Gas bubble disease'. *Proc. Workshop, Richland, Washington*, 8–9 October, E.R.D.A. Tech. Inf. Centre, Oak Ridge, Tennessee. Conf. 741033.

383. FIELDS, P. E. (1966). *Migrant salmon light-guiding studies at Columbia River dams.* Third Progress Report on Fisheries Eng. Research Prog. U.S. Army Corps of Engineers.

384. ——, et al. (1964). 'Migration of adult salmon through lighted fish ladders', *Research Fish. Seattle*, 1963, Contrib. no. 166, 36–7.

385. FILLION, D. B. (1967). 'The abundance and distribution of benthic fauna of three mountain reservoirs on the Kananaskis River in Alberta', *J. Applied Ecol.*, **4**, 1–11.

386. FJERDINGSTAD, E. (1975). 'Bacteria and fungi', pp. 129–53 in ref. 1140.

387. FLEMER, D. A., et al. (1971). 'The effects of steam electric station operation on entrained organisms', in Mihursky, J. A., and McErlean, A. J. (Co-principal investigators), *Postoperative Assessments of the Effects of Estuarine Power Plants*. Natural Resources Institute, University of Maryland, ref. no. 712b.

388. FLEMING, J. M. (1970). *An investigation into the growth of periphyton in the heated effluent from Trawsfynydd Nuclear Power Station: 1969*. C.E.R.L. Note. RD/L/N 199/70, Leatherhead, Surrey.

389. ——, and COUGHLAN, J. (1978). 'Preservation of vitally stained zooplankton for live/dead sorting', *Ches. Sci.*, **19**(2), 135–7.

390. FLIERMANS, C. B., et al. (1975). 'Direct measurement of bacterial stratification in Minnesota lakes', *Arch. Hydrobiol.*, **76**(2), 248–55.

391. FLOOK, R. A. (1978). 'Problems associated with the re-use of purified sewage effluents for power station cooling purposes', *Prog. Wat. Technol.*, **10**(1/2), 105–11.

392. FOERSTER, J. W., et al. (1974). 'Thermal effects on the Connecticut river. Phycology and chemistry', *J. Water Pollut. Contr. Fed.*, **46**, 2138–52.

393. FOX, D. L., and CORCORAN, E. F. (1957). 'Thermal and osmotic counter measures against some typical marine fouling organisms', *Corrosion*, **14**, 31–32.

394. FOX, J. L., and MOYER, M. S. (1973). 'Some effects of a power plant on marine microbiota', *Ches. Sci.*, **14**(1), 1–10.

395. FRASER, J. C. (1972). 'Regulated discharge and the stream environment', pp. 263–87 in ref. 843.

396. FREEMAN, R. F., and SHARMA, R. K. (1977). *Survey of Fish Impingement at Power Plants in the United States*, vol. II, *Inland Waters*. Rep. no. ANL/ES-56, Argonne, Ill.: Argonne National Laboratory.

397. FROST, W. E. (1939). 'River Liffey Survey. II. The food consumed by the brown trout (*Salmo trutta*. Linn) in acid and alkaline waters', *Proc. R. Ir. Acad.*, **45B**, 139–206.

398. FRY, F. E. J., et al. (1946). 'Lethal temperature relations for a sample of young speckled trout (*Salvelinus fontinalis*), *Univ. Toronto Stud. Biol. Series*, **54**, 1–35.

399. —— (1967). 'Responses of vertebrate poikilotherms to temperature', pp. 375–410 in ref. 938.

400. FRY, P. F., et al. (1971). 'Pumped storage as standby for nuclear plants', pp. 393–405 in ref. 588.

401. GALLAGHER, B. J. (ed.) (1974). *Energy Production and Thermal Effects*. Michigan: Ann Arbor Science Publishers.

402. GALLAWAY, B. J., and STRAWN, K. (1975). 'Seasonal and areal comparisons of fish diversity indices at hot-water discharge in Galveston Bay, Texas', *Contributions in Marine Science*, 79–89.

403. GAMESON, A. L. H. (1957). 'Weirs and the aeration of rivers', *J. Inst. Water Res. Engrs.*, **11**(6), 477–90, 72.

404. ——, and TRUESDALE, G. A. (1959). 'Some oxygen studies in streams', ibid., **9**, 571–94.

405. ——, et al. (1959). 'A preliminary temperature survey of a heated river', *J. Wat. and Wat. Engng.*, **63**, 13.

406. GAMMON, K. M. (1969). 'Planning cooling water for power stations', *Proceedings 4th Int. Conf. on Wat. Pollut. Research, Prague*, 927–36.

407. GANAPATI, S. V., and SCREENIVASAN, A. (1968). 'Aspects of limno-biology, primary production and fisheries in the Stanley reservoir; Madras State', *Hydrobiologia*, **32**(3–4), 551–69.

408. GARDNER, K., and SKULBERG, O. (1965). 'Radionuclide accumulation by *Anodonta piscinalis* Nilsson (Lamellibranchiata) in a continuous flow system', ibid., **26**(1–2), 151–69.

409. GARSIDE, E. T. (1970). 'Structural responses (Temperature)', pp. 561–615 in ref. 602.

410. GARTRELL, F. E., *et al.* (1971). 'Environmental quality protection. Large steam electric power stations', pp. 453–72 in ref. 546.

411. GAUFIN, A. R., and HERN, S. (1971). 'Laboratory studies on the tolerance of aquatic insects to heated water', *J. Kansas Entomol. Soc.*, **44**, 240–5.

412. GEEN, G. H. (1974). 'Effects of hydro-electric development in western Canada on aquatic ecosystems', *J. Fish. Res. Board Can.*, **31**, 913–27.

413. —— (1975), 'Ecological consequences of the proposed Moran Dam on the Fraser river', ibid., **32**, 126–35.

414. GENTILE, J. H., *et al.* (1976). *Power plants, chlorine and estuaries*. Environ-mental Research Laboratory, Office of Research and Development, U.S. Environmental Protection Agency, Narragansett, Rhode Island, EPA-600/3-76-055.

415. GENTRY, J. B., *et al.* (1975). 'Thermal ecology of dragonflies in habitats receiving reactor effluent', pp. 563–74 in ref. 551.

416. GESSNER, F. (1970). 'Plants', pp. 363–406 in ref. 602.

417. GIBBONS, J. W., and SHARITZ, R. R. (eds.) (1974). *Thermal Ecology*, Tech. Inf. Centre, U.S. Atomic Energy Commission. Conf. 730505.

418. ——, *et al.* (1975). 'Ecology of artificially heated streams, swamps and reservoirs on the Savannah River Plant; The Thermal Studies program of the Savannah River Ecology Laboratory', pp. 389–99 in ref. 551.

419. GIBSON, R. J., and GALBRAITH, D. (1975). 'The relationships between invertebrate drift and salmonid populations in the Matamek River Quebec, below a lake', *Trans. Amer. Fish. Soc.*, **104**(3), 529–35.

420. GILLHAM, E. W. F., and SIMPSON, D. T. (1974). 'Land restoration with pulverised fuel ash. Part 2. Comparisons between grass and winter wheat', *J. Br. Grassld Soc.*, **29**, 207–12.

421. GINN, T. C., and O'CONNOR, J. M. (1976). 'Response of the estuarine amphipod *Gammarus daiberi* to chlorinated power plant effluent', *Estuarine and Coastal Mar. Sci.*, **6**, 459–69.

422. GINN, C., WALLER, W. T., and LAUER, G. J. (1976). 'Survival and reproduction of *Gammarus* spp. (Amphipoda) following short-term exposure to elevated temperatures', *Ches. Sci.*, **17**(1), 8–14.

423. GLENDENNING, IAN (1977). 'Energy from the sea', *Chemistry and Industry*, 16 July, 588–99.

424. GLYMPH, L. M. (1973). 'Summary: Sedimentation of reservoirs', pp. 342–48 in ref. 5.

425. GOLDBERG, R. (1978). 'Some effects of gas-supersaturated seawater on *Spisula solidissima* and *Argopecten irradians*', *Aquaculture*, **4**, 282–7.

426. GOLDEN, D. M. (1975). 'Coastal power plant heat disposal considerations', *J. Env. Eng. Division: Proc. Am. Soc. Eng.*, **101**(EE 3), 365–79.

427. GOLDMAN, J. C., *et al.* (1978). 'Biological and chemical effects of chlorination at coastal power plants', pp. 291–306 in ref. 579.

428. ——, and QUIMBY, H. L. (1979). 'Phytoplankton recovery after power plant entrainment'. *J. Water Pollut. Contr. Fed.*, **51**(7), 1816–23.

429. GOLTERMAN, H. L. (1975). 'Chemistry', pp. 39–80 in ref. 1140.

430. GONZALEZ, J. G., and YEVICH, P. (1976). 'Response of an estuarine population of the blue mussel, *Mytilus edulis* to heated water from a steam generating plant', *Marine Biology*, **34**, 177–89.

431. GOODYEAR, C. P. (1977). 'Assessing the impact of power plant mortality on the compensatory reserve of fish populations', pp. 186–95 in ref. 1100.

432. GORDON, D. C., and LONGHURST, A. R. (1979). 'The environmental aspects of a tidal power project in the upper reaches of the Bay of Fundy', *Mar. Pollut. Bull.*, **10**, 38–45.

433. GORHAM, E. (1976). 'Acid precipitation and its influence upon aquatic ecosystems—an overview', *J. Water Air Soil Pollut.*, **6**, 457–82.

434. GORYAJNOVA, L. I. (1975). 'The effect of warm effluent from the Novorossiysk thermal power plant on zooplankton', *Hydrobiol. J.*, **11**(6), 19–23.

435. GOSS, L. B., and BUNTING, D. L. (1976). 'Thermal tolerance of zooplankton', *Water Research*, **10**, 387–92.

436. ——, and CAIN, C. (1977). 'Power plant condenser and service water system fouling by *Corbicula*, The Asiatic clam', pp. 11–17 in ref. 567.

437. GRABOW, W. O. K., *et al.* (1975). 'Behaviour in a river and a dam of non-transferable drug resistance', *Water Research*, **9**, 777–82.

438. GRAHAM, T. P. (1974). 'Chronic malnutrition in four species of sunfish in a thermally loaded impoundment', pp. 151–7 in ref. 417.

439. GRAHAM, T. R., and JONES, J. W. (1962). 'The biology of Llyn Tegid trout, 1960', *Proc. Zool. Soc. Lond.*, **139**, 657–83.

440. GRAUBY, A. (1978). 'Thermal release from nuclear power plants. Their utilization for plant and fish production', *5th Int. Fair and Tech. Meeting of Nuclear Industries*. Vol. D. Paper 3/8. Basel (3–8 Oct).

441. GRAY, R. H., and HAYNES, J. M. (1977). 'Depth distribution of adult Chinook salmon (*Oncorhynchus tsawytscha*) in relation to season and gas supersaturated water', *Trans. Amer. Fish. Soc.*, **106**(6), 617–620.

442. GRAYUM, M. M. (1973). 'Effects of thermal shock and ionizing radiation on primary productivity', pp. 639–44 in Nelson D. J. (ed.) *Radionuclides in Ecosystems*. International Symposium on Radioecology. U.S.A.E.C., Washington, U.S.

443. GREGORY, M. (1974). *Angling and the Law*. London and Tonbridge: Charles Knight.

444. GRIEVE, J. A., *et al.* (1978). 'A program to introduce site-specific chlorination regimes at Ontario hydro generating stations', pp. 77–84 in ref. 579.

445. GRIMAS, U. (1961). 'The bottom fauna of natural and impounded lakes in northern Sweden (Ankarvattnet and Blasjön)', *Rep. Inst. Freshw. Res. Drottingholm*, **42**, 183–237.

446. —— (1964). 'Studies on the bottom fauna of impounded lakes in southern Norway'. ibid., **45**, 94–104.

447. GRIMES, C. B. (1975). 'Entrapment of fishes on intake water screens at a steam electric generating station', *Ches. Sci.*, **16**(3), 172–7.

448. ——, and MOUNTAIN, J. A. (1971). *Effects of thermal effluent upon marine fishes near the Crystal River Steam Electric Station*. Florida Department of Natural Resources. Marine Research Lab. Professional Papers Series No. 17.

449. GROBERG, W. J., *et al.* (1978). 'Relation of water temperature to infections of Coho salmon (*Oncorhynchus kisutch*), Chinook salmon (*O. tshawytscha*) and

steelhead trout (*Salmo gairdneri*) with *Aeromonas salmonicida* and *A. hydro-phila.*, *J. Fish. Res. Board Can.*, **35**(1), 1–7.

450. GROSS, M. G., *et al.* (1978). 'Suspended sediment discharge of the Susquehanna River to northern Chesapeake Bay, 1966 to 1976', *Estuaries*, **1**(2), 106–10.

451. GROTBECK, L. M., and BECHTHOLD, J. L. (1975). 'Fish impingement at Monticello Nuclear Plant', *Proc. of the A.S.C.E. Journal of the Power Division.* **101**(P.01), 69–83.

452. GUMA'A, S. A. (1978). 'The effects of temperature on the development and mortality of eggs of perch. *Perca fluviatilis*', *Freshwat. Biol.*, **8**, 221–7.

453. GUNDERSEN, K., and BIENFANG, P. (1972). 'Thermal pollution: Use of deep, cold, nutrient-rich sea-water for power plant cooling and subsequent aquaculture in Hawaii', *Marine Pollut. and Sea Life.* F.A.O. Rome, 513–16.

454. GURNEY, J. D., and COTTER, I. A. (1966). *Cooling Towers.* London: Maclaren and Sons Ltd.

455. GUTHRIE, R. K., CHERRY, D. S., and FEREBEE, R. N. (1974). 'A comparison of thermal loading effect on bacterial populations in polluted and non-polluted aquatic systems', *Water Research*, **8**, 143–8.

456. ——, ——, and RODGERS, J. H. (1974). 'The impact of ash basin effluent on biota in the drainage system', *Proc. 7th Mid Atlantic Ind. Waste Conf. 12–14 November, Philadelphia, Pa: Drexel University.*

457. ——, *et al.* (1978). 'The effects of coal ash basin effluent and thermal loading on bacterial populations of flowing streams', *Environ. Pollut.*, **17**, 297–302.

458. HAGG, R. W., and GORHAM, P. R. (1977). 'Effects of thermal effluent on standing crop and net productions of *Elodea canadensis* and other submerged macrophytes in Lake Wabamum, Alberta', *J. Applied Ecol.*, **14**, 835–51.

459. HADDERINGH, R. H. (in press). 'Mortality of young fish in the cooling water system of Bergum power station', *Verh. int. Verh. Limnol.*, **20**.

460. HAIR, J. R. (1971). 'Upper lethal temperature and thermal shock tolerances of the Opossum shrimp *Neomysis awatschensis* from the Sacramento–San Joa-quim estuary, California', *Calif. Fish Game*, **57**, 17–27.

461. HALL, G. E. (ed.) (1971). *Reservoir Fisheries and Limnology*, Spec. Pub. no. 8. Washington: D.C. American Fish. Soc.

462. HAMILTON, D. H. (1978). 'Chlorine application for the control of condenser fouling', pp. 687–94 in ref. 579.

463. HAMMERTON, D. (1972). 'The Nile River—A case history', pp. 177–214 in ref. 843.

464. HANNAH, L. (1979). *Electricity before Nationalisation.* London: Macmillan Press.

465. HANNON, E. H. (1978). 'A Document Collection of Electricity Utility Studies related to steam—Electric Power Station Cooling System Effects on Water Quality and Aquatic Biota', *Atomic Ind. Forum Research Project, EA-872, (877-1) (Interim Report)*, Palo Alto, California: Elec. Power Research Inst.

466. HARAM, O. J., and JONES, J. W. (1971). 'Some observations on the food of the qwyniad, *Coregonus clupeoides pennantii*. Valenciennes of Llyn Tegid (Lake Bala). North Wales', *J. Fish. Biol.*, **3**(3), 287–96.

467. HARDEN-JONES, F. R. (1968). *Fish Migration.* London: Edward Arnold.

468. HARDING, D. (1966). 'Lake Kariba. The hydrology and development of fisheries', pp. 7–18 in ref. 689.

469. HARDISTY, M. W., *et al.* (1974). 'Dietary habits and heavy metal concentra-tions in fish from the Severn Estuary and the Bristol Channel', *Mar. Pollut. Bull.*, **5**(4), 61–63.

470. ——, HUGGINS, R. J., *et al.* (1974). 'Ecological implications of heavy metal in fish from the Severn Estuary, ibid., **5**(1), 12–15.

471. ——, —— (1975). 'A survey of the fish populations of the middle Severn Estuary based on power station sampling', *Internat. J. Environ. Studies*, **7**, 227–42.

472. HARMSWORTH, R. V. (1974). 'Artificial cooling lakes as unique aquatic ecosystems', pp. 56–66 in ref. 401.

473. HARVEY, H. W. (1945). *Recent Advances in the Chemistry and Physics of Sea Water.* Cambridge: Cambridge University Press.

474. HARVEY, R. S. (1974). 'Temperature effects on the sorption of radionuclides by aquatic organisms', pp. 28–42 in ref. 417.

475. HATHAWAY, E. S. (1927). 'The relation of temperature to the quantity of food consumed by fishes', *Ecology*, **8**(4), 428–34.

476. HATLE, S., and LAMPAR, M. (1975). 'Exploitation of waste heat', pp. 731–40 in ref. 551.

477. HAVEY, K. A. (1974). 'Effect of regulated flows on standing crops of juvenile salmon and other fishes at Barrows Stream, Maine', *Trans. Amer. Fish. Soc.*, **103**(1), 1–9.

478. HAWES, F. B. (1970). *Thermal problems—old hat in Britain. C.E.G.B. Newsletter no. 83.* London: Central Electricity Generating Board.

479. —— (1976). 'North Wales Hydro-electric Power Act 1973', Paper 1, pp. 1–5, in *Dinorwic Power Station: Seminar on Aquatic Environmental Studies.* 7 November 1975. Llanberis, C.E.G.B., North Wales.

480. ——, *et al.* (1975). 'Environmental effects of the heated discharges from Bradwell Nuclear Power Station and of the cooling systems of other power stations', pp. 423–47 in ref. 551.

481. HAWKES, H. A. (1962). 'Biological aspects', pp. 311–432 in ref. 607.

482. —— (1969). 'Ecological changes of applied significance from waste heat', pp. 15–53 in ref. 624.

483. —— (1975). 'River zonation and classification', chap. 14 in ref. 1140.

484. HAZEN, T. C., and FLIERMANS, C. B. (1969). 'Distribution of *Aeromonas Hydrophila* in natural and man-made thermal effluents', *Appl. and Environ. Microbiol.*, **38**(1), 166–8.

485. HEATH, A. G. (1977). 'Toxicity of intermittent chlorination to freshwater fish, influence of temperature and chlorine form', *Hydrobiologia*, **56**(1), 39–47.

486. HECHTEL, G. J., *et al.* (1970). 'Biological effects of thermal pollution. Northport'. *Stony Brook Technical Report.* New York: Marine Sciences Research Center, State University of New York.

487. HEINLE, D. R. (1969). 'Temperature and zooplankton', *Ches. Sci.*, **10**, 186–209.

488. ——, *et al.* (1974). *Zooplankton investigations at the Morgantown power plant.* University of Maryland, Natural Resources Institute, ref. no. 74, 10.

489. HELA, I., and LAEVASTU, T. (1961). *Fisheries Hydrography.* London: Fishing News (Books) Ltd.

490. HELLAWELL, J. M. (1978). *Biological Surveillance of Rivers—A Biological Monitoring Handbook.* Stevenage, U.K.: N.E.R.C. and Water Research Centre.

491. ——, *et al.* (1974). 'The upstream migratory behaviour of salmonids in the River Frome, Dorset', *J. Fish. Biol.*, **6**, 729–44.

492. HENRIKSEN, A., and JOHANNESSEN, M. (1975). *Deacidification of acid water—A survey of the literature and general appraisal.* S.N.S.F. Project on the effect of acid rainfall on forests and fish (Dept. of the Environment, English translation) (cyclostyled).

493. HERBERT, D. W. M., *et al.* (1961). 'The effect of china-clay wastes on trout streams', *Int. J. Air and Water Poll.*, **5**, 56–74.

494. HERGOTT, S. J., *et al.* (1978). 'Power plant cooling-water chlorination in northern California', *J. Water Pollut. Contr. Fed.*, **50**(11), 2590–2601.

495. HERRICKS, E. E., and CAIRNS, J. (1974–6). 'The recovery of stream macrobenthos from low pH stress', *Rivista de Biologia*, **10**(1–4), 1–11.

496. HERSHBERGER, W. K., *et al.* (1978). 'Chronic exposure of Chinook salmon eggs and alevins to gamma irradiation: Effects on their return to freshwater as adults', *Trans. Amer. Fish Soc.*, **107**(4), 622–31.

497. HETHERINGTON, J. A. (1976). *Radioactivity in surface and coastal waters of the British Isles (1974)*. Fish. Radiobiol. Lab. Tech. Rep. FRL 11. M.A.F.F., Lowestoft.

498. HEWETT, C. J., and JEFFERIES, D. F. (1978). 'The accumulation of radioactive caesium from food by the plaice (*Pleuronectes platessa*) and the brown trout (*Salmo trutta*)', *J. Fish. Biol.*, **13**, 143–53.

499. HICKMAN, M. (1974). 'Effects of the discharge of thermal effluent from a power station on Lake Wabumum, Alberta, Canada—The epipelic and epipsammic algal communities', *Hydrobiologia*, **45**(2–3), 199–215.

500. ——, and KLARER, D. M. (1975). 'The effect of the discharge of thermal effluent from a power station on the primary productivity of epiphytic algal community', *Brit. Phycol. J.*, **10**, 81–91.

501. HILL, B. J. (1977). 'The effect of heated effluent on egg production in the estuarine prawn *Upogebia africana (Ortmann)*', *J. exp. mar. Biol. Ecol.*, **29**, 291–302.

502. HILLMAN, R. E. (1977). 'Techniques for monitoring reproduction and growth of fouling organisms at power plant intakes', pp. 5–9 in ref. 567.

503. ——, *et al.* (1977). 'Abundance, diversity and stability in shore-zone fish communities in an area of Long Island Sound affected by the thermal discharge of a nuclear power station', *Estuarine and Coastal Mar. Sci.*, **5**, 355–81.

504. HILSENOFF, W. L. (1971). 'Changes in the downstream insect and amphipod fauna caused by an impoundment with a hypolimnion drain', *Ann. Entomol. Soc. Amer.*, **64**, 743–6.

505. HINDLEY, P. D., and MINER, R. M. (1972). 'Evaluating water surface heat exchange co-efficients, *Journal of the Hydraulics Division, Proc. Amer. Soc. Civil Engineers*, HY 8, 1411–26.

506. HINES, A. H. (1979). 'Effects of a thermal discharge on reproductive cycles in *Mytilus edulis* and *Mytilus californianus* (Mollusca; Bivalvia)', *Fishery Bulletin*, **77**(2), 498–503.

507. HIRAYAMA, K., and HIRANO, R. (1970a). 'Influence of high temperature and residual chlorine on marine phytoplankton', *Marine Biol.*, **7**, 205–13.

508. ——, —— (1970b). 'Influences of high temperature and residual chlorine on the marine planktonic larvae', *Bull. Fac. Fish. Nagasaki Univ.*, **29**, 83–90.

509. HOADLEY, A. W., *et al.* (1975). 'Preliminary studies of fluorescent pseudomonads capable of growth at 41°C in swimming pool water', *Appl. Microbiol.*, **29**(4), 527–31.

510. HOCUTT, C. H. (1973). 'Swimming performance of three warmwater fishes exposed to a rapid temperature change', *Ches. Sci.*, **14**(1), 11–16.

511. HOCKLEY, A. R. (1965). 'Population changes in *Limnoria* in relation to temperature', pp. 457–64 in Becker G., and Liese, W. *Holz und Organismen*. Berlin: Duncker and Humblot.

512. HOFFMAN, G. L., and BAUER, O. N. (1971). 'Fish parasitology in reservoirs—A review', pp. 495–512 in ref. 461.

513. HOLCIK, J. (1977). 'Changes in fish community of Klicava Reservoir with particular reference to Eurasian perch (*Perca fluviatilis*) 1957–72', *J. Fish. Res. Board Can.*, **34**, 1734–47.

514. HOLME, N. A., and McINTYRE, A. D. (1971). *Methods for Study of Marine Benthos.* I.B.P. Handbook no. 16. Oxford and Edinburgh: Blackwell Scientific Publications.

515. HOLMES, D. W., *et al.* (1974). 'Pond and cage culture of channel catfish in Virginia', *J. Tenn. Acad. Sci.*, **49**, 74–78.

516. HOLMES, N. J. (1970a). 'Marine fouling in power stations', *Mar. Pollut. Bull.*, **1**(7), 105–6.

517. —— (1970b). *The design of chlorination schedules for reducing mussel fouling in power station cooling systems.* C.E.R.L. Lab Report RD/L/R 1686, Leatherhead, Surrey.

518. HOLMES, R. H. A. (1975). *Fish and weed on Fawley power station screens.* C.E.R.L. Lab. note RD/L/N 129/75, Leatherhead, Surrey.

519. HOMER, M. (1976). 'Seasonal abundance, biomass, diversity and trophic structure of fish in a salt-marsh tidal creek affected by a coastal power plant', pp. 259–67 in ref. 368.

520. HOOD, D. W. (1963). 'Chemical oceanography', in Barnes, H. (ed.), *Oceanogr. Mar. Biol. Ann. Rev.*, **1**, 129–55. London: Allen and Unwin.

521. HOOPER, W. C., *et al.* (1978). ''Rearing of brook trout and lake trout in the thermal effluent of a coal-fired generating station', *Trans. Can. Electr. Assoc. (Eng. Oper. Div.)* **17**(4), paper 78-1-223.

522. HORTON, P. A. (1961). 'The bionomics of brown trout in a Dartmoor stream', *J. Anim. Ecol.*, **30**, 311–38.

523. HOSS, D. E., *et al.* (1975). 'Effects of temperature, copper and chlorine on fish during simulated entrainment in power plant condenser cooling systems', pp. 519–27 in ref. 551.

524. HOSTGAARD-JENSEN, *et al.* (1977). 'Chlorine decay in cooling water and discharge into seawater', *J. Water Pollut. Contr. Fed.*, 1832–41.

525. HOUGHTON, G. U. (1966). 'Maintaining and safety and quality of water supplies', pp. 173–82 in ref. 689.

526. HOUSE OF LORDS (1977). E.E.C. Environment Policy Select Committee on the European Communities. R/2005/76. *Water Standards for Freshwater Fish.* London: H.M.S.O.

527. HOWELL, F. G., and GENTRY, J. B. (1974). 'Effects of thermal effluents from nuclear reactors on species diversity of aquatic insects', pp. 562–71 in ref. 417.

528. HOWELLS G., and HOLDEN, A. V. (1978). *Effects of acid waters on fish.* C.E.R.L. Lab. Note RD/L/N 142/79, Leatherhead, Surrey.

529. HOWELLS, G. D. (1977). 'In and out of hot water', *Mar. Pollut. Bull.*, **8**(11), 245–8.

530. HUBBERT, M. K. (1971). 'Energy resources for power production', pp. 13–43 in ref. 546.

531. HUGENIN, J. E. (1976). 'Heat exchangers for use in the culturing of marine organisms', *Ches. Sci.*, **17**(1), 61–64.

532. HUGHES, S. (1973). Summary Report. *Supersaturation levels in Kootenay River downstream from Libby Dam and in the Columbia River south of the border.* B.C. Fish Wildl. Branch (mimeo).

533. HULTBERG, H. (1975). 'Thermally stratified acid water in late winter: a key factor inducing self-accelerating processes which increase acidification', *Proc. Int. Symp. on Acid Precipitation and the Forest Ecosystem.* Columbus, Ohio (May).

534. HUMPHRIS, T. H. (1977). 'The use of sewage effluent as power station cooling water', *Water Research*, **11**, 217–23.

535. ——, and RIPPON, J. E. (1978). *The effect of chlorine on nitrifying bacteria,* C.E.R.L. Lab. Note RD/L/N 164/77, Leatherhead, Surrey.

536. HUNT, F. R. (1971). 'Power station site selection in England and Wales', pp. 647–57 in ref. 546.

537. HUNT, P. C., and JONES, J. W. (1972). 'The effect of water level fluctuations on a littoral fauna', *J. Fish. Biol.*, **4**(3), 385–94.

538. HUTCHINSON, G. E. (1957). *A Treatise on Limnology*, vol. 1. (New York: Wiley).

539. HUTCHINSON, V. H. (1976). 'Factors influencing thermal tolerances of individual organisms', pp. 10–26 in ref. 368.

540. HYNES, H. B. N. (1958). 'The effect of drought on the fauna of a small Welsh mountain stream', *Verh. int. Ver. Limnol.*, **13**, 826–33.

541. —— (1960). *The Biology of Polluted Waters*. Liverpool: Liverpool University Press.

542. —— (1961). 'The effect of water level fluctuations on littoral fauna', *Verh. int. Ver. Limnol.*, **14**, 652–6.

543. —— (1970). *The Ecology of Running Waters*, Liverpool: Liverpool University Press.

544. I.A.E.A. (1966). *Proc. Disposal of Radioactive Wastes into Seas, Oceans and Surface Waters*. Vienna: International Atomic Energy Agency.

545. —— (1969). *Environmental Contamination by Radioactive Materials*. Vienna: International Atomic Energy Agency.

546. —— (1971). *Environmental Aspects of Nuclear Power Stations*. Vienna: International Atomic Energy Agency.

547. —— (1972). *Thermal Discharges at Nuclear Power Stations: Their management and Environmental Impacts*, West, P. J. (ed.). Vienna: International Atomic Energy Agency.

548. —— (1974a). *Environmental Surveillance around Nuclear Installations*, vol. 1. Vienna: International Atomic Energy Agency.

549. —— (1974b). ibid., vol. 2.

550. —— (1974c). *Population Dose Evaluation and Standards for Man and his Environment*. Vienna: International Atomic Energy Agency.

551. —— (1975a). *Environmental Effects of Cooling Systems at Nuclear Power Stations*. Vienna: International Atomic Energy Agency.

552. —— (1975b). *Impact of Nuclear Releases into the Aquatic Environment*. Vienna: International Atomic Energy Agency.

553. —— (1979). *Methodology for assessing impacts of radioactivity on aquatic ecosystems*. Vienna: International Atomic Energy Agency. Technical Report Series. No. 190.

554. ICANBERRY, J., and ADAMS, J. R. (1974). 'Zooplankton survival in cooling water systems of four thermal power plants on the California coast. Interim report March 1971–Jan. 1972', in ref. 417.

555. IDLER, D. R., and CLEMENS, W. R. (1959). 'The energy expenditures of Fraser River sockeye salmon during the spawning migration to Chilko and Stuart Lakes', *Int. Pacif. Salm. Fish. Comm. Prog. Rep.* 6.

556. ILES, R. (1963). 'Cultivating fish for food and sport in power station water', *New Scientist*, 31 January, 324.

557. ILLIES, J. (1952). 'Die Mölle. Faunistisch — ökologische Untersuchungen an eimen Forellenbach im Lipper Bergland', *Arch. Hydrobiol.*, **46**, 424–612.

558. IMHOFF, K. R. (1965). 'Uber die Reinigungsleistung der Ruhrstauseen', *Gas-u Wasserfach Wasser*, **106**, 1264–7.

559. I.C.S.U. (1972). *Man-made lakes as modified ecosystems*. Scope Report 2. Internat. Council Sci. Unions.

560. ISOM, B. G. (1971). 'Effects of storage and mainstream reservoirs on benthic macro-invertebrates in the Tennessee Valley', pp. 179–92 in ref. 461.

561. JACKSON, P. B. N. (1966). 'The establishment of fisheries in man-made lakes in the tropics', pp. 53–69 in ref. 689.

562. JASKE, R. T., and GOEBEL, J. B. (1967). 'Effects of dam construction on temperatures of Columbia River', *J. Amer. Water Works Assoc.*, **59**, 935.

563. JEFFRIES, D. S., COX, C. M., and DILLON, P. J. (1979). 'Depression of pH in lakes and streams in Central Ontario during snowmelt', *J. Fish. Res. Board Can.*, **36**, 640–6.

564. JENKINS, R. M. (1970). 'Reservoir Fish Management', pp. 173–82 in ref. 85.

565. ——, and MORAIS, D. I. (1971). 'Reservoir sport fishing effort and harvest in relation to environmental variables', pp. 371–84 in ref. 461.

566. JENSEN, L. D. (ed.) (1974). 'Entrainment and intake screening', *Proc. 2nd Entrainment and Screening Workshop*, Rep. no. 15. N.Y.: Edison Electric Institute.

567. —— (1977). 'Biofouling Control Procedures: Technology and Ecological Effects. *Pollution Engineering and Technology*, 5, p. 113. New York and Basel: Marcel Dekker.

568. —— *et al.* (1969). *The Effects of Elevated Temperature upon Aquatic Invertebrates.* Edison Electric Institute Publication no. 69-900, Report no. 4.

569. JENSEN, K. W., and SNEKVIK, E. (1972). 'Low pH levels wipe out salmon and trout populations in southernmost Norway', *Ambio*, **1**(6), 223–5.

570. JESTER, D. B. (1971). 'Effects of commercial fishing, species introductions and drawdown control on fish populations in Elephant Butte Reservoir, New Mexico', pp. 265–86 in ref. 461.

571. JETER, C. R. (1977). 'An approach to thermal water quality standards', Paper II-A-3 in ref. 663.

572. JOBSON, H. E. (1977). 'Bed conduction computation for thermal models', *J. Hydraul. Div. Am. Soc. Civil Engineers Tech. Notes*, Oct. 1977, Proc. paper 13251, 1213–17.

573. —— (1978). *Thermal model for evaporation from open channels.* International Association for Hydraulic Research (cyclostyled).

574. ——, and KEEFER, T. N. (1977). 'Thermal modelling of highly transient flows in the Chattahoochee River near Atlanta, Georgia', Paper to Special Symposium on River Quality Assessments, Amer. Water Res. Assoc. 2–3 November, Tucson, Arizona (preprint).

575. ——, *et al.* (1979). 'Chattahoochee River Thermal Alterations', *Journal of the Hydraulics Division, A.S.C.E.*, 105, no. HY4. Proc. Paper 14499, 295–311.

576. JOHNSON, P. B., and HASLER, A. D. (1977). 'Winter aggregations of carp (*Cyprinus carpio*) as revealed by ultrasonic tracking', *Trans. Amer. Fish. Soc.*, **106**(6), 556–9.

577. JOHNSON, R. G. (1965). 'Temperature variation in the infaunal environment of a sand flat', *Limnol. Oceanogr.*, **10**(1), 114–20.

578. JOHNSON, S. P. (1979). *The pollution control policy of the European Communities.* London: Graham and Trotman.

579. JOLLEY, R. L., *et al.* (eds.) (1978). *Water Chlorination: Environmental Impacts and Health Effects*, vol. 2. Michigan: Ann Arbor Science.

580. JONES, D. J. (1973). 'Variation in the trophic structure and species composition of some invertebrate communities in polluted kelp forests in the North Sea', *Marine Biology*, **20**, 351–65.

581. JONES, J. G. (1975). 'Heterotrophic micro-organisms', pp. 141–54 in ref. 1140.

582. JONES, J. R. E. (1964). *Fish and River Pollution.* London: Butterworth.

583. JONES, J. W. (1959). *The Salmon. The New Naturalist.* London: Collins.

584. JONES, R. E. (1978). *Heavy metals in the estuarine environment*. Water Research Centre. Technical Report TR 73, April. Stevenage, U.K.

585. JORDAN, H. M., and LLOYD, R. (1964). 'The resistance of rainbow trout (*Salmo gairdneri* Richardson) and roach (*Rutilus rutilus* (L)) to alkaline solutions', *Int. J. Air and Water Pollut.*, **8**, 405–9.

586. JUNK, W. J. (1977). 'The invertebrate fauna of the floating vegetation of Bung Borapet, a reservoir in Central Thailand', *Hydrobiologia*, **53**(3), 229–38.

587. KAMATH, P. R., *et al.* (1975). 'Seasonal features of thermal abatement of shoreline discharges at nuclear sites', pp. 217–27 in ref. 551.

588. KARADI, G. M., *et al.* (eds.) (1971). *Pumped Storage Development and its Environmental Effects*. American Water Resources Association, University of Wisconsin, Milwaukee (19–24 September 1971).

589. KASTER, J. L., and JACOBI, G. Z. (1978). 'Benthic macroinvertebrates of a fluctuating reservoir', *Freshwater Biol.*, **8**, 283–90.

590. KATZ, B. M. (1977). 'Chlorine dissipation and toxicity presence of nitrogenous compounds'. *J. Water Pollut. Contr. Fed.*, **49**(7), 1627–35.

591. KAYA, C. M. (1977). 'Reproductive biology of rainbow and brown trout in a geothermally heated stream, the Firehole River of Yellowstone National Park', *Trans. Amer. Fish. Soc.*, **16**(4), 354–61.

592. KELSO, J. R. M. (1974). 'Influence of a thermal effluent on movement of Brown bullhead (*Ictalurus nebulosus*) as determined by ultrasonic tracking', *J. Fish. Res. Board Can.*, **31**, 1507–13.

593. —— (1976a). 'Movement of yellow perch (*Perca flavescens*) and white sucker (*Catostomus commersoni*) in a nearshore Great Lakes habitat, subject to a thermal discharge', ibid., **33**, 42–53.

594. ——, and MINNS, C. K. (1975). 'Summer distribution of the nearshore fish community near a thermal genetating station as determined by acoustic census', ibid., **32**, 1409–18.

595. ——, and LESLIE, J. K. (1979). 'Entrainment of larval fish by the Douglas Point Generating Station, Lake Huron, in relation to seasonal succession and distribution, ibid., **36**, 37–41.

596. KENNEDY, V. S., and MIHURSKY, J. A. (1967). 'Bibliography on the effects of temperature in the aquatic environment', *University of Maryland Nat. Resources Inst., Contrib. no.* 326, May.

597. KERR, J. E. (1953). 'Studies of fish preservation at the Contra Costa Steam Plant of the Pacific Gas and Electric Company', *Fish. Bull.*, no. 92, California Fish and Game.

598. KERR, N. M. (1976). 'Farming marine flatfish using waste heat from sea-water cooling', *Energy World*, October, 2–10.

599. KERR, P. C., *et al.* (1970). *The Interrelationships of Carbon and Phosphorus in Regulating Heterotrophic and Autotrophic Populations in Aquatic Ecosystems*. Fed. Water Qual. Admin., U.S. Dept. Int. Water Poll. Cont. Res. Series 16050 FGS 07/70. Washington D.C.: U.S. Government Printing Office.

600. KHALANSKI, M. (1975). 'Etudes réalisées en France sur les conséquences écologiques de la refrigeration des centrales. Thermiques circuit ouvert', pp. 461–76 in ref. 551.

601. KIDD, G. J. A. (1953). *Fraser River Suspended Sediment Survey—Interim Report for period 1949–1952*. B.C. Dept. Lands, Forests, and Water Resources, Vancouver.

602. KINNE, O. (ed.) (1970). *Marine Ecology*, vol. I, *Environmental Factors*, Part 1, pp. 680. Wiley-Interscience.

603. —— (ed.) (1975) Ibid., vol. II, *Physiological Mechanisms*, Part 2, pp. 452–992. London, New York, Sydney, Toronto: John Wiley and Sons.

604. KING, J. R., and MANCINI, E. R. (1976). 'Effects of power plant cooling water entrainment on the drifting macroinvertebrates of the Wabash River (Indiana)', pp. 368–72 in ref. 368.

605. KIPELAINEN, J. E. (1971). 'Pumped storage prevents thermal pollution', in ref. 588.

606. KITITSINA, L. A. (1973). 'Effect of hot effluent from thermal and nuclear power plants on invertebrates in cooling ponds', *Hydrobiological J.*, **9**(5), 67–79.

607. KLEIN, L. (1962). *Aspects of River Pollution*, vol. 2, *Causes and Effects*. London: Butterworths.

608. KNIGHT-JONES, E. W., and MORGAN, E. (1966). 'Responses of marine animals to changes in hydrostatic pressure', pp. 267–300 in Barnes, H. (ed.), *Oceanography and Marine Biology*. London: George Allen and Unwin.

609. KNUTSON, K. M., *et al.* (1976). *Seasonal pumped entrainment of fish at the Monticello, Mn., Nuclear Power Installation*. St. Cloud, Minnesota: Dept. of Biol. Sci., St. Cloud State University.

610. KOBAYASI, H. (1961a). 'Chlorophyll content in sessile algal community of a Japanese mountain river', *Bot. Mag., Tokyo*, **74**, 228–35.

611. —— (1961b). 'Productivity in sessile algal community of a Japanese mountain river', ibid., **74**, 331–41.

612. KOLEHMAINEN, S. E., *et al.* (1975). 'Thermal studies on tropical marine ecosystems in Puerto Rico', pp. 409–21 in ref. 551.

613. KOOPS, F. B. J. (1972). *Report on plankton investigations in the cooling circuit of Flevo power station near Lelystad*. N.V. tot Keuring van Electrotechnische Materialen, Arnhem, Nederland, no. IV, 7984–72.

614. —— (1974). 'Investigation of hydrobiological effects due to cooling water discharges of electric power plants', *Hydrobiol. Bull.*, **7**(3), 86–95.

615. —— (1975). 'Plankton investigations near Flevo power station', *Verh. int. Ver. Limnol.*, **19**, 2207–13.

616. —— (1976). *Hydrobiological cooling water research in the Netherlands'*. Centre Belge d'Etude et de Documentation des Eaux, 318–20.

617. KOPPE, P. (1974). 'Water pollution and the capacity of waters for waste heat', Paper 2.2–2, in *Proc. 9th World Energy Conf., Detroit*. London: W.E.C.

618. KORN, L., and SMITH, E. M. (1971). 'Rearing juvenile salmon in Columbia River Basin storage reservoirs', pp. 287–98 in ref. 461.

619. KORNEEV, A. (1976). 'Warm water pond-fish culture in the U.S.S.R.', *Fishery Gazette*, 26.

620. KOSHELEVA, S. I., and TKACHENKO, V. M. (1973). 'Present and predicted hydrochemistry of the Kionka Sector of the Kakhova Reservoir in connection with the proposed construction of the Zaporozh'yl Regional Electric Power Plant', *Hydrobiol. J.*, **9**(4), 50–53.

621. KRAFT, M. E. (1972). 'Effect of controlled flow reduction on a trout stream', *J. Fish. Res. Board Can.*, **29**, 1405–11.

622. KREH, T. V., and DERWORT, J. E. (1976). 'Effects of entrainment through Oconee Nuclear Station on Carbon—14 assimilation rates of phytoplankton', pp. 331–5 in ref. 368.

623. KRENKEL, P. A., and PARKER, F. L. (1969a). 'Engineering aspects, sources and magnitude of thermal pollution', in ref. 624.

624. —— (eds.) (1969b). *Biological Aspects of Thermal Pollution*. Vanderbilt University Press.

625. KREOSKI, J. R. (1969). Benton Harbor Power Plant Limnological Studies. Part III. *Some Effects of Power Plant Waste Heat Discharge on the Ecology of Lake*

Michigan. Great Lakes Research Division, The University of Michigan, Report no. 44.

626. KROGER, R. L. (1974). 'Invertebrate drift in the Snake River, Wyoming', *Hydrobiologia* **44**, 369–80.

627. KUTTY, M. N., and SUKUMARAN, N. (1975). 'Influence of upper and lower temperature extremes on the swimming performance of *Tilapia Mossambica*', *Trans. Amer. Fish. Soc.*, **105**(4), 755–61.

628. KYSER, J. M., *et al.* (1975). 'Analysis of three years' complete field temperature data from different sites of heated surface discharges into Lake Michigan', pp. 249–309 in ref. 551.

629. LAGLER, K. F. (ed.) (1969). *Man-made Lakes—Planning and Development.* United Nations Development Programme. Rome, F.A.O.

630. LANDRY, A. M., and STRAWN, K. (1973). 'Annual cycle of sport fishing activity at a warmwater discharge into Galveston Bay, Texas'. *Trans. Amer. Fish Soc.*, **102**(3), 573–7.

631. —— (1974). 'Number of individuals and injury rates of fishes caught on revolving screens at the P. H. Robinson Generating Station', pp. 263–71 in ref. 566.

632. LANDY, M. (1979). *Environmental Impact Statement glossary.* N.Y., London. I.F.I./Plenum Press.

633. LANGFORD, T. E. (1963). *The coarse fishes of the River Ancholme, Lincs. Report to Lincolnshire River Board.* Boston, Lincs. (cyclostyled).

634. —— (1966). 'Fishery biology as applied by a River Board'. *Year Book of the River Authorities Association*, pp. 3–16.

635. —— (1967). *Preliminary observations on the macro-invertebrate faunas of rivers, in relation to heated effluents from power stations'.* C.E.R.L. Lab. Note RD/L/N 124/67, Leatherhead, Surrey.

636. —— (1970). 'The temperature of a British river upstream and downstream of a heated discharge from a power station', *Hydrobiologia*, **35**(3–4), 353–75.

637. —— (1971a). 'The distribution, abundance and life-histories of stoneflies (Plecoptera) and mayflies (Ephemeroptera) in a British river, warmed by cooling-water from a power station', ibid., **38**(2), 339–77.

638. —— (1971b). 'The biological assessment of thermal effects in some British rivers', pp. 1–33 in ref. 204.

639. —— (1972). 'A comparative assessment of thermal effects in some British and North American rivers', pp. 319–51, in ref. 843.

640. —— (1974). 'Ecology and cooling water from power stations—A review of recent biological research in Britain', Paper 2.2–4, in *Proc. 9th World Energy Conf., Detroit.* London: W.E.C.

641. —— (1975). 'The emergence of insects from a British river warmed by power station cooling-water. Part II. The emergence patterns of some species of Ephemeroptera, Trichoptera, and Megaloptera, in relation to water temperature and river flow upstream and downstream of the cooling-water outfalls'. *Hydrobiologia*, **47**(1), 91–133.

642. —— (1977). 'Biological problems with the use of sea-water for cooling'. *Chemistry and Industry*, 16 July, 612–16.

643. ——, and ASTON, R. J. (1972). 'The ecology of some British rivers in relation to warm water discharges from power stations', *Proc. Roy. Soc. Lond.*, **181**, 45–57.

644. ——, and BRAY, E. S. (1969). 'The distribution of stoneflies (Plecoptera) and mayflies (Ephemeroptera) in a lowland region of Britain (Lincolnshire)', *Hydrobiologia*, **34**(2), 243–71.

645. ——, and DAFFERN, J. R. (1975). 'The emergence of insects from a British river warmed by power station cooling-water. Part I. The use and performance of

insect emergence traps in a large spate river and the effects of various factors on total catches upstream and downstream of the cooling-water outfalls', ibid., **46**(1), 71–114.

646. ——, *et al.* (1977). *The tracking of salmonids in the Afon Seiont River system, using ultra-sonic tagging techniques.* C.E.R.L. Lab. Note RD/L/N 51/77, Leatherhead, Surrey.

647. ——, and HOWELLS, G. D. (1976). 'The use of biological monitoring in the freshwater environment by the electrical industry in the U.K.', pp. 115–24, in Alabaster, J. S. (ed.) *Use of Biological Monitoring in the Freshwater Environment.* London: Academic Press.

648. ——, *et al.* (1978). 'Factors affecting the impingement of fishes on power station cooling water intake screens', pp. 281–8, in *Proc. 12th European Symposium on Marine Biology*, Stirling (September 1977). Pergamon Press.

649. ——, *et al.* (1979). *The movements and distribution of some common bream (Abramis brama L.) in the vicinity of power station intakes and outfalls in British rivers as observed by ultra-sonic tracking.* C.E.R.L. Lab. Note RD/L/N 145/78, Leatherhead, Surrey.

650. LANGRIDGE, J., and McWILLIAM, J. R. (1967). 'Heat responses of higher plants', pp. 231–86 in ref. 938.

651. LANZA, G. R., *et al.* (1975). 'Biological effects of simulated discharge plume entrainment at Indian Point nuclear power station, Hudson River estuary, U.S.A., pp. 45–123, in *Combined Effects of Radioactive, Chemical and Thermal Releases to the Environment.* Vienna: International Atomic Energy Agency.

652. LARKIN, P. A. (ed.) (1958). *The Investigation of Fish Power Problems.* H. R. MacMillan Lectures in Fisheries, University of British Columbia, Vancouver, B.C.

653. —— (1979). 'Maybe you can't get there from here: A foreshortened history of research in relation to management of Pacific salmon', *J. Fish. Res. Board Can.*, **36**, 98–106.

654. LARSON, G. L., *et al.* (1977). 'Laboratory determination of acute and sublethal toxicities of inorganic chloramines to early life stages of Coho salmon (*Oncorhynchus kisutch*)', *Trans. Amer. Fish. Soc.*, **106**(3), 268–77.

655. LAUER, G. J., *et al.* (1974). 'Entrainment studies on Hudson River organisms', pp. 77–92 in ref. 566.

656. LAVIS, M. E., and SMITH, K. (1972). 'Reservoir storage and the thermal regime of rivers with special reference to the River Lune, Yorkshire'. *Sci. Total Environ.*, **1**, 81–90.

657. LAWLER, MATUSKY, and SKELLY—Engineers (1979). *Ecosystem effects of phytoplankton and zooplankton entrainment.* Electric Power Research Institute, EA-1038 Interim Report, April 1979. Palo Alto, Ca.

658. LAWRIE, J. P. (1976). 'An assessment of the heated water installation at the Furnace salmon hatchery (1971–75)', *Rep. Salm. Res. Trust. Ireland*, **20**, 58–65.

659. LEASON, D. B. (1974). 'Planning aspects of cooling-towers', *Atmospheric Environment*, **8**, 307–12.

660. —— (1975). 'The planning of coastal power stations', pp. 5–16 in ref. 205.

661. —— (1976). 'Future power stations—What will they be like?', Paper to Ann. Conf. of Royal Inst. Chartered Surveyors, Edinburgh, 1975. London: C.E.G.B.

662. LEE, G. F. (1979). 'Persistence of chlorine in cooling-water from an electric generating station', *Proc. A.S.C.E. J. Environ. Eng. Div.*, **105** (E.E.4), 757–73.

663. LEE, S. S., and SENGUPTA, S. (1977). *Waste Heat Management and Utilisation*, vols. I, II, and III. Proceedings of a Conference, 9–11 May 1976. Miami Beach, Florida, Department of Engineering, University of Miami.

664. LEENTVAAR, P. (1966). 'The Brokopondo Research Project'. Surinam, pp. 33–41 in ref. 689.
665. —— (1974). 'Inundation of a tropical forest in Surinam (Dutch Guiana), South America', *Proc. 1st Int. Congress Ecol.*, 348–54. Pudoc, Wageningen, Netherlands.
666. LEGGETT, W. C. (1976). 'The American Shad (*Alosa sapidissima*) with special reference to its migration and population dynamics in the Connecticut River', pp. 169–226 in ref. 745.
667. LEIVESTAD, H., *et al.* (1976). *Effects of acid precipitation on freshwater organisms* (cyclostyled).
668. LELAND, *et al.* (1977). 'Heavy metals', *J. Water Pollut. Contr. Fed.*, June, 1340–68.
669. LEMKUHL, D. M. (1972). 'Change in thermal regime as a cause of reduction of benthic fauna downstream of a reservoir', *J. Fish. Res. Board Can.*, **29**(9), 1329–32.
670. LENAT, D. R. (1978). 'Effects of power plant operation on the littoral benthos of Belews lake, North Carolina', pp. 580–96 in ref. 1067.
671. LEOPOLD, L. B. *et al.* (1964). *Fluvial Processes in Geomorphology*. San Francisco, Freeman.
672. LEVADNAYA, G. D., and KUZ'MINA, A. (1972). *Effect of Impoundment on Phytobenthos of the Ob' and Yenisei*, pp. 75–7. Central Siberian Botanical Garden. Siberian Division, U.S.S.R.: Novosibirsk Academy of Sciences.
673. LEWIS, B. G. (1961). *Biological observations made during trials of chlorination for mussel control at East Yelland, 1959–1960*. C.E.R.L. Lab. Report RD/L/R1052, Leatherhead, Surrey.
674. —— (1964). *Water flow and marine fouling in culverts: A review of literature up to 1962*. C.E.R.L. Lab. Note RD/L/M60, Leatherhead, Surrey.
675. LEWIS, K. (1973). 'The effect of suspended coal particles on the life forms of the aquatic moss *Eurhynchium riparioides* (Hedw)', *Freshwater. Biol.*, **3**, 251–7.
676. LEWIS, W. M., *et al.* (1968). 'Loss of fishes over the drop box spillway of a lake', *Trans. Amer. Fish. Soc.*, **97**(4), 492–4.
677. LEYNAUD, G., and ALLARDI, J. (1975). 'Incidences d'un rejet thermique en milieu fluivial sur les mouvements des populations ictyologiques', pp. 401–7 in ref. 551.
678. LIEPOLT, R. (1972). 'Uses of the Danube River', pp. 233–50 in ref. 843.
679. LIKENS, G. E., *et al.* (1972). 'Acid rain', *Environment*, **14**(2), 33–39.
680. ——, and BORMANN, F. H. (1974). 'Acid rain. A serious regional environmental problem', *Science*, 14 June, **184**, 1176–9.
681. —— (1975). 'Acidity in rain water: Has an explanation been presented?', ibid., 30 May, **188**, 958.
682. LISCOM, K. L. (1971). 'Orifice placement in gatewells of turbine intakes for bypassing juvenile fish around dams', *Trans. Amer. Fish. Soc.*, **100**(2), 319–24.
683. LITTLE, E. C. S. (1966). 'The invasion of man-made lakes by plants', pp. 75–84 in ref. 689.
684. LITTLE, M. G., and JONES, H. R. (1979). *The uses of herbaceous vegetation in the drawdown zone of reservoir margins*. Tech. Rep. TR. 105. Water Research Centre, Medmenham, U.K.
685. LLOYD, R., and JORDAN, D. H. M. (1964). 'Some factors affecting the resistance of rainbow trout (*Salmo gairdneri* Richardson) to acid waters', *Int. J. Air and Water Pollut.*, **8**, 393–403.
686. LOCAN, D. T., and MAURER, D. (1975). 'Diversity of marine invertebrates in a thermal effluent', *J. Wat. Pollut. Contr. Fed.*, **47**(3), 515–23.

687. LOOSANOFF, V. L. (1962). 'Effect of turbidity on some larval and adult bivalves', in Hidman, J. B. (ed.) *Proc. Gulf Caribb. Fish. Int. 14th Annual Session*, pp. 80–95. Institute of Marine Sciences of the University of Miami.

688. LOUKASHKIN, A. S., and GRANT, N. (1959). 'Behaviour and reactions of the Pacific sardine, *Sardinops caerulea* (Girard), under the influence of white and colored lights and darkness'. *Proc. Calif. Acad. of Sciences*, **24**(15), 509–48.

689. LOWE-McCONNELL, R. H. (ed.) (1966). *Man Made Lakes*. Inst. Biol. Symposia no 15. London and New York: Academic Press.

690. —— (1973). 'Summary: Reservoirs in relation to man-made lake fisheries', pp. 641–54 in ref. 5.

691. LYNAM, S., and BROWN, D. J. A. (1978). *Bioassay of coal and ash leachate.* C.E.R.L. Lab. Memo. LM/BIOL/043, Leatherhead, Surrey.

692. MACAN, T. T. (1957). 'The Ephemeroptera of a stony stream', *J. Anim. Ecol.*, **26**, 317–42.

693. —— (1958). 'The temperature of a stony stream', *Hydrobiologia*, **12**, 89–106.

694. —— (1960). *A Key to the British Fresh- and Brackish-water Gastropods.* Sci. Publ. Freshwater Biol. Ass. no. 13. Ferry House, Ambleside.

695. —— (1963). *Freshwater Ecology*. London: Longmans, Green and Co., Ltd.

696. —— (1970). *Biological Studies of the English Lakes*. London: Longmans.

697. —— (1974). 'Freshwater invertebrates', pp. 143–55 in Hawksworth, D. L. (ed.) *The Changing Flora and Fauna of Britain*. London, New York: Academic Press.

698. ——, and WORTHINGTON, E. B. (1951). *Life in Lakes and Rivers. The New Naturalist*. London: Collins.

699. MACKICHAN, K. A. (1967). 'Diurnal temperature variations of three Nebraska streams', *U.S. Geol. Surv. Prof. Papers, 575.B*, 223–34.

700. MACQUEEN, J. F. (1978). *On the calculation of contaminant concentrations in various types of cooling-water sources.* C.E.R.L. Lab. Note RD/L/N 129/78. Leatherhead, Surrey.

701. ——, and HOWELLS, G. (1978). 'Waste-heat disposal—A cool look at warm water', *C.E.G.B. Research*, **7**, 32–44.

702. MADDOCK, L., and SWANN, C. L. (1977). 'A statistical analysis of some trends in sea temperature and climate in the Plymouth area in the last 70 years', *J. Mar. Biol. Ass. U.K.*, **57**, 317–38.

703. M.A.F.F. and A.R.A. (1972). *Interim Report. Study Group on the fisheries implications of water transfers between catchments*. 30 pp. London: Min. of Ag. Fish. Food and Assn. River Authorities.

704. MAINS, E. M. (1977). 'Corps of Engineers responsibilities and actions to maintain Columbia Basin anadromous fish runs', pp. 40–3 in ref. 973.

705. MAJOR, R. L., and MIGNELL, J. L. (1966). 'Influence of Rocky Reach Dam and the temperature of the Okagon River on the upstream migration of sockeye salmon', *Fishery Bull.*, **66**(1), 131–47.

706. MANGARELLA, P. A., and VAN DUSEN, E. S. (1973). *Submerged thermal discharges from ocean-sited power plants*, 8 pp. A.S.M.E. 73-WA/OCT-13.

707. MANGUM, D. C., *et al.* (1973). 'Methods for controlling marine fouling in intake systems'. U.S. Dept. Interior, June, N.T.I.S. Accessions no. PB 221 909, Washington, U.S.A.

708. MANN, K. H. (1965). 'Heated effluents and their effects on the invertebrate fauna of rivers', *Proc. Soc. Water Treat. Exam.*, **14**, 45–50.

709. MARCIAK, Z. (1977). 'Influence of thermal effluents from an electric power plant on the growth of bream in the Konin Lake complex', *Rocz. Nauk. roln* (H), **97**, 41–43.

710. MARCY, B. C. (1971). 'Survival of young fish in the discharge canal of a nuclear power plant', *J. Fish. Res. Board Can.*, **28**, 1057–60.

711. —— (1976a). 'Fishes of the Lower Connecticut river and the effects of the Connecticut Yankee Plant', pp. 61–114 in ref. 745.

712. —— (1976b). 'Plankton, fish eggs and larvae of the Lower Connecticut River and the effects of the Connecticut Yankee Plant including entrainment', pp. 115–40 in ref. 745.

713. ——, *et al.* (1978). 'Effects and impacts of physical stress on entrained organisms', pp. 135–88 in ref. 970.

714. MARKOWSKI, S. (1959). 'The cooling water of power stations: A new factor in the environment of marine and freshwater invertebrates', *J. Anim. Ecol.*, **28**, 243–58.

715. —— (1960). 'Observations on the response of some benthonic organisms to power station cooling water', ibid., **29**, 349–57.

716. MARMER, G. J., *et al.* (1975). 'Comparison of thermal scanning and *in situ* techniques for monitoring thermal discharges', *Water Resources Bull.*, **11**(6), 1157–80.

717. MARTIN, M., *et al.* (1977). 'Copper toxicity experiments in relation to abalone deaths observed in a power plant's cooling-waters', *Calif. Fish. Game*, **63**(2), 95–100.

718. MARTIN, D. B., and ARNESON, R. D. (1978). 'Comparative limnology of a deep-discharge reservoir and surface-discharge lake on the Madison River, Montana'. *Freshwat. Biol.*, **8**, 33–42.

719. MARTIN MARIETTA CORPORATION (1977). Summary of Current Findings: *Calvert Cliffs Nuclear Power Plant Aquatic Monitoring Program*. Maryland Power Plant Siting Program.

720. MASSENGILL, R. R. (1976a). 'Benthic fauna. 1965–1967 versus 1968–1972', pp. 39–54 in ref. 745.

721. —— (1976b). 'Entrainment of zooplankton at the Connecticut Yankee Plant', pp. 55–60 in ref. 745.

722. MATHUR, D., *et al.* (1977). 'Impingement of fishes at Peach Bottom Atomic Power Station, Pennsylvania', *Trans. Amer. Fish. Soc.*, **106**(3), 258–67.

723. MATTICE, J. S., and ZITTEL, H. E. (1976). 'Site-specific evaluation of power plant chlorination', *J. Water Pollut. Contr.*, **48**(10). 2284–308.

724. McCAIN, J. (1975), 'Fouling community changes induced by the thermal discharge of a Hawaiian power plant', *Environ. Pollut.*, **9**, 63–70.

725. McCAULEY, R., and HUGGINS, N. (1976). 'Behavioural thermoregulation by rainbow trout in a temperature gradient', pp. 171–5 in ref. 368.

726. McERLEAN, A. J., *et al.* (1973). 'Abundance, diversity and seasonal patterns of estuarine fish populations', *Estuarine and Coastal Mar. Sci.*, **1**, 19–36.

727. McFADDEN, J. T. (1977). 'An argument supporting the reality of compensation in fish populations and a plea to let them exercise it', pp. 153–79 in ref. 1100.

728. McFARLANE, R. W. (1976). 'Fish diversity in adjacent ambient, thermal, and post-thermal freshwater streams', pp. 268–71 in ref. 368.

729. McFIE, H. H. (1973). 'Biological, chemical and related engineering problems in large storage lakes of Tasmania', pp. 56–62 in ref. 5.

730. McGUIRE, H. E. (1977). *The effect of liquid waste discharges from steam generating facilities*. Richland, Battelle Pacific N.W. Laboratories, B.N.W.L. 2393. Washington, U.S.A.

731. McK BARY, B. (1956). *The Effect of Electric Fields on Marine Fishes*. Scottish Home Dept. Marine Research Rep. No. 1. 32 pp. Edinburgh: H.M.S.O.

732. McKELVEY, K. K., and BROOKE, M. (1959). *The Industrial Cooling Tower.* Amsterdam, London, New York, Princetown, Elsevier Publishing Co.

733. McLAREN, I. A. (1963). 'Effects of temperature on growth of zooplankton and the adaptive value of vertical migration', *J. Fish. Res. Board Can.*, **20**(3), 685–727.

734. McLARNEY, W. O., *et al.* (1974). 'Effects of increasing temperature on social behaviour in groups of bullheads (*Ictalurus natalis*)', *Environ. Pollut.*, **7**, 111–19.

735. McLEAN, R. I. (1973). 'Chlorine and temperature stress on estuarine invertebrates', *J. Water Pollut. Contr.*, **45**(5), 837–41.

736. McLOUGHLIN, J. (1976). *The Law and Practice relating to Pollution Control in the Member States of the European Communities: A Comparative Survey.* London: Graham and Trotman Ltd.

737. McMAHON, J. W., and DOCHERTY, A. E. (1975). 'Effects of heat enrichment on species succession and primary production in freshwater plankton', pp. 529–45 in ref. 551.

738. McMAHON, R. F. (1975). 'Effects of artificially elevated water temperatures on the growth, reproduction and life-cycle of a natural population of *Physa virgata* Gould', *Ecology*, **56**, 1167–72.

739. McNEELY, D. L., and PEARSON, D. (1974). 'Distribution and condition of fishes in a small reservoir receiving heated waters', *Trans. Amer. Fish. Soc.*, **3**, 518–30.

740. McWHINNIE, M. A. (1967). 'The heat responses of invertebrates (exclusive of insects)', pp. 353–74 in ref. 938.

741. MELDRIM, J. W., and GIFT, J. J. (1970). 'An experimental study of temperature preference and avoidance of the white perch'. *Ichthyological Associates Report.* 15 pp. New York (cyclostyled).

742. MENON, N. R., *et al.* (1977). 'Biology of marine fouling in Mangalore Waters', *Marine Biol.*, **41**, 127–40.

743. MENZIES, W. J. M. (1939). 'Conference on salmon problems', pp. 100–1 in Moulton, F. R. (ed.), *Pubs. Amer. Assn. Advmt. Sci.*, **8.**

744. MERRIMAN, D. (1973). 'Some biological aspects of aquatic thermal discharges'. Paper to First World Congress on Water Resources. Symposium on Thermal Pollution, Chicago, Illinois (September) (cyclostyled).

745. ——, and THORPE L. M. (1976). 'The Connecticut River Ecological Study, The Impact of a Nuclear Power Plant', *Amer. Fish. Soc. Monograph*, no. 1.

746. MESAROVIC, M. M. (1975). 'Waste heat disposal from steam-electric plants with reference to the stochastic nature of some environmental conditions and to thermal pollution control regulations', pp. 311–29 in ref. 551.

747. MIDDAUGH, D. P., *et al.* (1977). 'Response of early life history stages of the striped bass, *Morone saxatilis*, to chlorination', *Ches. Sci.*, **18**, 141–53.

748. ——, and DEAN, J. M., *et al.* (1978). 'Effect of thermal stress and total residual chlorine on early life stages of the Mummichog, *Fundulus heteroclitus*', *Marine Biol.*, **46**, 1–8.

749. MIGACHEV, V. F., *et al.* (1974). 'Ash and slag yielded by the steam power plants and a feasibility of their utilization for obtaining useful products' Paper 4. 1–7, in *Proc. 9th World Energy Conf., Detroit.* London: W.E.C.

750. MIHURSKY, J. A., *et al.* (1970). 'Thermal pollution, aquaculture and pathobiology in aquatic systems', *J. Wildl. Diseases*, 6 Oct., Proc. Annual Conference, 347–55.

751. MILANOV, T. (1973). 'Cooling problems in thermally polluted recipients', *Nordic Hydrology*, **4**, 237–55.

752. MILLER, M. C., *et al.* (1976). 'Effects of a power plant operation on the biota of a thermal discharge channel', pp. 251–8 in ref. 368.

753. MILLER, R. W. (1974). 'Incidence and cause of gas-bubble disease in a heated effluent', pp. 79–93 in ref. 368.
754. MILLS, D. H. (1964). *The ecology of the young stages of the Atlantic Salmon in the River Bran, Ross-shire*, 58 pp. Scottish Freshwater and Salmon Fisheries Research, 32. Edinburgh: H.M.S.O.
755. —— (1966), 'Smolt transport', *Salm. Trout Mag.*, 138–41.
756. ——, and SHACKLEY, P. E. (1971). *Salmon smolt transportation experiments on the Conon River system, Ross-shire*, 8 pp. Scottish Freshwater and Salmon Fisheries Research, 40. Edinburgh: H.M.S.O.
757. MINCKLEY, W. L. (1963). 'The Ecology of a Spring Stream, Doe Run, Meade County, Kentucky', *Wildl. Monog. Chestertown*, 11.
758. MINISTRY OF HEALTH (1949). *Prevention of River Pollution* (Hobday). London: H.M.S.O.
759. MINNS, C. K., *et al.* (1978). 'Spatial distribution of nearshore fish in the vicinity of two thermal generating stations, Nanticoke and Douglas Point, on the Great Lakes', *J. Fish. Res. Board Can.*, **35**, 885–92.
760. MINSHALL, G. W., and WINGER, P. V. (1968). 'The effect of reduction in stream flow on invertebrate drift', *Ecology*, **49**, 580–2.
761. MITCHELL, D. S. (1969). 'The ecology of vascular hydrophytes on Lake Kariba', *Hydrobiologia*, **34**(3–4), 448–64.
762. —— (1973). 'Aquatic weeds in man-made lakes', pp. 606–12 in ref. 5.
763. MITCHELL, N. T. (1977a). *Radioactivity in surface and coastal waters of the British Isles. 1976. Part 1. The Irish Sea and its environs.* Fisheries Radiobiol. Lab. M.A.F.F. Tech. Rep. FRL. 13, Lowestoft.
764. —— (1977b). *Radioactivity in surface and coastal waters of the British Isles. 1976. Part 2. Areas other than the Irish Sea and its environs.* ibid., FRL. 14.
765. MITCHELL, R. (ed.) (1972). *Water Pollution. Micro-biology.* New York: Wiley-Interscience.
766. MITTON, J. B., and KEOHN, R. (1975). 'Genetic organisation and adaptive response of allozymes to ecological variables in *Fundulus heteroclitus*', *Genetics*, **79**, 97–111.
767. MITZINGER, W. (1974). 'Energetics and landscaping, exemplified by the development of one district of the German Democratic Republic.' Paper 2.3–2. *Proc. 9th World Energy Conf., Detroit.* London: W.E.C.
768. MOAZZAM, M., and RIZVI, S. H. N. (1980). 'Fish entrapment in the seawater intake of a power plant at Karachi coast', *Env. Biol. Fish.*, **5**(1), 49–57.
769. *Modern Power Station Practice* (1971). Vol. 1. *Planning and Layout.* London: C.E.G.B. and Pergamon Press.
770. MÖLLER, H. (1978a). *Effects of Power Plant Cooling on Aquatic Biota—An Indexed Bibliography.* Institüt für Meereskunde an der Christian-Abrechts Universität Kiel, no. 58.
771. —— (1978b). 'Ecological effects of cooling water of a power plant at Kiel Fjord'. *Sonderdruck aus Bd.* 26(1977/8). H. 3–4, S. 117–30.
772. MONN, M. G., *et al.* (1979). 'A survey of capital costs of closed-cycle cooling-systems for steam electric power plants. *Proc. 41st. Ann. Mtg. American Power Conf.* Chicago, Illinois. (Cyclostyled).
773. MOORE, C. J., and FRISBIE, C. M. (1972). 'A winter sport fishing survey in a warm-water discharge of a steam-electric station on the Patuxent River, Maryland', *Ches. Sci.*, **13**(12), 110–15.
774. ——, *et al.* (1973). 'A sport fishing survey in the vicinity of a steam-electric station on the Patuxent estuary, Maryland', ibid., **14**(3), 160–70.
775. ——, *et al.* (1975). 'A comparison of food habits of white perch (*Morone*

americana) in the heated effluent canal of a steam electric station and in an adjacent river system', *Environmental Letters*, **8**(4), 315–23.

776. MOORE, D. J., and JAMES, K. W. (1973). 'Water temperature surveys in the vicinity of power stations with special reference to infra-red techniques', *Water Research*, **7**, 807–20.

777. MOORE, R. H. (1978). 'Variations in the diversity of summer estuarine fish populations in Aransas Bay, Texas. 1966–1973', *Estuarine and Coastal Mar. Sci.*, **6**, 495–501.

778. MORDUKHAI-BOLTOVSKOI, P. D. (ed.) (1979). 'The River Volga and its life'. *Monographia Biologica*. 33. Junk-P-V, The Hague.

779. MORGAN, R. P., and CARPENTER, E. J. (1978). 'Biocides', pp. 95–134 in ref. 970.

780. ——, and PRINCE, R. D. (1977). 'Chlorine toxicity to eggs and larvae of five Chesapeake Bay fishes', *Trans. Amer. Fish. Soc.*, **106**(4), 380–5.

781. ——, —— (1978). 'Chlorine effects on larval development of striped bass (*Morone saxatilis*), white perch (*M. americana*) and blueback herring (*Alosa aestivalis*)', ibid., **107**(4), 636–41.

782. MORRIS, A. W., and BALE, A. J. (1975). 'The accumulation of cadmium, copper, manganese and zinc by *Fucus vesiculosus* in the Bristol Channel', *Estuarine and Coastal Mar. Sci.*, **3**, 153–63.

783. MORRIS, L. A., *et al.* (1968). 'Effects of main stem impoundments and channelization upon the limnology of the Missouri river, Nebraska', *Trans. Amer. Fish. Soc.*, **97**(4), 380–8.

784. MORTIMER, C. H. (1956). 'The oxygen content of air-saturated fresh waters and aids in calculating percentage saturation', *Mitt. int. Ver. Limnol.*, **6**, 20.

785. MOUNT, D. I. (1971). 'Thermal standards in the United States of America', pp. 195–99 in ref. 546.

786. MÜLLER, K. (1963). 'Diurnal rhythm in "organic drift" of *Gammarus pulex*', *Nature* (Lond.), **198**, 806–7.

787. MUIR, N. (1976). 'Swedish study looks at district heating for small towns', *Energy International*, **13**(4), 30–32.

788. MURARKA, I. P., *et al.* (1978). *Validation and Software Documentation of the ANL Fish Impingement Model.* Rep. ANL/ES-62, 124 pp. Argonne Nat. Lab., Argonne, Illinois.

789. MURPHY, T. M., and BRISBIN, I. L. (1974). 'Distribution of alligators in response to thermal gradients in a reactor cooling reservoir', pp. 313–21 in ref. 417.

790. NAKATANI, R. E. (1969). 'Effects of heated discharges on anadromous fishes', pp. 294–317 in ref. 843.

791. ——, *et al.* (1971). 'Thermal effects and nuclear power stations in the U.S.A.', pp. 561–72 in ref. 546.

792. NASH, C. E. (1968). 'Power stations as sea farms', *New Scientist*, 14 November, 367–9.

793. —— (1969). 'Thermal aquaculture', *Sea Frontier*, **15**(5), 268–76.

794. —— (1970). 'Marine fish farming', *Mar. Pollut. Bull.*, **1**(2), 28–30.

795. —— (1974). 'Residual chlorine retention and power plant fish farms', *Prog. Fish. Cult.*, **36**(2), 92–95.

796. NATIONAL ACADEMY OF SCIENCES (1969). *Eutrophication. Causes, Consequences, Correctives. Proceedings of a Symposium.* Washington D.C.: National Academy of Sciences.

797. NAUDASCHER, E., and ZIMMERMANN, C. (1972). 'Modellversuche über die Wärmebelastung eines Flusses durch ein Kernkraftwerk', *Energie*, **24**(1), 10–15.

798. NAUMAN, S. W., and CORY, R. L. (1969). 'Thermal additions and epifaunal organisms at Chalk Point, Maryland', *Ches. Sci.*, **10**, 218.
799. NAYLOR, E. (1959). 'The fauna of a warm dock', *Proc. XVth Int. Congr. Zool. Sect. 3*, 259–62.
800. —— (1965a). 'Biological effects of a heated effluent in docks at Swansea, S. Wales', *Proc. Zool. Lond.*, **144**(2), 253–68.
801. —— (1965b). 'Effects of heated effluents upon marine and estuarine organisms' in Russell, F. S., and Yonge, M. (eds.), *Advances in Marine Biology*, 3, pp. 63–103. London, New York: Academic Press.
802. NEBEKER, A. V. (1971a). 'Effect of temperature at different altitudes on the emergence of aquatic insects from a single stream', *J. Kans. Entomol. Soc.*, **44**, 26–35.
803. —— (1971b). 'Effect of high winter water temperatures on adult emergence of aquatic insects', *Water Research*, **5**, 777–83.
804. ——, *et al.* (1976). 'Carbon dioxide and oxygen-nitrogen ratios as factors affecting salmon survival in air-supersaturated water', *Trans. Amer. Fish. Soc.*, **10.6**(3), 425–9.
805. ——, *et al.* (1979). 'Temperature and oxygen–nitrogen gas ratios affect fish survival in air-supersaturated water', *Water Research*, **13**, 299–303.
806. NEEDHAM, P. R., and JONES, A. C. (1959). 'Flow, temperature, solar radiation and ice in relation to the activities of fishes in Sagehen Creek, California', *Ecology*, **40**, 465–74.
807. ——, and USINGER, R. L. (1956). 'Variability in the macro-fauna of a single riffle in Prosser Creek, California as indicated by the Surber Sampler', *Hilgardia*, **24**(14), 383–409.
808. NEEL, J. K. (1951). 'Interrelations of certain physical and chemical features in a head-water limestone stream', *Ecology*, **32**, 368–91.
809. —— (1963). 'Impact of reservoirs', pp. 575–93, in Frey, D. G. (ed.), *Limnology in North America*. Madison, Wisconsin.
810. NEGUS, C. L. (1966). 'A quantitative study of growth and production of unionid mussels in the River Thames at Reading', *J. Anim. Ecol.*, **35**, 513–32.
811. NEILL, W. H., and BRAUER, G. (1970). 'Ecological responses of fishes and fish-food organisms to heated effluents: Case study of Lake Monona, Wisconsin'. *Annual Report to the Wisconsin Utilities Association and Office of Water Resources Research*. Department of the Interior Laboratory of Limnology, University of Wisconsin, Madison, Wisconsin.
812. ——, and MAGNUSSON, J. J. (1974). 'Distributional ecology and behavioural thermoregulation of fishes in relation to heated effluent from a power plant at Lake Monona, Wisconsin', *Trans. Amer. Fish. Soc.*, **10.3**(4), 663–710.
813. NELSON, D. H. (1974). 'Growth and developmental responses of larval toad populations to heated effluent in a South Carolina Reservoir', pp. 264–76 in ref. 417.
814. NELSON, D. J., and EVANS, F. C. (eds) (1973). 'Radionuclides in ecosystems'. *Proc. 3rd Nat. Symp. on Radioecology, 10–12 May 1971, vol. 2. U.S.A.E.C. Conf. 670503*, Oak Ridge, Tennessee.
815. ——, *et al.* (1972). 'Radio-nuclides in river systems', pp. 367–88 in ref. 843.
816. NELSON, J. S. (1965). 'Effects of fish introductions and hydro-electric developments on fishes in the Kananaskis River System, Alberta', *J. Fish. Res. Board Can.*, **22**, 721–53.
817. NELSON, W. R., and WALBURG, C. H. (1977). 'Population dynamics of Yellow Perch (*Perch flavescens*), Sauger (*Stizostedion canadense*) and Walleye (*S. vitreum, vitreum*) in four main stem Missouri reservoirs,

818. NELSON-SMITH, A. (1970). 'The problem of oil pollution of the sea', pp. 215–306, in Russell, F. S., and Yonge, M. (eds.), *Advances in Marine Biology*, vol. 8. London, New York: Academic Press.

819. NETBOY, A. (1968). *The Atlantic Salmon—a vanishing species*. London: Faber and Faber.

820. NETSCH, N. F., *et al.* (1971). 'Distribution of young gizzard and threadfin shad in Beaver reservoir', pp. 95–105 in ref. 461.

821. NEWELL, R. C. (ed.) (1970). *The Biology of Intertidal Animals*. Logos Press Ltd.

822. NEWMAN, L. (1975). 'Acidity in rainwater: Has an explanation been presented', *Science*, 30 May, **188**, 957–8.

823. NEWTON, M. E., and FETTERHOLF, C. M. (1966). 'Limnological data from Ten Lakes, Genesee and Livingston Counties, Michigan, 1965'. *Report to Water Resources Commission*, Lansing, Michigan: Michigan Department of Natural Resources.

824. NEY, J. J., and SCHUMACHER, P. D. (1978). 'Assessment of damage to fish larvae by entrainment sampling with submersible pumps', *Environ. Sci. and Technol.*, **12**(6), 715–6.

825. NIKOLSKY, G. V. (1963). *The Ecology of Fishes*. London and New York: Academic Press.

826. NILSSON, C. (1978). 'Changes in the aquatic flora along a stretch of the River Umealven, N. Sweden following hydro-electric exploitation', *Hydrobiologia*, **61**(3), 229–36.

827. NILSSON, N. A. (1965). 'Food segregation between salmonid species in North Sweden', *Rep. Inst. Freshw. Res. Drottingholm*, **46**, 58–78.

828. NISBET, M. (1961). 'Un example de pollution de rivière par vidage d'une retenue hydroelectrique', *Verh. int. Vere. Limnol.*, **14**, 678–80.

829. NORTH, W. J. (1969). 'Biological effects of heated water discharge at Morro Bay, California', *Proc. Int. Seaweed Symp. 6*.

830. ——, and ADAMS, J. R. (1968). 'The status of thermal discharges on the Pacific coast'. *Prepared for Second I.B.P. Workshop on the effects of Thermal Additions in the Marine Environment*, Solomons, Maryland, November 1968 (cyclostyled).

831. NORTH OF SCOTLAND HYDRO-ELECTRIC BOARD (1973). *Power from the Glens*.

832. N.T.I.S. (1973). 'A Review of Thermal Power Plant Intake Structure Designs and Related Environmental Considerations'. Springfield, Virginia: U.S. Dept. Commerce. National Technical Information Service.

833. —— (1977). 'The Environmental Effects of Using Coal for Generating Electricity'. Washington D.C.: U.S. Dept. Commerce. National Technical Information Service. Report No. PB-267 237.

834. NURSALL, J. R. (1952). 'The early development of bottom fauna in a new power reservoir in the Rocky Mountains of Alberta', *Can. J. Zool.*, **30**, 387–409.

835. NUTTALL, P. M. (1972). 'The effects of sand deposition upon the macroinvertebrate fauna of the River Camel, Cornwall', *Freshwat. Biol.*, **2**, 181–6.

836. NYMAN, L. (1975). 'Behaviour of fish influenced by hotwater effluents as observed by ultrasonic tracking', *Rep. Inst. Freshw. Res. (Swed.)*, **54**, 63–75.

837. OBENG, L. E. (ed.) (1969). *Man-made lakes: The Accra Symposium*. Ghana Universities Press.

838. O'CONNELL, T. R., and CAMPBELL, R. S. (1953). 'The benthos of the Black River and Clearwater Lake, Missouri', *Univ. Mo. Stud.*, **26**(2), 25–41.

839. ODEN, B. J. (1979). 'The freshwater littoral meiofauna in a South Carolina reservoir receiving thermal effluents', *Freshwat. Biol.*, **9**, 291–304.

840. OEHME, F. W. (ed.) (1978). *Toxicity of heavy metals in the environment. Parts I and II.* New York: Marcel Dekker.

841. OGAWA, H. (1979). 'Modelling of power plant impacts on fish populations', *Environ. Management*, **3**(4), 321–30.

842. OGLESBY, G. B., *et al.* (1978). 'Toxic substances in discharges of hypolimnetic waters from a seasonally stratified impoundment', *Environ. Conservat.*, **5**(4), 287–93.

843. OGLESBY, R. T., *et al.* (eds) (1972). *River Ecology and Man.* New York and London: Academic Press.

844. OPPENHEIMER, C. H. (1970). 'Bacteria, fungi and blue-green algae', pp. 347–62 in ref. 602.

845. OPRESKO, D. M., and HANNON, E. H. (1979). *Chemical effects of power plant cooling-waters: an annotated bibliography.* Palo Alto, Electric Power Research Institute. EA-1072 (ORNL/EIS-134), California, U.S.A.

846. ORLOB, G. T. (1969). Discussion, pp. 53–61 in ref. 624.

847. O.R.R.C. (1962). *Sport Fishing Today and Tomorrow. O.R.R.C. Study Report 7.* Washington D.C.

848. OSEID, D., and SMITH Jr, L. L. (1972). 'Swimming endurance and resistance to copper and Malathion of bluegills treated by long-term exposure to sublethal levels of hydrogen sulphide', *Trans. Amer. Fish. Soc.*, **101**(4), 620–5.

849. O'SULLIVAN, D. (1974). 'International European water clean-up is under way', *Env. Sci. and Technol.*, **8**(7), 602–4.

850. OTTENDORFER, L. J. (1975). 'Impact of thermal discharges on the ecosytem of the Austrian Danube. Present situation and future development', pp. 785–92 in ref. 551.

851. OTTO, R. G. (1976). 'Thermal effluents, fish and gas-bubble disease in southwestern Lake Michigan', pp. 121–9 in ref. 368.

852. ——, *et al.* (1976). 'Lethal and preferred temperatures of the Alewife (*Alosa pseudoharengus*) in Lake Michigan', *Trans. Amer. Fish. Soc.*, **1**, 97–106.

853. OVERHOLZ, W. J., *et al.* (1977). 'Hypolimnion oxygenation and its effects on the depth distribution of rainbow trout (*Salmo gairdneri*) and gizzard shad (*Dorosoma cepedianum*)', *Trans. Amer. Fish. Soc.*, **106**(4), 371–5.

854. OWENS, M., and EDWARDS, R. W. (1961). 'The effects of plants on river conditions. II. Further studies and estimates of net productivity of macrophytes in a chalk stream', *J. Ecol.*, **49**, 119–26.

855. ——, —— (1964). 'A chemical survey of some English rivers', *Proc. Soc. Water Treat. Exam.*, **13**, 134–44.

856. ——, *et al.* (1969). 'The prediction of the distribution of dissolved oxygen in rivers', *Advances in Water Pollution Research. Proc. 4th Int. Conf. Prague*, 125–37.

857. PAGE-JONES, R. M. (1971). 'The automatic detection of low-levels of dissolved free chlorine in fish farming experiments using seawater effluents', *Prog. Fish. Cult.*, **33**, 99–102.

858. PALMER, C. J., *et al.* (1979). 'A further occurrence of *Atherina boyeri* Risso. 1810 in North-Eastern Atlantic waters', *Env. Biol. Fish.*, **4**(1), 71–75.

859. PANNELL, J. P. M., *et al.* (1962). 'An investigation into the effects of warmed water from Marchwood power station into Southampton Water', *Proc. Inst. Civil Engs.*, **23**, 35–62.

860. PARK, D. L., and FARR, W. E. (1972). 'Collection of juvenile salmon and steelhead trout passing through orifices in gatewells of turbine intakes at Ice Harbour Dam', *Trans. Amer. Fish. Soc.*, **2**, 381–4.

861. PARKER, E.D., *et al.* (1973). 'Ecological comparisons of thermally affected aquatic environments', *J. Water Pollut. Contr. Fed.*, **45**(4), 726–33.
862. PARKER, F. L., and KRENKEL, P. A. (eds.) (1969). *Engineering Aspects of Thermal Pollution.* Portland, Oregon: Vanderbilt University Press.
863. PARKHURST, J. D., *et al.* (1962). 'Effect of wind, tide and weather on nearshore ocean conditions', *Proc. Internat. Conf. on Wat. Pollut. Research.* Paper 3.40. (September) London.
864. PATEL, B. (1959). 'The influence of temperature on the reproduction and moulting of *Lepas anatifera* L. under laboratory conditions', *J. Mar. Biol. Ass. U.K.*, **38**, 589–97.
865. PATRIARCHE, M. U., and CAMPBELL, R. S. (1958). 'The development of the fish population in a new flood-control reservoir in Missouri, 1948–1954', *Trans. Amer. Fish. Soc.*, **87**, 240–58.
866. PATRICK, R. (1969). 'Some effects of temperature on freshwater algae', pp. 161–85 in ref. 624.
867. —— (1974). 'Effects of abnormal temperatures on algal communities', pp. 335–49 in ref. 417.
868. PAUL, J. F., and LICK, W. J. (1974). 'A numerical model for thermal plumes and river discharges', *Proc. 17th Conf. Great Lakes*, 445–55. Ann Arbor, Michigan.
869. PAULEY, G. B., and NAKATANI, R. E. (1967). 'Histopathology of "gas-bubble" disease in salmon fingerlings', *J. Fish. Res. Board Can.*, **24**, 867–71.
870. PEARSON, W. D., and FRANKLIN, D. R. (1968). 'Some factors affecting drift rates of *Baetis* and *Simuliidae* in a large river', *Ecology*, **49**, 75–81.
871. PELTIER, W. H., and WELCH, E. B. (1969). 'Factors affecting growth of rooted aquatic plants in a river', *Weed Sci.*, **17**, 412–16.
872. ——, —— (1970). 'Factors affecting growth of rooted aquatic plants in a reservoir', ibid., **18**, 7–9.
873. PENN, A. F. (1975). 'Development of James Bay: The role of environmental impact assessment in determining the legal right to an interlocutory injunction', *J. Fish. Res. Board Can.*, **32**(1), 136–60.
874. PENTREATH, R. J. (1977). 'The accumulation of ^{110}Ag. by the plaice, *Pleuronectes platessa* L. and the Thornback Ray, *Raja clavata* L., *J. exp. mar. Biol. Ecol.*, **29**, 315–25.
875. —— (1978a). '^{237}Pu experiments with the plaice, *Pleuronectes platessa*', *Marine Biology*, **48**, 327–35.
876. —— (1978b). '^{237}Pu experiments with the Thornback Ray, *Raja Clavata*', ibid., **48**, 337–42.
877. PERKINS, E. J. (1974). *The Biology of Estuaries and Coastal Waters.* London, New York and San Francisco: Academic Press.
878. PETERS, D. S., *et al.* (1974). 'The effect of temperature on food evacuation rate in the pinfish (*Lagodon rhomboides*), spot (*Lepistomus zanthurus*) and silverside (*Menidia menidia*)', *Proc. 26th. Ann. Conf. of the S.E. Assoc. of Game and Fish. Commissioners 1972*, 637–43. U.S.A.
879. PETERSON, R. H., *et al.* (1979). 'Temperature preference of several species of *Salmo* and *Salvelinus* and some of their hybrids', *J. Fish. Res. Board Can.*, **36**, 1137–40.
880. PETR, T. (1968). 'Distribution, abundance and food of commercial fish in the Black Volta and the Volta man-made lake in Ghana during its first period of filling (1964–1966). I. Mormyridae', *Hydrobiologia*, **32**(3–4), 417–48.
881. PIPE, E. J. (1972). 'The Central Electricity Generating Board and regional water authorities', *Proc. Conf. on Management of National and Regional Water Resources.* Inst. Civil. Engs. Hilton Hotel, London, 3–4 October, C.E.G.B., London.

882. PITA, F. W., and HYNE, N. J. (1975). 'The depositional environment of zinc, lead and cadmium in reservoir sediments', *Water Research*, **9**, 701–6.

883. PIVNICKA, K., and SVATORA, M. (1977). 'Factors affecting shift in predominance from Eurasion perch (*Perca fluviatilis*) to roach (*Rutilus rutilus*) in the Klicava Reservoir, Czechoslovakia', *J. Fish. Res. Board Can.*, **34**, 1571–5.

884. POLIKARPOV, G. G. (1966). *Radioecology of Aquatic Organisms. The Accumulation and Biological Effect of Radioactive Substances*. Amsterdam: North-Holland Publishing Co., Reinhold Book Division.

885. POLIVANNAYA, M. F. (1974). 'Food of perch and ruffe in the zone of effluent from the Kurakhovka State Regional Electric Power Station', *Hydrobiol. J.* (U.S.S.R.), **10**, 84–94.

886. ——, and SERGEYEVA, O. A. (1971a). 'Zooplankton dynamics and biology of some dominant species in cooling ponds of the southern Ukraine', *in Simpoz. povliyan. podogr. vod TEPP. (copy 2)* Borok.

887. ——, —— (1971b). 'Biology of common species of Cladocera in the Kurakhovka cooling pond', *Hydrobiol. J.* (U.S.S.R.), **7**, 6.

888. PORTMANN, J. E. (1970). 'The effect of china clay on the sediments of St. Austell and Mevagissey Bays', *J. Mar. Biol. Ass. U.K.*, **50**, 577–91.

889. POWER, G. (1978). 'Problems of fisheries management in Northern Québec'. (Preprint). *Inst. Fisheries Management Ann. Study Course*. University of Lancaster (18–22 September).

890. POWER, M. E., and TODD, J. H. (1976). 'Effects of increasing temperature on social behaviour in territorial groups of pumpkinseed sunfish (*Lepomis gibbosus*)', *Environ. Pollut.*, **10**, 217–23.

891. PRECHT, J., *et al.* (eds.) (1973). *Temperature and Life*. Berlin, Heidelberg, New York: Springer-Verlag.

892. PRENTICE, E. F. (1969). 'Gull predation in a reactor discharge plume', in *Biological Effects of Thermal Discharges. Annual Prog. Rept. for 1968*. AEC Research and Development Rept, BNWL-1050 Battelle Northwest.

893. PRESTON, A. (1971). 'The radiological consequences of releases from nuclear facilities to the aquatic environment', pp. 3–23 in ref. 546.

894. PRICE, A. H., *et al.* (1976). 'Observations of *Crassostrea virginica*, cultured in the heated effluent and discharged radionuclides of a nuclear power reactor', *Proc. Nat. Shellfisheries Ass.*, **66**.

895. PRINGLE, B. H., *et al.* (1968). 'Trace metal accumulation in estuarine molluscs. A.S.C.E.', *J. Sanit. Eng. Div.*, **94**, 455–75.

896. PROFITT, M. A. (1969). 'Effects of Heated Discharges upon Aquatic Resources of White River, at Petersburg, Indiana', *Indiana Water Resources Center, Rep. no. 3*.

897. PRUTER, A. T. (1972). 'Review of commercial fisheries in the Columbia River and in contiguous ocean waters', pp. 81–122 in ref. 898.

898. ——, and ALVERSON, D. L. (eds.) (1972). *The Columbia River Estuary and Adjacent Ocean Waters*. University of Washington Press: Seattle and London.

899. P.S.E.G. REPORT (1977). 'Integration of Thermal and Food Processing Residuals into a System for Commercial Culture of Freshwater Shrimp (Power Plant Waste Heat Utilization in Aquaculture)', *Pub. Serv. Elec. and Gas Comp. Rep. Final Report*, January, vol. 1.

900. PYEFINCH, K. A. (1966). 'Hydro-electric schemes in Scotland. Biological problems and effects on salmonid fisheries', in ref. 689.

901. QUENNERSTEDT, N. (1958). 'Effect of water level fluctuation on lake vegetation', *Verh. int. Ver. Limnol.*, **13**, 901–6.

902. RACHYUNAS, L. A. (1973). 'Feeding of fishes in the cooling ponds of the

Lithuanian State Power Plant', *Hydrobiol. J.*, **9**, 13–18. (U.S.S.R.), *Hydrobiol. J.*, **9**, 21.

903. RADFORD, D. S. (1972). 'Some effects of hydro-electric power installations on aquatic invertebrates and fish in the Kananaskis River system', *Alberta Conservat*, Summer, 19–21.

904. ——, and HARTLAND-ROWE, R. (1971). 'A preliminary investigation of bottom fauna and invertebrate drift in an unregulated and regulated stream in Alberta', *J. Appl. Ecol.*, **8**, 883–903.

905. RANDOLPH, K. N. (1975). 'The daily routine of channel catfish *Ictalurus punctatus* in culture ponds'. PhD. Thesis. University of Oklahoma. Norman.

906. RANEY, E. C., and MENZEL, B. W. (1969). 'Heated Effluents and Effects on Aquatic Life with Emphasis on Fishes (A bibliography)', Cornell Univ. Water Res. and Mar. Sci. Centre Philad. Elect. Co., and Ichthyol. Ass. BLU. no. 2.

907. RANKIN, J. S., *et al.* (1974). 'Thermal effects on the microbiology and chemistry of the Connecticut River—A summary', pp. 350–5 in ref. 417.

908. RAY, D. L. (ed.) (1959). *Marine Boring and Fouling Organisms*, 539 pp. Seattle: University of Washington Press.

909. RAY, N. J. (1965). 'Chemical changes in cooling water', *C.E.G.B. Midland Region Research Rep.* 103/65.

910. RAY, S. S., *et al.* (1976). 'A State-of-the-Art, Report on Intake Technologies'. Rept. to Office of Res. and Dev. U.S. Env. Protect. Agency, Washington D.C. EPA-60 0/7-76-02 0.

911. RAYMOND, H. L. (1968). 'Migration rates of yearling Chinook salmon in relation to flows and impoundments in the Columbia and Snake Rivers', *Trans. Amer. Fish. Soc.*, **97**(4), 356–9.

912. —— (1969). 'Effect of John Day Reservoir on the migration rate of juvenile Chinook Salmon in the Columbia River', ibid., **3**, 513–14.

913. RAYMONT, J. E. G., and CARRIE. B. G. A. (1964). 'The production of zooplankton in Southampton Water', *Int. Revue ges Hydrobiologia*, **49**(2), 185–232.

914. REEVE, M. R., and COSPER, E. (1970). 'The acute thermal effects of heated effluents on the copepod *Acartia tonsa* from a sub-tropical bay and some problems of assessment', *F.A.O. Technical Conference on Marine Pollution and its Effects on Living Resources and Fishing*, pp. 1–5.

915. REISH, D. J., *et al.* (1977). 'Marine and estuarine pollution', *J. Water Pollut. Contr. Fed.*, June, 1316–40.

916. REUTTER, J. M., and HERDENDORF, C. E. (1974). 'Laboratory estimates of the seasonal final temperature preferenda of some Lake Erie fish', *Proc. 17th Conf. Great Lakes Res.*, 59.

917. REYMEYSEN, J., *et al.* (1979). 'Dry cooling towers'. U.N.I.P.E.D.E., Warsaw. Paper 2 O/D.2., June 1979.

918. RICE, T. R. (1965). 'The role of plants and animals in the cycling of radionuclides in the marine environment', *Health Physics*, **11**, 953–64.

919. ——, and BAPTIST, J. P. (1974). 'Ecological effects of radioactive emissions from nuclear power plants', in *Human and Ecologic Effects of Nuclear Power Plants*, 373–439.

920. RICHARDS, B. R. (1977). 'The use of artificial substrates for studies of the biofouling community', pp. 1–3 in ref. 567.

921. RICHARDS, F. P., *et al.* (1977). 'Temperature preference studies in environmental impact assessments: An overview with procedural recommendations', *J. Fish. Res. Board Can.*, **34**, 728–61.

922. RICHKUS, W. A. (1975). 'The response of juvenile alewives to water currents in an experimental chamber', *Trans. Amer. Fish. Soc.*, **104**(3), 494–8.

923. —— (1977). 'Aquatic impact assessment at Calvert Cliffs', *Record of the Maryland Power Plant Siting Act*, **6**(1), 1–7.

924. RIGGS, C. D. (1958). 'Unexploited potentials of our freshwater commercial fisheries', *Trans. Amer. Fish. Soc.*, **87**, 299–308.

925. RIPPON, J. (1971). 'The bacteria of aquatic environments in power stations', pp. 95–114 in ref. 205.

926. RIPPON, J. E., and WOOD, M. J. (1975). *Micro-organisms in water treatment plants*. C.E.R.L. Lab. Note RD/L/N 88/75, Leatherhead, Surrey.

927. RITCHIE, J. (1927). 'Report on prevention of growth of mussels in sub-marine shafts and tunnels at Westbank Electric Station, Portobello', *Trans. R. Scottish Soc. of Arts*.

928. ROBBINS, T. W., and MATHUR, D. (1976). 'The Muddy Run pumped storage project. A case history', *Trans. Amer. Fish. Soc.*, **105**(1), 165–72.

929. ROBERTS, D. (1977). 'Mussels and pollution', pp. 67–80 in Bayne, B. L. (ed.), *Marine Mussels; Their Ecology and Physiology. Int. Biol. Prog. 10.* Cambridge University Press.

930. ROBERTS, M. H., and GLEESON, R. A. (1978). 'Acute toxicity of bromochlorinated seawater to selected estuarine species with a comparison to chlorinated seawater toxicity', *Marine Environ. Research*, **1**(1), 19–30.

931. RODHE, W. (1964). 'Effects of impoundment on water chemistry and plankton in Lake Ransaren (Swedish Lappland)', *Verh. int. Ver. Limnol.*, **15**, 437–43.

932. ROESSLER, M. A. (1971). 'Environmental changes associated with a Florida power plant', *Mar. Pollut. bull.*, **2**(6), 87–90.

933. ROGERS, A. (1977). *Dinorwic pumped-storage scheme. Aquatic environmental studies programme. 1. Introduction and general appraisal.* Manchester: C.E.G.B. NW/SSD/SR 82/77.

934. ROMERIL, M. G. (1972). *Trace metals in the common cockle*, Cerastoderma edule. C.E.R.L. Lab. Note RD/L/N 179/72, Leatherhead, Surrey.

935. —— (1974). 'Trace metals in sediments and bivalve mollusca in Southampton Water and the Solent', *Rev. Int. Oceanogr. Med.* Tome, **xxxiii**, 31–47.

936. ——, and DAVIS, M. H. (1976). 'Trace metal levels in eels grown in power station cooling-water', *Aquaculture*, **8**, 139–49.

937. ROOSENBURG, W. J. (1969). 'Greening and copper accmulation in the American oyster, *Crassostrea virginica*, in the vicinity of a steam electric generating station, *Ches. Sci.* **10**, 241–52.

938. ROSE, A. H. (ed.) (1967). *Thermobiology*. London and New York: Academic Press.

939. ROSENQVIST, I. TH. (1977). *Sur jord-Surt vann (Acid soil-Acid water)*. Oslo: Ingenirforlaget.

940. —— (1978). 'Alternative sources for acidification of river water in Norway', *Sci. Tot. Env.*, **10**, 39–49.

941. ROSS, F. F. (1959). 'The operation of thermal power stations in relation to streams', *J. Proc. Inst. Sewage Purif.*, **16**, 2–11.

942. ——, and WHITEHOUSE, J. W. (1973). 'Cooling-towers and water quality', *Water Research*, **7**, 623–31.

943. ROTHWELL, R. (1971). 'The calefaction of Lake Trawsfynydd', pp. 115–30 in ref. 204.

944. ROUND, F. E. (1973). *The Biology of the Algae*, 275 pp. London: Edward Arnold.

945. RUCKER, R. R. (1972). *Gas bubble disease of salmonids—a critical review*. U.S. Fish Wildl. Service Tech. Paper 58.

946. —— (1976) 'Gas bubble disease: mortalities of coho salmon, *Oncorhynchus*

kisutch, in water with constant total gas pressure and different oxygen–nitrogen ratios', *Fish. Bull.*, **73**(4), 915–18.

947. RUGGLES, C. P., and WATT, W. D. (1975). 'Ecological changes due to hydroelectric development on the St. John River', *J. Fish. Res. Board Can.*, **32**, 161–70.

948. RUHR, C. E. (1957). 'Effect of stream impoundment in Tennessee on the fish populations of tributary streams', *Trans. Amer. Fish. Soc.*, **86**, 144–57.

949. RULIFSON, R. A. (1977). 'Temperature and water velocity effects on the swimming performances of young-of-the-year striped mullet (*Mugil cephalus*), spot (*Leiostomus xanthurus*), and pinfish (*Lagodon rhomboides*)', *J. Fish. Res. Board. Can.*, **34**, 2316–22.

950. RUTTNER, F. (1963). *Fundamentals of Limnology*, (Trans.) by Frey, D. G., and Frey, F. E. J. University of Toronto Press.

951. RYDER, R. A., and HENDERSON, H. F. (1975). 'Estimates of potential fish yield for the Nasser Reservoir, Arab Republic of Egypt', *J. Fish. Res. Board Can.*, **32**(11), 2137–51.

952. RYLAND, J. S. (1960). 'The British species of *Bugula* (Polyzoa)', *Proc. Zool. Soc. Lond.*, **134**, 65–105.

953. RZOSKA, J. (1966). 'The biology of reservoirs in the U.S.S.R.', pp. 149–53 in ref. 689.

954. —— (ed.) (1976). *The Nile: Biology of an ancient river*. Dr W. Junk, The Hague.

955. SADLER, K. (1979). *Effect of the warm-water discharge from Castle Donington Power Station on fish populations in the River Trent*. C.E.R.L. Lab. Note. RD/L/N/11/79, Leatherhead, Surrey.

956. SAILA, S. B., and LORDA, E. (1977). 'Sensitivity analysis applied to matrix model of the Hudson River striped bass population', pp. 311–32 in ref. 1100.

957. *Salmon and freshwater fisheries Act (1923)*. London: H.M.S.O.

958. *Salmon and freshwater fisheries Act (1975)*. London: H.M.S.O.

959. SANDERS, J. E., *et al.* (1978). 'Relation of water temperature to bacterial kidney disease in Coho salmon (*Oncorhynchus kisutch*), sockeye salmon (*O. nerka*) and steelhead trout (*Salmo gairdneri*)', *J. Fish. Res. Board Can.*, **35**, 8–11.

960. SANDHOLM, M., *et al.* (1973). 'Uptake of selenium by aquatic organisms', *Limnol. Oceanogr.*, **18**(3), 496–9.

961. SAPPO, G. B. (1975). 'Effect of hot effluent discharge from the Konakovo Power Plant on the growth of bream in the Ivan'kovo Reservoir', *Hydrobiol. J. Ichthyol.*, **11**(6), 43–7.

962. —— (1976). 'The formation of local population of bream (*Abramis brama orientalis*) in the heated water zone of the Konakovo Power Station', *ibid.* **16**(1), 35–45.

963. SAVAGE, P. D. V. (1975). 'Studies on algae in the vicinity of Fawley and Poole power stations', pp. 57–64 in ref. 205.

964. SCHIEWE, M. H. (1974). 'Influence of dissolved atmospheric gas on swimming performance of juvenile Chinook salmon', *Trans. Amer. Fish. Soc.*, **103**(4), 717–21.

965. SCHMIDT, W. (1915). 'Annelen der Hydrographic and Maritimen Meterologic', 111–24 and 169–78.

966. SCHOENEMANN, D. E., *et al.* (1961). 'Mortalities and downstream migrant salmon at McNary Dam', *Trans. Amer. Fish. Soc.*, **90**, 58–72.

967. SCHOFIELD, C. L. (1976). 'Acid precipitation: Effects on fish', *Ambio*, **5**(5–6), 228–30.

968. SCHOUMACHER, R. (1976). 'Biological considerations of pumped storage development', *Trans Amer. Fish. Soc.*, **105**(1), 155–7.

969. SCHUBEL, J. R., *et al.* (1978). 'Thermal effects of entrainment', pp. 19–94 in ref. 970.

970. ——, and MARCY, B. C. (eds.) (1978). *Power Plant Entrainment. A Biological Assessment.* London, New York, Academic Press.

971. SCHULER, V. J., and LARSON, L. E. (1974). 'Experimental studies evaluating aspects of fish behaviour as parameters in the design of generating station systems'. *Paper to Am. Soc. Civil Engineers Water Res. Eng. Prog.*, Los Angeles, California.

972. ——, —— (1975). 'Improved fish protection at intake systems', *J. Env. Eng. Div.* A.S.C.E. 101. No. EE6. Paper 11756. Dec. 897–910.

973. SCHWIEBERT, E. (ed.) (1977). 'Columbia River Salmon and Steelhead', *Special Publication no. 10. American Fisheries Society. Washington D.C.*

974. SCOTT, D. L. (1973). *Pollution in the Electric Power Industry. Its Control and Costs.* Lexington, Toronto, and London: Lexington Books.

975. SCOTTISH HOME DEPARTMENT (1957). *The Passage of Smolts and Kelts through Fish Passes.* London: H.M.S.O.

976. SCOTTON, L. N., and ANSON, D. T. (1977). 'Protecting aquatic life at plant intakes', *Power*, 74–76.

977. SCRIVEN, R. A., and HOWELLS, G. D. (1977). 'Stack Emissions and the Environment', *C.E.G.B. Research no. 5*, 28–40.

978. SEEGERT, G. L., and BROOKS, A. S. (1978a). 'The effects of intermittent chlorination on Coho salmon, alewife, spottail shiner and rainbow smelt', *Trans. Amer. Fish. Soc.*, **107**(2), 346–53.

979. ——, —— (1978b). 'Dechlorination of water for fish culture: Comparison of the activated carbon, sulfite reduction, and photochemical methods', *J. Fish. Res. Board Can.*, **35**, 88–92.

980. ——, ——, and LATIMER, D. L. (1977). 'The effects of a 30 minute exposure of selected Lake Michigan fishes and invertebrates to residual chlorine', pp. 91–99 in ref. 567.

981. SEIP, H. M., and TOLLAN, A. (1978). 'Acid precipitation and other possible sources for acidification of rivers and lakes', *Sci. and Total Environ.*, **10**, 253–70.

982. SEVERN ESTUARY STUDY GROUP (1977). *Environmental Effects of a Severn Estuary Barrage. An Appraisal of Present Knowledge and Future Research*, 2nd draft, London.

983. SEVERN–TRENT WATER AUTHORITY (1978). *Water Quality 1976–77.* Birmingham.

984. SHADIN, V. I. (1956). 'Life in Rivers' (Russian), *Jizni presnih Vod S.S.S.R.* 3, 113–256. Moscow (op. cit. in ref. 543).

985. SHARITZ, R. R., *et al.* (1974). 'Impact of production-reactor effluents on vegetation in a southeastern swamp forest', pp. 356–62 in ref. 417.

986. SHARMA, R. K., and FREEMAN, R. F. (1977). *Survey of Fish Impingement at Power Plants in the United States, vol. 1, The Great Lakes,* Rep ANL/ES-56, Argonne, Illinois: Argonne Nat. Lab.

987. SHAW, T. L. (1971). 'Low-head estuary pumped storage', in ref. 588.

988. —— (1975). 'Tidal power and the environment'. *New Scientist*, 23 October, 202–6.

989. SHEADER, M., and EVANS, F. (1975). 'Feeding and gut structure of *Parathemisto gaudichaudi* (Guerin) (Amphipoda, Hyperiidea)', *J. Mar. Biol. Ass. U.K.*, **55**, 641–52.

990. SHELBOURNE, J. E. (1970). 'Marine fish cultivation: priorities and progress in Britain', pp. 15–36, in McNeil, W. J. (ed.), *Marine Aquaculture.* Corvallis, Oregon: Oregon State University Press.

991. SHERBERGER, F. F., *et al.* (1977). 'Effects of thermal shocks on drifting aquatic insects: A laboratory simulation', *J. Fish. Res. Board Can.*, **34**(4), 529–36.

992. SHIELDS, J. T. (1958). 'Experimental control of carp reproduction through water drawdowns in Port Randall Reservoir, South Dakota', *Trans. Amer. Fish. Soc.*, **87**, 23–33.

993. SIMMONS, C. N., *et al.* (1974). 'An ecological extraction of heater water discharge on phytoplankton blooms in the Potomac River', *Hydrobiologia*, **45**(4), 441–66.

994. SINCLAIR, D. C. (1965). 'The effects of water level changes on the limnology of two British Columbia coastal lakes, with particular reference to the bottom fauna'. M.Sc. Thesis, University of British Columbia, Vancouver, B.C.

995. SINHA, N. R. P., and JONES, J. W. (1974). *The European Freshwater Eel.* Liverpool: Liverpool University Press.

996. SLATICK, E. (1970). *Passage of adult salmon and trout through pipes.* U.S. Dept. Int. Fish and Wildl. Serv. Spec. Sci. Report No. 592, 1–18.

997. SMIL, V. (1976). 'China opts for small-scale energy supplies', *Energy International*, **13**(2), 17–19.

998. SMITH, E. D. (1974). 'Electro-physiology of the electrical shark repellant', *Trans. Inst. Elec. Engrs.*, Aug, 1–15.

999. SMITH, K. (1972). 'River water temperatures—an environmental review', *Scot. Geog. Mag.*, **88**, 211–20.

1000. —— (1975). 'Water temperature variations within a major river system', *Hydrology*, **6**, 153–69.

1001. SNYDER, D. E. (1975). 'Passage of fish eggs and young through a pumped storage generating station', *J. Fish. Res. Board Can.*, **32**(8), 1259–65.

1002. SOLTERO, R. A., and WRIGHT, J. C. (1975). 'Primary production on a new reservoir, Bighorn Lake—Yellowtail Dam, Montana, U.S.A.', *Freshwat. Biol.*, **5**(5), 407–22.

1003. SOUTHWOOD, J. R. E. (1966). *Ecological Methods.* London, Chapman and Hall.

1004. SPANOVSKAYA, V. D., and GRYGORASH, V. A. (1977). 'Development of food of age 0, Eurasian perch (*Perca fluviatilis*) in reservoirs near Moscow, U.S.S.R.', *J. Fish. Res. Board Can.*, **34**(10), 1551–8.

1005. SPEECE, R. E. (1971). 'Hypolymnion aeration', *J. Amer. Water Works Assoc.* **64**(1), 6–9.

1006. SPENCE, J. A., and HYNES, H. B. N. (1971). 'Differences in benthos, upstream and downstream of an impoundment', *J. Fish. Res. Board Can.*, **28**, 35–43.

1007. SPENCER, J. F. (1970). 'Diurnal and seasonal temperature changes in the littoral soil on Pwllcrochan Flats, Milford Haven, in relation to Pembroke Power Station'. *Effects of Industry on the Environment*, Field Studies, 11–27.

1008. —— (1977). *Temperature studies in the Blackwater Estuary, 1974–75.* C.E.R.L. Lab. Rep. RD/L/R/ 1962, Leatherhead, Surrey.

1009. SPIGARELLI, S. A. (1975). 'Behavioural responses of Lake Michigan fishes to a nuclear power plant discharge', pp. 479–98 in ref. 551.

1010. ——, *et al.* (1977). 'The influence of body weight on heating and cooling of selected Lake Michigan fishes', *Comp. Biochem. Physiol.*, **56A**, 51–7, 222.

1011. ——, and THOMMES, M. M. (1979). 'Temperature selection and estimated thermal acclimation by rainbow trout (*Salmo gairdneri*) in a thermal plume', *J. Fish. Res. Board Can.*, **36**, 366–76.

1012. SPRAGUE, J. B. (1963). 'Resistance of four freshwater crustaceans to lethal high temperatures and low oxygen', *J. Fish. Res. Board Can.*, **20**, 387.

1013. ——, and DRURY, D. E. (1969). 'Avoidance reactions of salmonid fish to representative pollutants', pp. 169–79 in *Advances Water Poll. Res., Proc. 4th Int. Conf.*, Oxford: Pergamon Press.

1014. SPRULES, W. M. (1947) 'An ecological investigation of stream insects in Algonquin Park, Ontario', *University of Toronto, Stud. Biol. Ser.*, **56**, 1–81.

1015. SPURR, G., and SCRIVEN, R. A. (1975). 'United Kingdom experience of the physical behaviour of headed effluents in the atmosphere and in various types of aquatic systems', pp. 227–47 in ref. 551.

1016. SQUIRES, L. E., *et al.* (1979). 'Algal response to a thermal effluent. Study of a power station on the Provo River, Utah, U.S.A.', *Hydrobiologia*, **63**, 17–32.

1017. STANGENBERG, M., and PAWLACZYK, M. (1961). 'The influence of warm water influx from a power station upon the formation of biocoenotic communities in a river', *Zesz-nauk. Politech. Wr. Wroclaw. No. 40, Inzyn Sanit.*, **1**, 67–106; *Water Poll. Abst.*, **35**(3), 579.

1018. STARRETT, W. C. (1972). 'Man and the Illinois River', pp. 131–70 in ref. 843.

1019. STAUFFER, J. R., *et al.* (1974). 'A field evaluation of the effects of heated discharges on fish distribution', *Water Resources Bull.*, **10**(5), 860–75.

1020. ——, *et al.* (1975a). 'Summer distribution of fish species in the vicinity of a thermal discharge, New River, Virginia', *Arch. Hydrobiol.*, **76**(3), 287–301.

1021. ——, *et al.* (1975b). 'Body temperature change of bluegill sunfish subjected to thermal shock', *Prog. Fish. Cult.*, **37**(2), 90–92.

1022. STEWART, L. (1968). *The movements of salmon in relation to variations in air and water temperatures.* Lancaster: Lancashire River Authority Fisheries Department Report.

1023. STILES, C. D., and BLAKE, N. J. (1976). 'Seasonal distribution of a podocopid ostracod in a thermally altered area of Tampa Bay, Florida', pp. 227–34 in ref. 368.

1024. STOBER, Q. J. (1964). 'Some limnological effects of Tiber Reservoir on the Marias River, Montana', *Proc. Mont. Acad. Sci.*, **23**, 111–37.

1025. ——, and HANSON, C. H. (1974). 'Toxicity of chlorine and heat to pink (*Oncorhynchus gorbuscha*) and Chinook salmon (*O. tshawytscha*)', *Trans. Amer. Fish. Soc.*, **3**, 569–75.

1026. STOCK, J. N., and STRACHAN, A. R. (1977). 'Heat as a marine fouling control process at coastal electric generating stations', pp. 55–62 in ref. 567.

1027. STORCK, T. W., *et al.* (1978). 'The distribution of limnetic fish larvae in a flood control reservoir in central Illinois', *Trans. Amer. Fish. Soc.*, **107**(3), 419–24.

1028. STORR, J. F. (1974). 'Plankton entrainment by the condenser systems of nuclear power stations of Lake Ontario', pp. 291–5 in ref. 417.

1029. ——, and SCHLENKER, G. (1974). 'Response of perch and their forage to thermal discharges in Lake Ontario', pp. 363–370 in ref. 417.

1030. STOTT, B., *et al.* (1963). 'Homing behaviour in gudgeon (*Gobio gobio* (L)'. *Anim. Behaviour*, **11**, 93–96.

1031. ——, and CROSS, D. G. (1973). 'The reactions of roach (*Rutilus rutilus L.*) to changes in the concentration of dissolved oxygen and free carbon dioxide in a laboratory channel', *Water Research*, **7**, 793–805.

1032. STRASKRABA, M., and STRASKRABOVA, V. (1969). 'Eastern European Lakes', pp. 65–7 in *Eutrophication. Causes, Consequences, Correctives.* Washington D.C.: National Academy of Science.

1033. STRASKRABOVA, V. (1975). 'Self purification capacity of impoundments', *Water Research*, **9**, 1171–77.

1034. STRATTON, C. L., and LEE, G. F. (1975). 'Cooling towers and water quality', *J. Water Pollut. Contr. Fed.*, **47**, 1901–12.

1035. STRAWN, K. (1970). 'Beneficial uses of warm water discharges in surface waters', in *Electric Power and Thermal Discharges*, Eisenbud, M., and Gleason, G. (eds.). New York: Gordon and Breach.

1036. STRICKLAND, J. B. (1969). 'Remarks on the effects of heated discharges on marine zooplankton', pp. 73–7 in ref. 624.

1037. STROUD, R. H. (1966). 'American experience in recreational use of artificial waters in man-made lakes', pp. 189–208 in ref. 689.

1038. ——, and DOUGLAS, P. A. (1968). 'Thermal pollution of water', *Bull. Sport. Fish. Inst.*, **191**, 1–8.

1039. STUART, T. A. (1962). 'The Leaping Behaviour of Salmon and Trout at Falls and Obstructions', *Freshwater and Salmon Fisheries Research 28*. Edinburgh: H.M.S.O.

1040. STUART, T. J., and STANFORD, J. A. (1978). 'A case of thermal pollution limited primary productivity in a southwestern U.S.A. reservoir', *Hydrobiologia*, **58**(3), 199–211.

1041. STUPKA, R. C., and SHARMA, R. K. (1977). *Survey of Fish Impingement at Power Plants in the United States*, vol. iii, *Estuaries and Coastal Waters*. Argonne, Illinois: Argonne National Laboratory, ANL/ES-56.

1042. *Sunday Times* (1978). 'Row over salmon barriers', 23 July, p. 5. London, Times Newspapers.

1043. SUTCLIFFE, D. W., and CARRICK, T. R. (1973). 'Studies on mountain streams in the English Lake District. pH, calcium and the distribution of invertebrates in the River Duddon', *Freshwat. Biol.*, **3**, 437–62.

1044. SVERDRUP, H. U., *et al.* (1963). *The Oceans: Their Physics, Chemistry and Biology*. Englewood Cliffs. N.J.: Prentice Hall.

1045. SWAIN, A., and NEWMAN, O. F. (1957). 'Hydrographical survey of the River Usk (U.K.)', *Fish. Investig. Series 1*, **6**(1).

1046. SWALE, E. M. F. (1964). 'A study of the phytoplankton of a calcareus river', *J. Ecol.*, **52**, 433–66.

1047. SWARTZMANN, G. L., *et al.* (1978). *Comparison of simulation models used in assessing the effects of power plant induced mortality on fish populations*. U.S.N.R.C. Report No. Nureg./CR-0474. Washington.

1048. SWEDEBERG, D. V. (1968). 'Food and growth of the freshwater drum and Lewis and Clark Lake, South Dakota', *Trans. Amer. Fish. Soc.*, **97**(4), 442–7.

1049. SWEERS, DOOR H. E. (1974). 'Two methods to measure the heat dissipation of discharged cooling water; a phenomenological approach', *Electrotechniek*, **52**, 11, 615–18.

1050. SYLVESTER, J. R. (1975a). 'Critical thermal maxima of three species of Hawaiian estuarine fish: a comparative study', *J. Fish. Biol.*, **7**, 257–62.

1051. —— (1975b). 'Biological considerations on the use of thermal effluents for finfish aquaculture', *Aquaculture*, **6**, 1–10.

1052. TALLING, J. F., and RZOSKA, J. (1967). 'The development of plankton in relation to hydrological regime in the Blue Nile', *J. Ecol.*, **55**, 637–62.

1053. TALMAGE, S. S., and COUTANT, C. C. (1980). 'Thermal Effects', *J. Wat. Pollut. Contr. Fed.*, **52**(6), 1575–1616.

1054. TANSEY, M. R., and FLIERMANS, C. B. (1978). 'Pathogenic species of thermophilic and thermotolerant fungi in reactor effluents of the Savannah River Plant', pp. 633–90 in ref. 1067.

1055. TARGETT, T. E., and McLEAVE, J. D. (1974). 'Summer abundance of fishes in a Maine tidal cove with special reference to temperature', *Trans. Amer. Fish. Soc.*, **103**(2), 325–30.

1056. TARZWELL, C. M. (1970). 'Thermal requirements to protect aquatic life', *J. Water Pollut. Contr. Fed.*, **42**(5), part 1, 823–8.

1057. TEMPLETON, R. G. (1971). 'An investigation of the advantages of autumn and spring stocking with brown trout, *Salmo trutta* L. in a Yorkshire reservoir', *J. Fish. Biol.*, **3**(3), 303–24.

1058. TEPPEN, T. C., and GAMMON, J. R. (1976). 'Distribution and abundance of fish populations in the Middle Wabash River', pp. 272–83 in ref. 368.

1059. THATCHER, T. O. (1978). 'The relative sensitivity of Pacific northwest fishes and invertebrates to chlorinated sea water', pp. 341–50 in ref. 579.

1060. ——, *et al.* (1976). 'Bioassays on the combined effects of chlorine, heavy metals and temperature on fishes and fish food organisms. Part 1. Effects of chlorine and temperature on juvenile brook trout (*Salvelinus fontinalis*)'. *Bulletin of Environmental Contamination and Toxicology*, **15**(1), 40–48.

1061. THOMANN, R. V. (1972). 'The Delaware River. A study in water quality management', pp. 130 in ref. 843.

1062. THOMPSON, T. J. (1971). 'Role of nuclear power in the United States of America', pp. 91–118 in ref. 546.

1063. THORHAUG, A. (1974). 'Effect of thermal effluents on the marine biology of southeastern Florida', pp. 518–31 in ref. 419.

1064. ——, *et al.* (1974). 'Impact of a power plant on a subtropical estuarine environment', *Mar. Pollut. Bull.*, 166–9.

1065. ——, *et al.* (1978). 'The effect of heated effluents from power plants on seagrass (*Thalassia*) communities, quantitatively comparing estuaries in the subtropics to the tropics'. ibid., **9**, 181–7.

1066. ——, *et al.* (1979). 'Biological effects of power plant thermal effluents in Card Sound, Florida', *Environ. Conservat.*, **6**(2), 127–37.

1067. THORP, J. H., and GIBBONS, J. W. (eds.) (1978). *Energy and Environmental Stress in Aquatic Systems*. Tech. Inf. Centre., U.S. Department of Energy.

1068. TILL, J. E. (1978). 'The effect of chronic exposure to plutonium-238 (IV) citrate on the embryonic development of carp and fathead minnow eggs', *Health Physics*, **34**(4), 333–43.

1069. TILLY, L. J. (1974). 'Respiration and net productivity of the plankton community in a reactor cooling reservoir', pp. 462–74 in ref. 417.

1070. ——, *et al.* (1978). 'Response of the Asiatic Clam, *Corbicula fluminea*, to gamma radiation', *Health Physics*, **35**(11), 704–7.

1071. TOPPING, G. (1973). 'Heavy metals in fish from Scottish waters', *Aquaculture*, **1**, 373–7.

1072. TREFETHEN, P. (1972). 'Man's impact on the Columbia River', pp. 77–98 in ref. 843.

1073. TREMBLEY, F. J. (1960). 'Research Projects on Effects of Condenser Discharge Water on Aquatic Life'. *Progress Report, 1956–1959*. Bethlehem, Pennsylvania: Institute of Research, Lehigh University.

1074. —— (1965). 'Effects of cooling-water from steam-electric power plants on stream biota', pp. 334–45, in *Biological Problems in Water Pollution, U.S. Dept. Health Education and Welfare*, 999 WP-25, Washington, D.C.: Government Printing Office.

1075. TRENT RIVER AUTHORITY. *Annual Reports 1964–1974*. Trent River Authority, Nottingham.

1076. TROTZKY, H. M., and GREGORY, R. W. (1974). 'The effects of water flow manipulation below a hydro-electric power dam on the bottom fauna of the Upper Kennebec river, Maine', *Trans. Amer. Fish. Soc.*, **103**(2), 318–24.

1077. TRUCHAN, J. G. (1977). 'Toxicity of residual chlorine to freshwater fish: Michigan's experience', pp. 79–89 in ref. 567.

1078. TUCKER, W. S. (1978). 'Power plant siting under the N.E.P.A. process. Success or failure', pp. 251–63 in ref. 81.

1079. TURNER, W. R. (1971). 'Sport fish harvest from Rough River, Kentucky, before and after impoundment', pp. 321–9 in ref. 461.

1080. TURNPENNY, A. W. H. (1981). *The use of multiple regression analysis in formulating operating strategies at power stations subject to fish impingement problems*. C.E.R.L. Lab. Note RD/C/2043/N81. Leatherhead, Surrey.

1081. ——, and UTTING, N. J. (1980). *Winter diel patterns of fish impingement at Dungeness 'A' power station with special reference to the sprat* (Sprattus, sprattus L.). C.E.R.L. Lab. Note RD/L/N 188/79, Leatherhead, Surrey.

1082. TUROBOYSKI, L. (1973). 'Investigations on influence of heated waters from the power station at Skawina on Skawinka and Vistula Rivers, Poland', *Arch. Hydrobiol.*, **20**(3), 443–60.

1083. UNITED NATIONS GENERAL ASSEMBLY (1962). *Report of the U.N. Scientific Committee on the Effects of Atomic Radiation. U.N. General Assembly Office Records, 17th Session, Supplement no. 16 (A/5216).*

1084. U.N.I.P.E.D.E. (1977). 'Expert Group on cooling water problems. Survey of legislation and rules on water quality in common use in a number of European countries'. *Study Document 20/D.4.*

1085. U.S.A.E.C. (1971). *Thermal Effects and U.S. Nuclear Power Stations.* Washington, D.C.: U.S.A.E.C. Division of Reactor Development and Technology.

1086. U.S. DEPARTMENT OF THE INTERIOR (1953). *Laboratory Investigations of 81 Fly Ashes.* Bureau of Reclamation. Eng. Lab. Branch. Concrete Lab. Report No. C.680 (11th September).

1087. UTTING, N. J. (1975a). *Bradwell Oyster Survey 1974 (Parts 1 and 2).* C.E.R.L. Lab. Rep. RD/L/R.1916, Leatherhead, Surrey.

1088. —— (1975b). *An experiment to determine the effect of entrainment through Bradwell Nuclear Power Station on the larvae of* Ostrea edulis. C.E.R.L. Lab. Note RD/L/N 45/75, Leatherhead, Surrey.

1089. ——, and MILLICAN, P. F. (1977). *Further experiments to determine the effect of entrainment through Bradwell Nuclear Power Station on the larvae of* Ostrea Edulis L. C.E.R.L. Lab. Note RD/L/N 130/77, Leatherhead, Surrey.

1090. UZIEL, M. S. (1980). 'Entrainment and Impingement at cooling-water intakes', *J. Wat. Pollut. Contr. Fed.*, **52**(6) 1616–1630.

1091. ——, and HANNON, E. H. (1979). *Impingement: an annotated bibliography.* Palo Alto. Electric Power Research Institute. EA-1050 Interim Report, April 1979.

1092. VADAS, R. L., KESER, M., and RUSANOWSKI, P. C. (1976). 'Influence of thermal loading on the ecology of intertidal algae', pp. 202–12 in ref. 368.

1093. ——, ——, ——, and LARSON, B. R. (1976). 'The effects of thermal loading on the growth and ecology of a northern population of *Spartina alterniflora*', pp. 54–63 in ref. 368.

1094. ——, *et al.* (1978). 'Effects of reduced temperatures on previously stressed populations of an intertidal alga', pp. 434–51 in ref. 1067.

1095. VAN DEN BROEK, W. L. F. (1978). 'Dietary habits of fish populations in the Lower Medway Estuary', *J. Fish. Biol.*, **13**, 645–54.

1096. —— (1979). 'A seasonal survey of fish populations in the Lower Medway Estuary, Kent based on power stations screen samples', *Estuarine and Coastal Mar. Sci.*, **9**, 1–15.

1097. VAN DER LEEDEN, F. (1975). *Water Resources of the World.* Washington, New York: Water Info. Centre. Inc. Port.

1098. VAN DER WALKER, J. G. (1966). 'Response of salmonids to low frequency sound', pp. 45–58, in Lairolga, W. N. (ed.) *Marine Biological Acoustics*, vol. 2. Pergamon Press.

1099. VAN VELSON, R. C. (1974). 'Self-sustaining rainbow trout (*Salmo gairdneri*) population in McConaughy reservoir, Nebraska', *Trans. Amer. Fish. Soc.*, **103**(1), 59–64.

1100. VAN WINKLE, W. (ed.) (1977). *Proceedings of the Conference on Assessing the Effects of Power-Plant-Induced Mortality on Fish Populations*. Pergamon Press.

1101. VARLEY, J. D., *et al.* (1971). 'Growth of rainbow trout in Flaming Gorge Reservoir during the first six years of impoundment', pp. 121–36 in ref. 461.

1102. VELZ, C. J. (1971). 'Environmental aspects of pumped storage development—reason or emotion', pp. 478–9 in ref. 588.

1103. VERSTRAETE, W., *et al.* (1975). 'Shifts in microbial groups of river water upon passage through cooling systems', *Environ. Pollut.*, **8**, 275–81.

1104. VIBERT, R. (1967). *Fishing with Electricity*. London: Fishing News (Books) Ltd.

1105. VIELVOYE, R. (1977). 'Russian claim of more efficient design for power stations'. *The Times,* Thursday, 22 September.

1106. VINER, A. B. (1970a). 'Hydrobiology of Lake Volta, Ghana. I. Stratification and Circulation of Water', *Hydrobiologia*, **35**(2), 209–29.

1107. ——, (1970b). 'Hydrobiology of Lake Volta, Ghana. II. Some observations on biological features associated with the morphology and water stratification, ibid., **35**(2), 230–48.

1108. WALBURG, C. H. (1971). 'Loss of young fish in reservoir discharge, and year class survival, Lewis and Clark Lake, Missouri River', pp. 441–48 in ref. 461.

1109. ——, *et al.* (1971). 'Lewis and Clark Tailwater biota and some relations of the tailwater and reservoir fish populations,' pp. 449–67 in ref. 461.

1110. WALKER, J. H., and LAWSON, J. D. (1977) . 'Natural stream temperature variations in a catchment', *Water Research*, **2**, 373–7.

1111. WALLACE, D. N. (1978). 'Two anomalies of fish larvae transport and their importance in environmental assessment', *New York Fish and Game Journal*, **25**(1), 59–71.

1112. WALLEN, I. E. (1951). 'The direct effect of turbidity on fishes', *Bull. Okla. Agric. Exp. Stn.*, **48**(2) 1–27.

1113. WANDESFORD-SMITH, G. (1978). 'American–European assessment of environmental impact assessment', *Ambio*, **7**(4), 181–2.

1114. WARD, D. V. (1978). *Biological Environmental Impact Studies: Theory and Methods*. N.Y., London: Academic Press.

1115. WARD, J. V. (1974). 'A temperature stressed stream ecosystem below a hypolimnial release mountain reservoir', *Arch. Hydrobiol.*, **74**(2), 247–75.'

1116. ——, and STANFORD, J. A. (eds.) (1979). *The Ecology of Regulated Streams.* New York: Plenum Press.

1117. WARDLE, C. S. (1976). 'Fish reactions to fishing gears', *Scot. Fish. Bull.*, **43**, 16–20.

1118. WARINNER, J. E., and BREHMER, M. L. (1966). 'The effects of thermal effluents on marine organisms', *Int. J. Air and Water Poll.*, **10**(4), 277–89.

1119. WATER RESEARCH CENTRE (1976). *Thermal Pollution. A Literature Survey covering the period 1968–1975*. W.R.C. Reading List no. 34, Stevenage, Herts.

1120. WATER RESOURCES BOARD (1973a). *Water Resources in England and Wales*, vol. 1. London: H.M.S.O.

1121. —— (1973b). ibid., vol. 2.

1122. WATERS, T. F. (1961). 'Standing crop and drift of stream bottom organisms', *Ecology*, **42**, 532–42.

1123. —— (1962). 'Diurnal periodicity in the drift of stream invertebrates', ibid., **43**, 316–20.

1124. —— (1969). 'Invertebrate drift, ecology and significance to stream fishes', pp. 121–34 in *Symposium on Salmon and Trout in Streams*. Vancouver: University of British Columbia.

1125. —— (1972). 'The drift of stream insects', *Ann. Rev. Entomol.*, **17**, 253–72.

1126. WATT, J. D., and THORNE, D. J. (1965). 'Composition and pozzolanic properties of pulverised fuel ashes. I. Composition of fly ashes from some British power stations and properties of their component particles', *J. Appl. Chem.*, **15**, 585–604.

1127. WAUGH, G. D. (1964). 'Observations on the effects of chlorine on the larvae of oysters (*Ostrea edulis* (L)) and barnacles (*Elminius modestus* (Darwin))', *Ann. Appl. Biol.*, **54**, 423–40.

1128. WORLD ENERGY CONFERENCE (1974). *Proceedings of the 9th World Energy Conference, Detroit, 22–27 September, 1974*, Divisions 1–4 (4 volumes). London: W.E.C.

1129. WELCH, M. O., and WARD, C. H. (1978). 'Primary productivity: Analysis of variance in a thermally enriched aquatic system', pp. 381–91 in ref. 1067.

1130. WELCH, P. S. (1952). *Limnology*, 538 pp. McGraw-Hill Book Co. Inc.

1131. WESCHE, T. A. (1976). 'Minimum stream flows for trout'. *Wyoming Wildlife*, **40**(1), 8–11, 35.

1132. WESTGUARD, R. L. (1964). 'Physical and biological aspects of gas-bubble disease in impounded adult Chinook salmon at McNary spawning channel', *Trans. Amer. Fish. Soc.*, **94**(3), 306–9.

1133. WHITE, G. C. (1972). *Handbook of Chlorination*. New York, Cincinnati, Toronto, London, Melbourne: van Nostrand Reinhold Co.

1134. WHITE, J. W., *et al.* (1977). 'A study of the fish community in the vicinity of a thermal discharge in the James River, Virginia', *Ches. Sci.*, **18**(2), 161–171.

1135. WHITEHOUSE, J. W. (1971b). 'Some aspects of the biology of Lake Trawsfynydd; a power station cooling pond', *Hydrobiologia*, **38**(2), 253–88.

1136. —— (1975). *Chlorination of cooling water: A review of literature on the effects of chlorine on aquatic organisms*, C.E.R.L. Lab. Report RD/L/M 496, Leatherhead, Surrey.

1137. ——, and ASTON, R. J. (1964). *A survey of the fauna above and below nine power station outfalls*. C.E.R.L. Lab. Note RD/L/N 12/64, Leatherhead, Surrey.

1138. WHITING, D., *et al.* (1976). *Analyses of angling competition catch-data from the River Trent—with special reference to the location of power stations*, C.E.G.B. SSD/MID/N13/76, Leatherhead, Surrey.

1139. WHITTON, B. A. (1975a). 'Algae', pp. 81–105 in ref. 1140.

1140. —— (ed.) (1975b) *River Ecology*. London: Blackwell Scientific.

1141. WICKETT, W. P. (1959). 'Damage to the Qualicum River stream bed by a flood in January, 1958', *Fish. Res. Board Can. Prog. Rep. Pacif. Coast. Stns*, **113**, 16–17.

1142. WILCOX, J. R. (1977). 'Waste heat utilisation in aquaculture. Some futuristic and plausible schemes', *Paper X-A-35*, in ref. 663.

1143. WILKONSKA, H., and ZUROMSKA, H. (1977). 'Changes in the species composition of fry in the shallow littoral of heated lakes of the Konin lakes complex', *Rocz. Nauk. roln.* (H)., **97**, 132–4.

1144. WILLIAMS, D. D., and HYNES, H. B. N. (1976). 'The recolonization mechanisms of stream benthos', *Oikos*, **27**, 265–72.

1145. WILLIAMS, R. B., and MURDOCH, M. B. (1973). *Effects of continuous low-level gamma radiation on sessile marine invertebrates.* IAEA-SM-158/34, pp. 551–63. Vienna: International Atomic Energy Agency.

1146. WILLS, D. (1972). *Physical Aspects of Cooling Water Discharges. 1963*—Oct. 1971. C.E. Bib. 217, London: C.E.G.B.

1147. WISCONSIN ELECTRIC POWER COMPANY (1976). Point Beach Nuclear Plant: *Final Report W.P.D.E.S., Intake Monitoring Studies Permit Number W1-0000957* Madison, Wisconson: Wisconsin Electric Power Company.

1148. WISDOM A. S. (1956). *The Law on the Pollution of Waters.* London.

1149. WITHLER, F. (1952). 'Sockeye reproduction in a tributary of Babine Lake, 1950–51', *Fish. Res. Bd. Can. Prog. Rep. Pacif. Coast. Stns*, **91**, 13–17.

1150. WITT, A., Jr. (1971). *The evaluation of environmental alteration by thermal loading and acid pollution in the cooling reservoir of a steam electric station.* NTIS, PB-197, Missouri Water Resources Research Abstract 4.W71-04019.

1151. WOJTALIK, T. A., and WATERS, T. F. (1970). 'Some effects of heated water on the drift of two species of stream invertebrates', *Trans. Amer. Fish. Soc.*, **99**(4),782–8.

1152. WOLTERS, W. R., and COUTANT, C. C. (1976). 'The effect of cold shock on the vulnerability of young bluegill to predation', pp. 162–4 in ref. 368.

1153. WONG, G. T. F., and DAVIDSON, J. A. (1977). 'The fate of chlorine in seawater', *Water Research*, **11**, 971–8.

1154. WONG, S. L., *et al.* (1978). 'Water temperature fluctuations and seasonal periodicity of *Cladophora* and *Potamogeton* in shallow rivers', *J. Fish Res. Board Can.*, **35**, 866–870.

1155. WOODHEAD, P. M. J. (1966). 'The behaviour of fish in relation to light in the sea', pp. 337–404, in Barnes, H. (ed.), *Oceanography and Marine Biology. An Annual Review*, vol. 4. London: Allen and Unwin.

1156. ——, and WOODHEAD, A. D. (1955). 'Reactions of herring larvae to light: a mechanism of vertical migration', *Nature* (Lond.), **176**(4477), 349–50.

1157. WOODS HOLE OCEANOGRAPHIC INSTITUTE (1952). *Marine Fouling and its Prevention*, Annapolis: U.S. Naval Institute.

1158. WREN, W. B. (1976). 'Temperature preference and movement of fish in relation to a long, heated discharge channel', pp. 191–4 in ref. 368.

1159. WÜNDERLICH, W. O. (1971). 'The dynamics of density-stratified reservoirs', pp. 219–32 in ref. 461.

1160. WURTZ, C. B., and DOLAN, T. (1960). 'A biological method used in the evaluation of effects of thermal discharge in the Schuylkill River', *Proc. 15th Ind. Waste Conf. Purdue*, pp. 461–72.

1161. ——, and RENN, C. E. (1965). 'Water temperatures and aquatic life', *Publ. no. 65–901. Edison Electric Institute* (June). Baltimore, Maryland: Johns Hopkins University.

1162. YEE, W. C. (1972). 'Thermal aquaculture: engineering and economics', *Environ. Sci. and Technol.*, **6**(3), 232–6.

1163. YODER, C. O., and GAMMON, J. R. (1976). 'Seasonal distribution and abundance of Ohio River fishes at the J. M. Stuart Electric Generating Station', pp. 284–95 in ref. 368.

1164. YONGE, C. M. (1949). *The Sea Shore.* London: Collins.

1165. YOSHIOKA, T., *et al.* (1971). 'Management of effluents from J.A.P.C.'s nuclear power stations', pp. 439–50 in ref. 546.

1166. YOUNG, D. L. (1974). 'Studies of Florida Gulf Coast salt marshes receiving thermal discharges', 532–50 in ref. 419.

1167. YOUNG, J. S., and FRAME, A. B. (1976). 'Some effects of a power plant effluent on estuarine epibenthic organisms', *Int. Rev. ges. Hydrobiol.*, **61**(1), 37–61.

1168. ——, and GIBSON, C. I. (1973). 'Effect of thermal effluent on migrating menhaden', *Mar. Pollut. Bull.*, **4**(6), June, 94–95.

1169. YOUNG, G. K., *et al.* (1971). 'Estuary water temperature sensitivity to meterologic conditions', *Water Resources Research* (5), 1173–81.

1170. YOURINOV, D. M., *et al.* (1974). 'Powerful hydroelectric stations and the role they play in a comprehensive utilisation of hydraulic resources in the U.S.S.R.', Paper 4. 2–19, *Proc. 9th World Energy Conf., Detroit.* London: W.E.C.

1171. ZEITOUN, I. H. (1978). 'The recovery and haematological rehabilitation of chlorine stressed adult rainbow trout (*Salmo gairdneri*)', *Environ. Biol. Fish.*, **3**(4), 355–9.

1172. ZILLICH, J. A. (1972). 'Toxicity of combine chlorine residuals to freshwater fish', *J. Water Pollut. Contr. Fed.*, **44**(2), 212–20.

1173. ZIMMERMAN, R. C., *et al.* (1974). 'Characteristics of non-thermal liquid effluents from two coal-fired power plants in Alberta, Canada', Paper 2.2–1, *Proc. 9th World Energy Conf., Detroit.* London: W.E.C.

Index

COMPILED BY HILARY DAVIES